Hydraulic Fracturing Wastewater

Treatment, Reuse, and Disposal

Hydraulic Fracturing Wastewater

Treatment, Reuse, and Disposal

Frank R. Spellman

CRC Press
Taylor & Francis Group
Boca Raton London New York

CRC Press is an imprint of the
Taylor & Francis Group, an **informa** business

CRC Press
Taylor & Francis Group
6000 Broken Sound Parkway NW, Suite 300
Boca Raton, FL 33487-2742

© 2017 by Taylor & Francis Group, LLC
CRC Press is an imprint of Taylor & Francis Group, an Informa business

No claim to original U.S. Government works

Printed on acid-free paper
Version Date: 20170313

International Standard Book Number-13: 978-1-1381-9792-3 (Hardback)

This book contains information obtained from authentic and highly regarded sources. Reasonable efforts have been made to publish reliable data and information, but the author and publisher cannot assume responsibility for the validity of all materials or the consequences of their use. The authors and publishers have attempted to trace the copyright holders of all material reproduced in this publication and apologize to copyright holders if permission to publish in this form has not been obtained. If any copyright material has not been acknowledged please write and let us know so we may rectify in any future reprint.

Except as permitted under U.S. Copyright Law, no part of this book may be reprinted, reproduced, transmitted, or utilized in any form by any electronic, mechanical, or other means, now known or hereafter invented, including photocopying, microfilming, and recording, or in any information storage or retrieval system, without written permission from the publishers.

For permission to photocopy or use material electronically from this work, please access www.copyright.com (http://www.copyright.com/) or contact the Copyright Clearance Center, Inc. (CCC), 222 Rosewood Drive, Danvers, MA 01923, 978-750-8400. CCC is a not-for-profit organization that provides licenses and registration for a variety of users. For organizations that have been granted a photocopy license by the CCC, a separate system of payment has been arranged.

Trademark Notice: Product or corporate names may be trademarks or registered trademarks, and are used only for identification and explanation without intent to infringe.

Visit the Taylor & Francis Web site at
http://www.taylorandfrancis.com

and the CRC Press Web site at
http://www.crcpress.com

Contents

Preface... xv
Acronyms and Abbreviations.. xvii
Author xix

SECTION I Introduction

Chapter 1 When an End Was a Beginning 3

End Before the Beginning.. 3
Nature's Alchemy 7
Mining Shale Gas, a Double-Edged Sword 8
Key Concepts and Definitions... 8
Purpose of Text.. 16
Thought-Provoking Questions 16
References and Recommended Reading 17

SECTION II Fracking Process and Produced Wastewater

Chapter 2 Hydraulic Fracking: The Process 21

Shale Gas Drilling Development Technology.. 21
Drilling... 21
    Well Casing Construction.. 25
    Drilling Fluids and Retention Pits.. 28
Aspects of Hydraulic Fracturing 28
    Fracture Design 30
    Description of the Hydraulic Fracturing Process.............................. 31
    Fracturing Fluids 33
Thought-Provoking Questions 37
References and Recommended Reading 37

Chapter 3 Fracking Water Supply 41

Fracking Water Use 41
Hydraulic Fracturing and Drinking Water Quality................................. 45
Thought-Provoking Questions 46
References and Recommended Reading 47

Chapter 4 Composition of Fracking Water 49

Water Chemistry 49
    Chemistry Concepts and Definitions.. 49
        Concepts 50
        Key Definitions 51

v

vi Contents

Water Solutions 52
Water Constituents 53
  Solids 53
  Turbidity 53
  Color 54
  Dissolved Oxygen 54
  Metals 54
  Organic Matter 55
  Inorganic Matter 55
  Acids 55
  Bases 55
  Salts 56
  pH 56
  Alkalinity 57
  Water Temperature 57
  Specific Conductance 57
  Hardness 58
Chemical Mixing 58
  Types of Fracturing Fluids and Additives 58
    Gelled Fluids 59
    Foamed Gels 61
    Water and Potassium Chloride Water Treatments 61
    Acids 61
    Use of Chemical Additives 61
Naturally Occurring Radioactive Material 65
Fracking Fluids and Their Constituents 66
  Commonly Used Chemical Components 67
  Toxic Chemicals 68
    Carcinogens 68
    Safe Drinking Water Act Chemicals 70
Trade Secrets and Proprietary Chemicals 70
Thought-Provoking Questions 71
References and Recommended Reading 71

SECTION III   *Fracking Water Management Choices*

Chapter 5 Fracking Wastewater Treatment 77

Fracking Water Treatment Options 77
Produced Wastewater Management Costs 78
  Transportation 78
  Water Sourcing 78
  Disposal 78
  Treatment 78
Produced Wastewater Quality and Constituents 79
  Salinity 79
  Suspended Solids 79
  Oil and Grease, Hydrocarbons, and Natural Organic Matter 79

Contents   vii

Dissolved Gas and Volatile Compounds 79
Iron and Manganese 79
Barium and Strontium 80
Boron and Bromide 80
Trace Metals 80
Radionuclides 80
Well Additives and Fracturing Chemicals 80
Thought-Provoking Questions 80
References and Recommended Reading 81

Chapter 6   Produced Wastewater Treatment Technology 83

Treatment of Produced Wastewater 83
Organic, Particulate, and Microbial Inactivation/Removal Technologies 84
Advanced Separators 84
Biological Aerated Filters 85
Hydrocyclone 88
Dissolved Air/Gas Flotation 89
Adsorption and Media Filtration 91
Oxidation 92
Settling Ponds or Basins 92
Air Stripping 93
Surfactant-Modified Zeolite/Vapor-Phase Bioreactor 94
Constructed Wetlands 94
Granular Activated Carbon 95
Ultraviolet Disinfection 96
Advantages 97
Disadvantages 97
Applicability 97
Microfiltration/Ultrafiltration 98
Common Filter Problems 98
Thought-Provoking Questions 99
References and Recommended Reading 99

Chapter 7   Desalination 101

Introduction 101
Desalination Technologies 101
Miscibility and Solubility 102
Suspension, Sediment, Particles, and Solids 102
Emulsion 102
Ion 102
Mass Concentration 103
Permeate 103
Concentrate, Reject, Retentate, Brine, or Residual Stream 103
Tonicity 103
Osmosis 103
Osmotic Pressure 104
Osmotic Gradient 104
Membrane 104

viii   Contents

Semipermeable Membrane 104
Reverse Osmosis System Flow Rating 104
Recovery or Conversion Rate 104
Concentration Factor 104
Rejection 104
Flux 105
Specific Flux (Permeability) 106
Concentration Polarization 106
Membrane Fouling 107
Membrane Scaling 107
Silt Density Index 107
Langelier Saturation Index 107
Antiscalants 107
Gas Laws 107
Boyle's Law 108
Charles's Law 108
Ideal Gas Law 109
Solutions 109
Solution Calculations 110
Concentrations 112
Moles 112
Predicting Solubility 113
Colligative Properties 114
Reverse Osmosis 114
Osmotic Pressure 115
Reverse Osmosis Process 116
Process Description 116
Reverse Osmosis Equipment 118
Membrane Materials 118
Membrane Modules 119
Plate-and-Frame Modules 119
Spiral-Wound Modules 120
Hollow-Fiber Modules 121
Tubular Modules 122
System Configuration 123
Nanofiltration 125
Electrodialysis and Electrodialysis Reversal 125
Forward Osmosis 126
Hybrid Membrane Processes 127
Two-Pass Nanofiltration 127
Dual RO with Chemical Precipitation 127
Dual RO with Softening Pretreatment and Operation at High pH 127
Dual RO with Seeded Slurry Precipitation and Recycling RO 128
High-Efficiency Electrodialysis 128
Electrodeionization 128
Thermal Desalination Technologies 129
Membrane Distillation 129
Multistage Flash Distillation 130
Multiple-Effect Distillation 130
Vapor Compression 130

Contents   ix

Alternative Desalination Processes 130
  Capacitive Deionization 130
  Softening 131
    Calculating Calcium Hardness as $CaCO_3$ 131
    Calculating Magnesium Hardness as $CaCO_3$ 132
    Calculating Total Hardness 132
    Calculating Carbonate and Noncarbonate Hardness 133
  Ion Exchange 134
Commercial Desalination Processes 135
  CDM Produced Wastewater Technology 135
Thought-Provoking Questions 136
References and Recommended Reading 136

SECTION IV   Produced Wastewater Disposal

Chapter 8   Surface Water Disposal Options 141

Introduction 141
Discharge to Surface Waters 142
Water Quality Regulations 143
  Genesis of Clean Water Reform 143
  Clean Water Act 144
  Safe Drinking Water Act 149
    SDWA Definitions 150
    SDWA Specific Provisions 152
    National Primary Drinking Water Regulations 153
    Nation Secondary Drinking Water Regulations 158
    1996 Amendments to SDWA 158
    Implementing SDWA 159
  Oil Pollution Act of 1990 159
Thought-Provoking Questions 164
References and Recommended Reading 164

Chapter 9   Disposal by Evaporation 167

Produced Wastewater Evaporation Ponds 167
Evaporation Ponds and Pits 169
Wastewater Stabilization Ponds 171
  Pond and Pit Morphometry Calculations 171
    Volume 171
    Shoreline Development Index 171
    Mean Depth 171
    Bottom Slope 172
    Volume Development 173
    Water Retention Time 173
    Ratio of Drainage Area to Water Body Capacity 173
  Impoundment Surface Evaporation 173
    Water Budget Model 173
    Energy Budget Model 174

x   Contents

Priestly–Taylor Equation 174
Penman Equation 174
DeBruin–Keijman Equation 175
Papadakis Equation 175
Thought-Provoking Questions 175
References and Recommended Reading 175

Chapter 10   Injection of Produced Wastewater 177

Define Your Terms, Please! 177
Symbols and Dimensions 177
Definitions 178
Aquiclude 178
Aquifer System 178
Aquitard 180
Coefficient of Volume Compressibility 180
Compaction 180
Compaction, Residual 180
Compaction, Specific 181
Compaction, Specific Unit 181
Compaction, Unit 181
Consolidation 181
Excess Pore Pressure 181
Expansion, Specific 181
Expansion, Specific Unit 181
Hydraulic Diffusivity 181
Hydrocompaction 182
Piezometric Surface 182
Stress, Applied 183
Stress, Effective 184
Stress, Geostatic 184
Stress, Gravitational 184
Stress, Neutral 184
Stress, Preconsolidation 185
Stress, Seepage 185
Subsidence 185
Subsidence/Head Decline Ratio 185
Unit Compaction/Head Decline Ratio 185
Wells 186
Conventional Water Wells 187
Shallow Wells 189
Deep Wells 189
Components of a Well 189
Well Casing 190
Grout 190
Well Pad 191
Sanitary Seal 191
Well Screen 191
Casing Vent 191
Drop Pipe 191
Miscellaneous Well Components 191

Contents   xi

Well Evaluation 191
Well Pumps... 192
Routine Operation and Recordkeeping Requirements...................................... 192
Well Maintenance... 193
  Troubleshooting Well Problems.. 194
Well Abandonment... 194
Injection Wells.. 195
  Class I Wells 195
  Class II Wells 197
  Class III Wells 198
  Class IV Wells 200
  Class V Wells...201
  Class VI Wells 201
Thought-Provoking Questions 202
References and Recommended Reading...202

Chapter 11 Offsite Treatment/Disposal 205

Offsite Commercial Disposal..205
Thought-Provoking Questions 206
References and Recommended Reading...206

SECTION V Safety and Health Impacts of Working with Produced Wastewater

Chapter 12 Safety and Health Considerations 211

Introduction 211
  A Network of Confusing and Constraining Rules and Standards.......................211
  Costly Modifications of Existing Installations to Meet New Legal Demands........212
  Inspections, Fines, or Time-Consuming Legal Hearings 212
  Above All, an Increasingly Burdensome
  Task of Recordkeeping and Paperwork...213
Safety and Health Regulations and Standards 214
OSHA Standards Applicable to Fracking Operations 215
  Occupational Safety and Health Act of 1970 (OSH Act) 215
  Industry Group 138—Oil and Gas Field Services: Process Description 217
OSHA Standards Most Often Cited for
Lack of Compliance or Willful Violation 219
  Subpart D, Walking–Working Surfaces 220
  Subpart E, Means of Egress 221
    Emergency Response...222
    Egress Requirements 225
  Subpart F, Powered Platforms, Manlifts,
  and Vehicle-Mounted Work Platforms...226
    What Is Fall Protection?...226
    Physical Factors at Work in a Fall 227
    Slips 227

xii   Contents

Trips	228
Stair Falls	228
Fall Protection Measures	229
Subpart G, Occupational and Environmental Control	229
Occupational Noise Exposure	230
Hearing Protection	230
Subpart H, Hazardous Materials	231
Hazardous Material	231
Flammable and Combustible Liquids	232
What Is a Hazardous Waste?	232
Subpart I, Personal Protective Equipment	234
Personal Protective Equipment	234
OSHA's PPE Standard	235
OSHA's PPE Requirements	235
Respiratory Protection	236
Subpart J, General Environmental Controls	237
Personal and Sanitation Facilities	237
Recommended Color Codes for Accident Prevention Tags	237
OSHA's Confined Space Entry Program	237
Permit-Required Confined Space Written Program	238
Pre-Entry Requirements	242
Permit System	243
Confined Space Training	243
Control of Hazardous Energy—Lockout/Tagout	244
Subpart K, Medical and First Aid	245
Subpart L, Fire Protection	246
Fire Safety	246
Fire Prevention and Control	247
Fire Protection Using Fire Extinguishers	248
Miscellaneous Fire Prevention Measures	249
Subpart N, Materials Handling and Storage	249
Rigging Safety Program	249
Safety: Ropes, Slings, and Chains	250
Subpart O, Machinery and Machine Guarding	252
Machine Guarding	252
Subpart P, Hand and Portable Powered Tools	253
Safe Work Practice for Hand Tools, Power Tools, and Portable Power Equipment	253
Subpart Q, Welding, Cutting, and Brazing	254
Hot Work Permit Procedure	254
Fire Watch Requirements	254
Welding and Cutting Safety	255
Subpart Z, Toxic and Hazardous Substances	259
Benefits of HazCom with GHS	259
Major Changes to the Hazard Communication Standard	260
Hazard Classification	261
Label Changes Under the Revised Hazard Communication Standard	261
Safety Data Sheet Changes Under the Revised Hazard Communication Standard	262
Worker Training	262

Labeling Requirements 262
Hazard Communication Program Audit Items 265
Thought-Provoking Discussion Questions 265
References and Recommended Reading 266

Glossary 271

Appendix. Chemicals Used in Hydraulic Fracturing 287

Index 303

Preface

With regard to gaining U.S. energy independence and a sustainable, reliable source of energy—and just when many of us thought that the United States was energy poor (having reached "peak oil" and thus running out of hydrocarbon supplies or lacking relatively easy access to potential hydrocarbon supplies)—technology is advancing to the point where, for example, natural gas supplies recently thought to be non-existent or too difficult to mine are now being mined and processed and are now available for both industrial and consumer use. This is important not only for U.S. future natural gas supplies but also because natural gas plays a key role in our nation's pursuit of a clean energy future.

Recent advances in drilling technologies (including horizontal drilling and hydraulic fracturing, or fracking) have made it more economical to recover massive supplies of natural gas in the United States. Fracking is not a new technology; it has been around for years. Indeed, hydraulic fracturing or well stimulation, as many in the industry like to call it, is a well-known practice that was developed by Halliburton 60 years ago.

The use of hydraulic fracturing is a double-edged sword. Making up one side of the sword is the fracturing process itself, known as a *frack job*, which involves the pressurized injection of fluids commonly made up of large quantities of water and chemical additives into a geologic formation (e.g., gas-bearing shale). The pressure exceeds the rock strength, and the fluid opens or enlarges fractures in the rock. As the formation is fractured, a propping agent, such as ceramic or sand beads (even peanut and walnut shells have been used), is pumped into the fractures to keep them from closing as the pumping pressure is released. The fracturing fluids (water and chemical additives) are then returned back to the surface. Natural gas will follow the produced wastewater and flow from pores and fractures in the rock into the well for subsequent extraction. Hydraulic fracturing has proven to be a viable way to access vital resources such as natural gas, oil, and geothermal energy. Simply, hydraulic fracturing has helped to expand natural gas production in the United States by unlocking large natural gas supplies in shale and other unconventional formations across the country. The results of hydraulic fracturing are startling, as natural gas production in 2010 reached the highest level in decades. According to estimates by the Energy Information Administration,[*] the United States possesses natural gas resources sufficient to supply the country for more than 110 years. Moreover, the technology has also been used successfully to stimulate water wells, whereby the fluid used is usually pure water (typically water and a chlorine-based disinfectant, such as bleach). Hydraulic fracturing has also been used to remediate waste spills by injecting air, bacteria, or other materials into a subsurface contaminated zone.

The other edge of the sword is occupied by critics of fracking who point out its environmental and public health impacts. A U.S. House of Representatives committee report[†] suggests that hydraulic fracturing raises myriad concerns, including risks to air quality, migration of gases and hydraulic fracturing chemicals to the surface, potential mishandling of wastes (especially produced wastewater), land subsidence (might be causal factor of earthquakes), and, most importantly, contamination of groundwater. Of these, the most significant to date has proven to be the fracking fluids used to fracture rock formations. These contain numerous chemicals that could harm human health and the environment, especially if they enter and contaminate drinking water supplies. Compounding the concerns of Congress, local politicians, and environmentalists is the opposition of many oil and gas companies to publically disclose the chemicals they use.

---

[*] Energy Information Administration, *U.S. Natural Gas Monthly Supply and Disposition Balance*, https://www.eia.gov/dnav/ng/ng_sum_sndm_s1_m.htm, 2017.

[†] USHR, *Chemicals Used in Hydraulic Fracturing*, U.S. House of Representatives Committee on Energy and Commerce, Washington, DC, 2011.

Hydraulic Fracturing Wastewater: Treatment, Reuse, and Disposal provides a fair and balanced discussion and comprehensive overview of the potential risks to drinking water supplies posed by fracking. This book is intended to serve as a reference book for administrators; legal professionals; research engineers; graduate students in chemical, natural gas, petroleum, or mechanical engineering; non-engineering professionals; and the general reader. For readers not familiar with my writings, I must warn you that I converse with you in my traditional conversational style. I do not, and will not, apologize for trying to communicate with the reader. Failure to do otherwise is not an option.

Acronyms and Abbreviations

API	American Petroleum Institute
bbl	Barrel, petroleum (42 gallons)
Bcf	Billion cubic feet
BLM	Bureau of Land Management
BMP	Best Management Practice
Btu	British thermal unit
CAA	Clean Air Act
CBNG	Coal bed natural gas
CEQ	Council on Environmental Quality
CERCLA	Comprehensive Environmental Response, Compensation, and Liability Act
CFR	Code of Federal Regulations
CH_4	Methane
CO	Carbon monoxide
CO_2	Carbon dioxide
COE	Coefficient of oil extraction
CWA	Clean Water Act
DRBC	Delaware River Basin Commission
EIA	Energy Information Administration
ELGs	Effluent Limitation Guidelines
EPA	Environmental Protection Agency
EPCRA	Emergency Planning and Community Right-to-Know Act
FR	*Federal Register*
ft	Foot/feet
FWS	Fish and Wildlife Service
gal	Gallon
GHG	Greenhouse gas
GWPC	Ground Water Protection Council
H_2S	Hydrogen sulfide
HAP	Hazardous air pollutant
HCl	Hydrochloric acid
IOGCC	Interstate Oil and Gas Compact Commission
IR	Infrared
Mcf	1,000,000 cubic feet (also, 1000 cubic feet)
mrem	Millirem
mrem/yr	Millirem per year
MSDS	Material Safety Data Sheet
NEPA	National Environmental Policy Act
NESHAPs	National Emission Standards for Hazardous Air Pollutants
NETL	National Energy Technology Laboratory
NORM	Naturally occurring radioactive material
NO_x	Nitrogen oxides
NPDES	National Pollution Discharge Elimination System
NYDEC	New York State Department of Environmental Conservation
O_3	Ozone
OPA	Oil Pollution Act
OSHA	Occupational Safety and Health Administration
PM	Particulate matter

ppm	Parts per million
RAPPS	Reasonable and Prudent Practices for Stabilization
RCRA	Resource Conservation and Recovery Act
RP	Recommended Practice
RQ	Reportable quantity
SARA	Superfund Amendments and Reauthorization Act
SCF	Standard cubic feet
SDWA	Safe Drinking Water Act
SO_2	Sulfur dioxide
SPCC	Spill Prevention, Control, and Countermeasures
SRBC	Susquehanna River Basin Commission
STRONGER	State Review of Oil and Natural Gas Environmental Regulations, Inc.
SWDA	Solid Waste Disposal Act
Tcf	Trillion cubic feet
TDS	Total dissolved solids
tpy	Tons per year
TRI	Toxic Release Inventory
UIC	Underground Injection Control
USC	United States Code
USDW	Underground source of drinking water
USGS	United States Geological Survey
VOC	Volatile organic compound
WQA	Water Quality Act
yr	Year

Author

Frank R. Spellman, PhD, is a retired adjunct assistant professor of environmental health at Old Dominion University, Norfolk, Virginia, and the author of more than 100 books covering topics ranging from concentrated animal feeding operations (CAFOs) to all areas of environmental science and occupational health. Many of his texts are readily available online, and several have been adopted for classroom use at major universities throughout the United States, Canada, Europe, and Russia; two have been translated into Spanish for South American markets. Dr. Spellman has been cited in more than 850 publications. He serves as a professional expert witness for three law groups and as an incident/accident investigator for the U.S. Department of Justice and a northern Virginia law firm. In addition, he consults on homeland security vulnerability assessments for critical infrastructures, including water/wastewater facilities, and conducts pre-Occupational Safety and Health Administration and Environmental Protection Agency audits throughout the country. Dr. Spellman receives frequent requests to co-author with well-recognized experts in several scientific fields; for example, he is a contributing author to the prestigious text *The Engineering Handbook*, 2nd ed. Dr. Spellman lectures on wastewater treatment, water treatment, and homeland security, as well as on safety topics, throughout the country and teaches water/wastewater operator short courses at Virginia Tech in Blacksburg. In 2011, he traced and documented the ancient water distribution system at Machu Picchu, Peru, and surveyed several drinking water resources in Amazonia, Ecuador. He has also studied and surveyed two separate potable water supplies in the Galapagos Islands, in addition to studying Darwin's finches while there. Dr. Spellman earned a BA in public administration, a BS in business management, an MBA, and both an MS and a PhD in environmental engineering.

Section I

Introduction

1 When an End Was a Beginning

The fascinating problem of the origin of liquid petroleum, with which we must associate natural gas, mineral waxes, and asphaltic materials, is primarily of interest to geologists and chemists, but its solution would have a wider impact, since it would throw some light on certain aspects of the early history of the earth and the first steps in the evolution of living forms.

—Sir Robert Robinson, organic chemist (1886–1975)

END BEFORE THE BEGINNING

It was an end that became a beginning. The end was final—sort of. Life did not end in a wink of an eye nor was it lights-out, totally. In an effort to ensure understanding, it could be said that the end was catastrophic, cataclysmic, calamitous, shattering, shocking, smothering—very smothering—but maybe a better descriptor would be tumultuous. Again, it was final, to a point. Sort of. All of this finality was caused by tectonic stretching, upheaval, land subsidence, and ocean inundation. Although tumultuous in all physical respects, it was a period of time (before its ending) when Earth experienced the first adaptive radiation of terrestrial life. A proliferation of new life forms such as free-sporing vascular plants (i.e., plants that have no pollen or seeds and are dispersed by spores) began to sprout and spread across the land, forming extensive woodlands, jungle-type environments, and forests that covered the continents. During the middle of this period, the Devonian Period (see Figure 1.1), several groups of plants developed leaves and true roots. Toward the end of the Devonian Period, the first seed-bearing plants appeared. The moss forests and bacterial and algal mats of the preceding period were joined by primitive root plants that created the first stable soil and harbored other forms of early life such as arthropods, which had appeared on land much earlier. The arthropods included mites, scorpions, trigonotarbids (an extinct group of arachnid arthropods that resembled spiders but lacked spinnerets to spin webs, thus dooming them to extinction), and myriapods (e.g., millipedes, centipedes). Fish achieved substantial diversity during this time, leading to this period being dubbed the "Age of Fishes." Fish types such as the early jawed placoderms dominated almost every known aquatic environment of the time. The first tetrapods (four-legged, land-living invertebrates), which evolved from lobe-finned fish, appeared in the middle of this period and gave rise to the first amphibians (Garwood and Dunlop, 2014; Garwood and Edgecombe, 2011; Niedzwiedzki, 2010). The ancestors of all tetrapods began adapting to walking on land, their strong pectoral and pelvic fins gradually evolving into legs.

For awhile, this period experienced a boom in ecosystems, both aquatic and terrestrial, the key words here being "for awhile." All booms come to an end eventually, and this was the case for this period of geologic time on Earth. The end was not sudden. No. The end took time—lots of time measured in millions of years. What happened to this beginning that turned it into an ending? Today what occurred is known as Earth's second Great Extinction (the second of five known mass extinctions), but this mass extinction differed from the other four. There was no massive spike in the dying out of species; that is, Mother Nature did not wield a giant axe and sever all life, nor did she send a giant asteroid, meteor, or bolide to end it all. What happened, especially in comparison with the other mass extinctions, can be described as strange, unusual, or even weird. The point is that not all life forms died, but why not? Well, that is a head-scratcher, for sure. Experts have various theories, but the fact is we do not know what we do not know about this and so many other historical events.

What *do* we know? We know that many species died off and that there was no apparent evolutionary speciation, which occurs when species evolve to replace ones that have died out. If you accept this point of view (which some might say is somewhat of an educated guess), the mass extinction

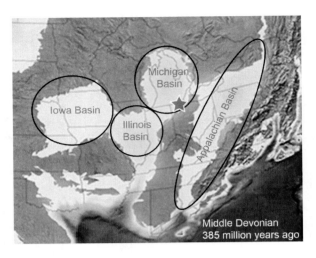

FIGURE 1.1 Middle Devonian Period. (From National Park Service, http://nature.nps.gov/geology/nationalfossilday/paleozoic_silica.cfm.)

may have occurred because life proliferated to the point where there were too many living within too limited of an environment. Not enough food, not enough good soil, and not enough nutrients were available to sustain the population, so the population stagnated and much of it died off.

At this point, the reader probably has a few questions:

Question: What period of time are we talking about?
Answer: The Devonian Period (see Figure 1.1), which occurred during the Paleozoic Era, approximately 416 to 359 million years ago (see Table 1.1).

Question: Why has so much time been spent describing past events that do not seem to have anything to do with today and the subject matter of this text?
Answer: The preceding was a basic introduction to the subject matter this text deals with, and it lays a foundation of understanding for readers and users of this text. Simply leaping into the topic of wastewater produced during hydraulic fracturing operations (fracking) would be one way to present the information, but to the author that would be the easy way out. Moreover, understanding a nebulous but essential characteristic is important for communication.

DID YOU KNOW?

The largest unit of geologic time is an *era*, and each era is divided into smaller time units that are called *periods*. A period of geologic time is divided into *epochs*, which in turn may be subdivided into still smaller units. The geologic column provides a standard by which we can discuss the relative age of rock formations and the rocks and the fossils that they contain. However, these time units are arbitrary and are of unequal duration. We are dealing with relative time, and we cannot be positive about the exact amount of time involved in each unit.

When an End Was a Beginning

TABLE 1.1
Geologic Time Scale

Erathem or Era	System, Subsystem or Period, Subperiod	Series or Epoch
Cenozoic (Age of Recent Life) 65 million years ago to present	**Quaternary** 1.8 million years to the present	**Holocene** 11,477 years ago (±85 yr) to the present Greek *holos* (entire) and *ceno* (new)
		Pleistocene (Great Ice Age) 1.8 million years ago to ~11,477 (±85 yr) ago Greek *pleistos* (most) and *ceno* (new)
	Tertiary 65.5 to 1.8 million years ago	**Pliocene** 5.3 to 1.8 million years ago Greek *pleion* (more) and *ceno* (new)
		Miocene 23.0 to 5.3 million years ago Greek *meion* (less) and *ceno* (new)
		Oligocene 33.9 to 23.0 million years ago Greek *oligos* (little, few) and *ceno* (new)
		Eocene 55.8 to 33.9 million years ago Greek *eos* (dawn) and *ceno* (new)
		Paleocene 65.5 to 58.8 million years ago Greek *palaois* (old) and *ceno* (new)
Mesozoic (Age of Medieval Life) 251.0 to 65.5 million years ago	**Cretaceous** (Age of Dinosaurs) 145.5 to 65.5 million years ago	Late or Upper Early or Lower
	Jurassic 199.6 to 145.5 million years ago	Late or Upper Middle Early or Lower
	Triassic 251.0 in 199.6 million years ago	Late or Upper Middle Early or Lower
Paleozoic (Age of Ancient Life) 542.0 to 251.0 million years ago	**Permian** 299.0 to 251.0 million years ago	Lopingian Guadalupian Cisuralian
	Pennsylvanian (Coal Age) 318.1 to 299.0 million years ago	Late or Upper Middle Early or Lower
	Mississippian 359.2 to 318.1 million years ago	Late or Upper Middle Early or Lower
	Devonian 416.0 to 359.2 million years ago	Late or Upper Middle Early or Lower

(continued)

TABLE 1.1 (continued)
Geologic Time Scale

Erathem or Era	System, Subsystem or Period, Subperiod	Series or Epoch
Paleozoic (continued)	**Silurian**	Pridoli
(Age of Ancient Life)	443.7 to 416.0 million years ago	Ludlow
542.0 to 251.0 million years ago		Wenlock
		Llandovery
	Ordovician	Late or Upper
	488.3 to 443.7 million years ago	Middle
		Early or Lower
	Cambrian	Late or Upper
	542.0 to 488.3 million years ago	Middle
		Early or Lower
Precambrian		
~4 billion years ago to 542.0 million years ago		

Question: Why is the Devonian Period important to this discussion?

Answer: The Devonian Period was a time when life on Earth evolved rapidly. As mentioned, the Devonian Period has been called the "Age of Fishes," and rightfully so because the seas were filled with rapidly diversifying fish that were unusually large and some plated in armor. This was a time when swimming in the seas with such large, armored fish would not have been advised. Their large teeth would certainly have been menacing. Enormous reef ecosystems grew and spread their winding curvature across ocean bottoms. Most of these sinuous reefs were dominated by *Porifera* (sponges), which made the coral (*Cnidaria*) reef systems rather smooth compared to today's reefs. Most important to this text and discussion is that these life-filled seas occupied continents; in fact, periodically massive inland oceans covered large parts of the Appalachian Basin (the geographic area that this discussion will focus on).

Ocean inundation was not the only action going on during the Devonian Period. The Devonian Period was also the age when trees evolved; plants thrived and grew into giants. As the flora spread across the Appalachian Basin (and elsewhere) their root systems broke up the soil and created huge amounts of nutrients, which ran off into the surrounding waters. Flora began to change not only

DID YOU KNOW?

We cannot study shale gas formation without referring to geologic time. Geologic time is often discussed in two forms:

- *Relative time*—A chronostratic arrangement (in geologic column) of geologic events and time periods in their proper order (displayed as the geologic time scale; see Table 1.1). This is done by using stratigraphic techniques (relative age relationships, or vertical/stratigraphic positions).
- *Absolute time*—A chronometric arrangement of numerical ages in millions of years or some other measurement; the time in years since the beginning or end of a period. These are commonly obtained via radiometric dating methods performed on appropriate rock types.

the environment but also the climate. This seems to be evidenced by the speculation that Devonian temperatures fluctuated dramatically, causing sea levels to rise and fall. Scientists theorize that the profusion of flora could have enriched the atmosphere with so much oxygen that temperatures dropped and caused a cooling event whereby ice locked up quantities of seawater. At that point, sea levels would have dropped, thus stranding and killing off many basin species. There were also several cycles of greenhouse conditions that increased ocean acidity, again killing off aquatic life-forms. Many of the resident life-forms in the basin that managed to survive the increased acidity ultimately became the victims of invasive species.

> **Question:** The Appalachian Basin and what might have occurred there in the Devonian Period are interesting, but what is this discussion about it leading to?
>
> **Answer:** The discussion to this point serves as an introduction to the Marcellus Formation (or simply the Marcellus Shale) which is a Middle Devonian Period unit of marine sedimentary rock found in eastern North America. It extends throughout the Appalachian Basin. The Marcellus Shale contains largely untapped natural gas reserves and is a hub for hydraulic fracturing operations to recover the gas for use by the high-demand markets nearby. And, because the shale gas is also near East Coast seaports, it is energy that is highly exportable.

The bottom line: The Devonian Period is characterized by its ending—its mass extinction event. However, because the beginning of oil and gas shale formation took place within areas such as the Appalachian Basin, it can be said that the Devonian Period was an end that was also a beginning.

NATURE'S ALCHEMY

When and if we think about alchemists, we might envision some dark, ancient laboratory with huge black pots boiling and some old, long-bearded codger pouring one goopy-looking mixture into another. The goals of such an operation might be to transform some base metal into a noble one such as gold, to create a steaming potion that when swallowed or topically applied will be a fountain of youth, to devise a magic formula that will cure any and all diseases, or to develop a universal solvent with the power to dissolve any substance.*

Mother Nature could do all of the above if she were persuaded to do so. She has provided cures for diseases and perhaps even the so-called fountain of youth, if we could only discover it. These are all out there for us to find; we just need to keep on keeping on in our scientific research and other pursuits. Every problem has a solution; it is just a matter of finding that solution.

Let's back up and talk about nature's alchemy as related to the Marcellus Shale formation. Marcellus Shale developed from the deposition and later compression of minute rock particles and organic matter at the bottom of a sea during the Middle Devonian era, about 383 to 392 million years ago. To understand the formation of shale gas, imagine a pressure cooker, because what occurred inside the Marcellus Shale formation is somewhat similar to pressure cooking. During the Middle Devonian, sediments eroding from the Acadian Mountains (eastern edge of the Appalachian Basin sea region) were washed down into the Catskill Delta. Coarser-grained sediments, including sand and gravel-sized particles, quickly settled near the shore. The sand and clay-size (finer-grained) fragments flowed as a slow underwater landslide, depositing and accumulating within the deepest part of the Appalachian Basin. Added to these fine rock fragments were organic materials such as algae and other aquatic microorganism. Nature's pressure cooker now held the right mix of ingredients, and with the reduction in oxygen level the organics did not decompose, instead becoming intermingled with the mineral fragments as a mucky ooze or mud flysch (a series of sedimentary layers). All this time, the pressure cooker was not only adding megatons of weight, translated to pressure, but was also increasing the temperature level. With time the pressure increased as the shale formation

---

* Water is just such a substance; given enough exposure and time, nothing on Earth is safe from it (Linden, 2003).

subsided, becoming buried deeper in the Earth and covered with thousands of additional sediment layers. Within the pressure cooker environment, temperatures ranged from 60°C to 100°C; thus, the perfect temperature range, proper pressure, and proper ingredients were in place for nature's alchemy to transform the entrapped shale and its contents: shale gas and oil. Interestingly, all of this deposition and the increases in pressure and temperature occurred at a time when geologic forces were causing the Appalachian Basin to become deeper, trapping shale contents and producing black shale, as the mountains rose up. This process continued for millions of years. This process continues today.

MINING SHALE GAS, A DOUBLE-EDGED SWORD

Many eons ago, Mother Nature, the only true alchemist, waved her magic wand and fashioned rock fragments, minerals, organic material, and we can only guess what else into an oozy, muddy mass that initially had some exposure to oxygen but then later no exposure to oxygen. She then increased the pressure-cooker pressure and temperature and eventually produced the end products of shale gas and shale oil. Later, humans discovered the gas and oil buried in the shale formations and developed technology enabling them to tap the deep, difficult-to-mine reserves of shale gas and shale oil. Improvements in hydraulic fracturing (horizontal drilling) techniques enabled shale gas and shale oil miners to obtain even more of the product they sought. Today hydraulic fracturing for shale gas and shale oil is not only ongoing in practice but also a booming industry.

Hydraulic fracturing of oil and gas shale in the United States has provided many benefits to the local regions where it is mined and to the country itself; however, as with all good things in life, there are drawbacks. It is difficult to characterize the process of hydraulic fracturing in strictly good, bad, and ugly terms. Instead, it is probably better to characterize hydraulic fracturing and its impact as having both a good side and bad side; in other words, hydraulic fracturing is the proverbial double-edged sword—something that has both advantages and disadvantages.

On the good edge of this double-edged sword is the fact that burning the resultant shale gas emits less carbon dioxide than does burning coal or even traditionally extracted natural gas. The smaller amount of carbon dioxide emitted means that the impact on global climate change is lessened. From the United States' point of view fracking also advances energy independence, and from a local perspective fracking is a direct benefit to local, county, and state economies because of the resulting growth in employment opportunities and an accompanying increase in disposable income for investment and other uses.

On the bad edge of this double-edged sword are a number of problems related to health and environmental impacts. For example, although burning natural gas from shale gas formations is less offensive to the environment than the burning of other fossil fuels, the drilling process unintentionally releases fugitive methane, a potent greenhouse gas, into the environment. Another problem is that fracking is a relatively unregulated industry. This means there is less pressure for operators to follow environmentally friendly protocols. Particularly alarming to local officials and residents is the localized environmental contamination caused by fracking operators who use proprietary slurries that can include hydrochloric acids, ethylene glycol, aluminum phosphate, and 2-butoxyethanol, among other chemicals, many of them listed in the appendix. These chemicals and produced waters (the term *produced wastewater*, or just *wastewater*, is emphasized throughout this text) unsurprisingly contribute to pollution of air, soil, groundwater, and local streams. As the title of this book indicates, the focus herein is on wastewater produced by hydraulic fracturing and its treatment, reuse, and disposal.

KEY CONCEPTS AND DEFINITIONS

Every branch of science, every profession, every engineering process has its own key concepts and language for communication. The topics of shale oil mining, hydraulic fracturing, and produced wastewater are no different. To work even at the edge of oil shale fracking, it is necessary to acquire a fundamental knowledge of the key concepts and vocabulary of the processes involved. As Voltaire said, "If you wish to converse with me, define your terms." This section introduces the key concepts and defines the terms used by fracking practitioners. Although an extensive glossary can be found at the end of the book, many concepts and terms are presented early in the text so readers can become familiar with them before the text addresses issues relevant to these concepts and terms. Fracking engineers and students of fracking should understand these concepts and terms; otherwise, it will be difficult (if not impossible) to practice or understand fracking, shale gas production, and the fate of produced wastewater. Hydraulic fracturing and shale gas drilling have an extensive and unique terminology that is generally well defined, but a few terms not only are poorly defined but are also defined from different and conflicting points of view.

Air quality—A measure of the amount of pollutants emitted into the atmosphere and the dispersion potential of an area to dilute those pollutants.

Aquifer—A body of rock that is sufficiently permeable to conduct groundwater and to yield economically significant quantities of water to wells and springs.

Basin—A closed geologic structure in which the beds dip toward a center location; the youngest rocks are at the center of a basin and are partly or completely ringed by progressively older rocks.

Bcf (billion cubic feet)—A gas measurement equal to 1,000,000,000 cubic feet.

Biogenic gas—Natural gas produced by living organisms or biological processes.

Btu (British thermal unit)—The amount of energy required to heat 1 pound of water by 1°F.

Casing—Steel piping positioned in a wellbore and cemented in place to prevent the soil or rock from caving in. It also serves to isolate fluids, such as water, gas, and oil, from the surrounding geologic formations.

Coalbed methane/coalbed natural gas (CBM/CBNG)—A clean-burning natural gas found deep inside and around coal seams. The gas has an affinity to coal and is held in place by pressure from groundwater. CBNG is produced by drilling a wellbore into the coal seams, pumping out large volumes of groundwater to reduce the hydrostatic pressure, and allowing the gas to dissociate from the coal and flow to the surface.

Completion—The activities and methods to prepare a well for production and following drilling include installation of equipment for production from a gas well.

Corridor—A strip of land through which one or more existing or potential utilities may be co-located.

Directional drilling—The technique of drilling at an angle from a surface location to reach a target formation not located directly underneath the well pad.

Disposal well—A well that injects produced water into an underground formation for disposal.

Drill rig—The mast, draw works, and attendant surface equipment of a drilling or workover unit.

Emission—Air pollution discharged into the atmosphere, usually specified by mass per unit time.

Endangered species—Species of plants or animals classified by the Secretary of the Interior or the Secretary of Commerce as endangered pursuant to Section 4 of the Endangered Species Act of 1973, as amended.

Exploration—The process of identifying a potential subsurface geologic target formation and the active drilling of a borehole designed to assess the natural gas or oil.

Flow line—A small-diameter pipeline that generally connects a well to the initial processing facility.

Flowback—The fracture fluids that return to the surface after a hydraulic fracture is completed; water used as a pressurized fluid during hydraulic fracturing that returns to the surface via the well (this occurs after the fracturing procedure is completed and pressure is released).

Formation (geologic)—A rock body distinguishable from other rock bodies and useful for mapping or description. Formations may be combined into groups or subdivided into members.

DID YOU KNOW?

One well can be fracked 10 or more times, and there can be up to 30 wells on one pad. An estimated 50% to 60% of the fracking fluid is returned to the surface during well completion and subsequent production, bringing with it toxic gases, liquids, and solid materials that are naturally present in underground gas deposits. Under some circumstances, none of the injected fluid is recovered (OGC, 2001).

Frack—Term for hydraulic fracturing, as adapted by the petroleum industry.

Fracturing fluids—A mixture of water and additives used to hydraulically induce cracks in the target formation.

Groundwater—Subsurface water that is in the zone of saturation and is the source of water for wells, seepage, and springs. The top surface of the groundwater is the *water table.*

Habitat—The area in which a particular species lives. In wildlife management, the major elements of a habitat are considered to be food, water, cover, breeding space, and living space.

Horizontal drilling—A drilling procedure in which the wellbore is drilled vertically to a kick-off depth above the target formation and then angled through a wide 90° arc such that the producing portion of the well extends horizontally through the target formation.

Hydraulic fracturing—Injecting fracturing fluids into the target formation at a force exceeding the parting pressure of the rock, thus inducing a network of fracture through which all oil or natural gas can flow to the wellbore.

Hydrostatic pressure—The pressure exerted by a fluid at rest due to its inherent physical properties and the amount of pressure being exerted on it from outside forces.

Injection well—A well used to inject fluids into an underground formation for either enhanced recovery or disposal.

Lease—A legal document that conveys to an operator the right to drill for oil and gas. Also, the tract of land on which a lease has been obtained where producing wells and production equipment are located.

Mcf—A natural gas measurement unit for either 1000 cubic feet or 1 million cubic feet.

MMcf—A natural gas measurement unit for 1 million cubic feet.

NORM (naturally occurring radioactive materials)—Includes naturally occurring uranium-235 and daughter products such as radium and radon.

Oil-equivalent gas (OEG)—The volume of natural gas needed to generate the equivalent amount of heat as a barrel of crude oil. An amount of approximately 6000 cubic feet of natural gas is equivalent to one barrel of crude oil.

Particulate matter (PM)—A small particle of solid or liquid matter (e.g., soot, dust, mist). PM_{10} refers to particulate matter having a size diameter of less than 10 micrometers (μm) and $PM_{2.5}$ to particulate matter less than 2.5 μm in diameter.

Permeability/porosity—Refers to the capacity of a rock to transmit a fluid, which is dependent on the size and shape of the pores and interconnecting pore throats. A rock may have significant porosity (many microscopic pores) but low permeability if the pores are not interconnected (see Figures 1.2 and 1.3). Permeability may also exist or be enhanced through fractures that connect the pores. Although shales may be as porous as other sedimentary rocks, their extremely small pore sizes make them relatively impermeable to gas flow, unless natural or artificial fractures occur. Porosity is the percent volume of the rock that is not occupied by solids. Again, permeability is a measure of the ease with which a fluid can flow through a rock—the greater the permeability of a rock, the easier it is for the fluid to flow through the rock. Permeability is measured in units of *darcies* (D) or *millidarcies* (mD). A darcy is the permeability that will allow a flow of 1 cm^3 per second of a fluid

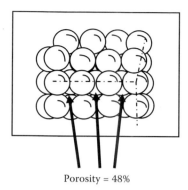

FIGURE 1.2 When spheres are stacked within a box, the empty space between each sphere equals 48% of the total volume. (Adapted from Raymond, M.D. and Leffler, W.L., *Oil and Gas Production in Nontechnical Language*, PennWell Corporation, Tulsa, OK, 2006.)

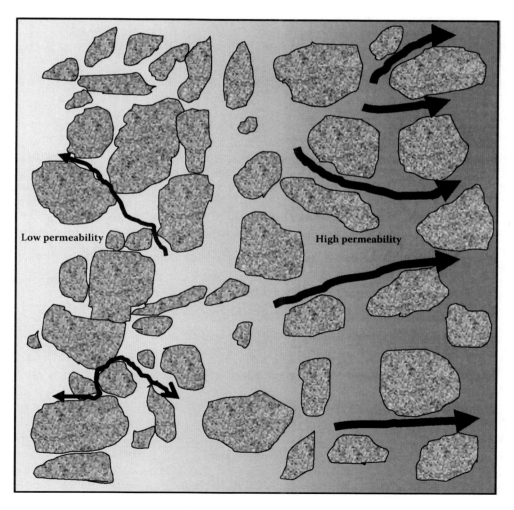

FIGURE 1.3 High- and low-permeability rock. (Adapted from Raymond, M.D. and Leffler, W.L., *Oil and Gas Production in Nontechnical Language*, PennWell Corporation, Tulsa, OK, 2006.)

DID YOU KNOW?

In 1850, Henri Darcy, the city water engineer for Dijon, France, received endless complaints about the filthy water coming from the city mains. Darcy installed sand filters to purify the system. While purifying water in the mains, Darcy experimented with fluid flow through porous materials and developed equations to describe it. For his efforts, Darcy earned immortality via the universally used *darcy*, a unit measuring how easily fluid flows through porous media.

with 1-centipoise (cP) viscosity (resistance to flow) through a distance of 1 cm through an area of 1 cm² under a differential pressure of 1 atmosphere (atm). In naturally occurring materials, permeability values range over many orders of magnitude.

Primacy—A right that can be granted to states by the federal government that allows state agencies to implement programs with federal oversight. Usually, the states develop their own set of regulations. By statute, states may adopt their own standards; however, these must be at least as protective as the federal standards they replace, and may be even more protective in order to address local conditions. When these state programs have been approved by the relevant federal agency (usually the USEPA), the state then has primary jurisdiction.

Produced water (wastewater)—Naturally occurring water found in shale formations; it generally flows to the surface during the entire lifespan of a well, often along with natural gas. Produced water and flowback from natural gas extraction may be reused in fracking operations, disposed of through underground injection, discharged to surface waters as long as it does not degrade water quality standards, or transferred to a treatment facility, if necessary, for processing and subsequent discharge into a receiving water body in compliance with effluent limits.

Propping agents/proppant—Silica sand or other particles pumped into a formation during a hydraulic fracturing operation to keep fractures open and maintain permeability.

Proved reserves—The portion of recoverable resources that is demonstrated by actual production or conclusive formation tests to be technically, economically, and legally producible under existing economic and operating conditions.

Reclamation—Rehabilitation of a disturbed area to make it acceptable for designated uses. This normally involves regrading, replacement of topsoil, re-vegetation, and other work necessary to restore it.

Setback—The distance that must be maintained between a well or other specified equipment and any protected structure or feature.

Shale gas—Natural gas produced from low-permeability shale formations.

Slickwater—Water-based fracking fluid mixed with friction-reducing agents, commonly potassium chloride.

Split estate—Condition that exists when the surface rights and mineral rights of a given area are owned by different persons or entities; also referred to as *severed estate*.

DID YOU KNOW?

Natural gas is generally priced and sold in units of 1000 cubic feet (abbreviated Mcf, using the Roman numeral for 1000). Units of a trillion cubic feet (Tcf) are often used to measure large quantities, such as resources or reserves in the ground, or annual nation energy consumption. A Tcf is 1 billion Mcf, which is enough natural gas to heat 15 million homes for one year, generate 100 billion kilowatt-hours of electricity, or fuel 12 million natural gas-fired vehicles for one year.

Stimulation—Any of several processes used to enhance near-wellbore permeability and reservoir permeability.

Stipulation—A condition or requirement attached to a lease or contract, usually dealing with protection of the environment or recovery of a mineral.

Sulfur dioxide (SO_2)—A colorless gas formed when sulfur oxidizes, often as a result of burning trace amounts of sulfur in fossil fuels.

Tcf—A natural gas measurement unit for 1 trillion cubic feet.

Technically recoverable resources—The total amount of a resource, discovered and undiscovered, that is thought to be recoverable with available technology, regardless of economics.

Thermogenic gas—Natural gas that is formed by the combined forces of high pressure and temperature found deep within the Earth's crust, resulting in natural cracking of the organic matter in the source rock matrix.

Thixotrophy—The property of a gel to become fluid when disturbed (as by shaking).

Threatened and endangered species—Plant or animal species that have been designated as being in danger of extinction.

Tight gas—Natural gas trapped in a hardrock, sandstone, or limestone formation that is relatively impermeable. The rock layers that hold the gas are very dense, preventing easy flow.

Tight sand—A very low or no permeability sandstone or carbonate.

Total dissolved solids (TDS)—The dry weight of dissolved material, organic and inorganic, contained in water; usually expressed in parts per million.

Underground Injection Control (UIC) Program—Program administered by the USEPA, primacy state, or Indian tribe under the Safe Drinking Water Act to ensure that subsurface emplacement of fluids does not endanger underground sources of drinking water.

Underground source of drinking water (USDW)—Defined in 40 CFR 144.3 as an aquifer or its portion: (a)(1) Which supplies any public water system; or (2) Which contains a sufficient quantity of groundwater to supply a public water system; and (i) Currently supplies drinking water for human consumption; or (ii) Contains fewer than 10,000 mg/L total dissolved solids; and (b) Which is not an exempted aquifer.

Wastewater (produced wastewater)—Term used by experts in the field to refer to fracked wastewater. It is the author's view that wastewater, contaminated water, and produced water are the same thing, and the terms are interchangeable. Think about it. Unless contaminated by natural processes, all wastewater is produced. All produced wastewater is human-produced wastewater.

Let's take a closer look at produced wastewater. No matter the label—produced, contaminated, dirty, filthy, grimy, greasy, spoiled, soiled, fouled, murky, polluted, sullied, slimy, skuzzy, yucky, unsanitary—to really understand wastewater it is necessary to understand the basic science involved in its production and to have some real understanding of what wastewater and produced water really are. Translating basic scientific principles to provide a basis for understanding the science involved in the makeup of wastewater would be an exercise without merit unless the reader understands what produced wastewater really is. The author's experience with college-level students studying environmental topics that address water and wastewater sciences and the operation and treatment of wastewater treatment facilities is that when they are asked to define wastewater they sort of stumble a bit. Consider some of their responses:

"Well, … you know … wastewater is poop … and the toothpaste and mouthwash we spit … and pee … all mixed together."

"Well, wastewater is whatever we flush down the toilet."

"It's poop and toilet paper and my leftover grease and kitchen waste that I flush."

"Well, if we dump it into a toilet and flush it or wash it down the drain in a sink, it's wastewater."

"Wastewater? Anything that goes down the toilet."

These responses are not unusual; they are typical responses from people who do not understand the science of wastewater. The mindset of flush it, drain it, or pour it out reflects an "out of sight, out of mind" mentality. It's just waste, right? Well, to a point it is.

So, what *is* wastewater? Simply, wastewater is any water that has been adversely affected in quality by an anthropogenic (i.e., human) influence. Municipal wastewater is usually conveyed in a combined sewer or sanitary sewer and treated at a wastewater treatment facility. Treated wastewater is discharged into receiving water via an effluent sewer (an outfall). In areas where centralized sewer systems are not accessible, wastewater is typically discharged to onsite systems. These onsite systems are typically comprised of a septic tank, drain field, outhouse, and optionally an onsite treatment unit. When referring to the management of wastewater, human excreta, solid waste, and stormwater (drainage), the all-encompassing term *sanitation* is generally used.

It is not unusual to hear people speak of sewage and wastewater as being one and the same thing. This is a misconception. Sewage is the subset of wastewater that is contaminated with feces or urine. Sewage includes domestic, municipal, or industrial liquid waste products that are disposed of, usually via a sanitary or combined sewer (pipe network) but sometimes in a cesspool or cesspit emptier. A *cesspool* is an underground holding tank (sealed at the bottom) or a soak pit (not sealed). It is used for the temporary collection and storage of feces, excreta, or fecal sludge as part of an onsite sanitation system and has similarities with septic tanks. Typically, it is a deep cylindrical chamber dug into the ground (like a hand-dug water well), with a diameter of approximately 1 meter and a depth of 2 to 3 meters.

Another term that is sometimes intermixed with wastewater and sewage is *sewerage*, which is the physical infrastructure, including pipes, pumps, screens, channels, and so forth, used to convey sewage from its origin to the point of treatment or disposal. With the exception of onsite septic systems, sewerage is found in all types of sewage treatment.

Getting back to wastewater, some of its sources include the following:

- Human waste
- Cesspool leakage
- Septic tank discharge
- Wastewater treatment plant discharge
- Washing water
- Rainfall collected (see Figure 1.4)
- Groundwater infiltration
- Surplus manufactured liquids
- Urban runoff
- Seawater ingress
- River water ingress
- Illegal ingress of pesticides, used oils, etc.
- Highway drainage
- Stormwater
- Industrial site drainage
- Organic and biodegradable waste
- pH waste
- Toxic waste
- Emulsion waste
- Agricultural drainage
- Hydraulic fracturing
- Produced wastewater (natural gas production)

The actual composition of wastewater varies widely; For example, it may contain the following:

FIGURE 1.4 Household rain barrel collection system. (Photographs by author.)

- Water (more than 95% used for flushing)
- Pathogens
- Non-pathogenic bacteria
- Organic particles
- Soluble organic material (e.g., urea, sugars, proteins, drugs)
- Inorganic particles, such as sand, grit, and metals
- Small fish and macroinvertebrates
- Soluble inorganic material, road salt, sea salt, cyanic, hydrogen sulfide, etc.
- Macro-solids, such as $100 bills (flushed when police knock on the door), bags of cocaine and heroin, guns, dead animals, plants, fetuses, sanitary napkins, etc.

- Gases such as methane, carbon dioxide, or hydrogen sulfide
- Emulsions such as adhesives, mayonnaise, hair colorants, or paints
- Pharmaceuticals and hormones
- Personal care products such as perfumes, body lotions, shampoos, tanning lotions, or lipstick
- Pesticides, poisons, herbicides, etc.

Produced wastewater is a complex mixture of dissolved and particulate organic and inorganic chemicals. The actual physical and chemical properties of produced wastewater vary widely depending on the geological age, depth, and geochemistry of the hydrocarbon-bearing formation, as well as the chemical composition of the oil and gas phases in the reservoir and process chemicals added during production (Neff et al., 2011).

Water quality—The chemical, physical, and biological characteristics of water with respect to its suitability for a particular use.

Watershed—All lands that are enclosed by a continuous hydrologic drainage divide and lay upslope from a specified point on a stream.

Whipstock—A wedge-shaped piece of metal placed downhole to deflect the drill bit.

Workover—To perform one or more remedial operations on a producing or injection well to increase production; deepening, plugging back, pulling, and resetting the line are examples of worker operations.

PURPOSE OF TEXT

Experience has shown that writing is one thing and communicating with the reader is another. Many times writing is easy but communicating is not. The fact is that scientific descriptions often are meaningless to those not trained in the language of science. Therefore, the topics in this book move naturally from the simple to the complex, and the science and engineering are presented in the context of their application; readers not versed in science and engineering can absorb the language and basic premises while reading about water use, treatment, reuse, and disposal in hydraulic fracturing. No one has to wade through a tedious exposé of science and engineering nomenclature. The bottom line is that this is a somewhat unorthodox science and engineering book. Although many science and engineering books can be frustrating, with the context being too facile and watered down to be interesting, this is a narrative science and engineering book presented in conversational style. I am confident that upon completion of the reader's journey through *Hydraulic Fracturing Wastewater: Treatment, Reuse, and Disposal* the reader should have a better feel for what produced wastewater is and what the related issues are. The issues are real.

THOUGHT-PROVOKING QUESTIONS

1. Is the environment just one of those things that will have to take a few bruises while we explore for and extract the hydrocarbons that we need to sustain our way of life?
2. Are environmental concerns real or just talking points for radical groups?
3. Can we always protect the environment from human-caused damage?
4. Should we focus our attention on renewable sources of energy instead of drilling for more oil and natural gas?
5. Is natural gas really as clean as the experts say it is?
6. Does fracking contribute to global warming? How?
7. Who do you think is more responsible for pollution: individuals, companies, or the government?
8. Do you think it is important for frackers to understand the language of fracking? Explain.

REFERENCES AND RECOMMENDED READING

Allen, D.T. et al. (2013). Measurements of methane emissions at natural gas production sites in the United States. *PNAS*, 110(44):17768–17773.

De Witt, Jr., W. et al. (1993). *Principal Oil and Gas Plays in the Appalachian Basin (Province 131): Middle Eocene Intrusive Igneous Rocks of the Central Appalachian Valley and Ridge Province—Setting, Chemistry, and Implications for Crucial Structure*, USGS Bulletin 1839. U.S. Geological Survey, Washington, DC.

Garwood, R. and Dunlop, J. (2014). The walking dead: blender as a tool for paleontologists with a case study on extinct arachnids. *Journal of Paleontology*, 88(4):735–746.

Garwood, R. and Edgecombe, G. (2011). Early terrestrial animals, evolution, and uncertainty. *Evolution: Education & Outreach*, 4(3):489–501.

Howarth, R.W. and Atkinson, D.R. (2011). *Assessment of the Greenhouse Gas Footprint of Natural Gas from Shale Formations Obtained by High-Volume, Slick-Water Hydraulic Fracturing*. Department of Ecology and Evolutionary Biology, Cornell University, Ithaca, NY.

Linden, S.J. (2003). *The Alchemy Reader: The Hermes Trismegistus to Isaac Newton*. Cambridge University Press, Cambridge, U.K.

Neff, J., Lee, K., and DeBlois, E.M. (2011). Produced water: overview of composition, fates, and effects. In *Produced Water: Environmental Risks and Advances in Mitigation Technologies* (Lee, K. and Neff, J., Eds.), pp. 3–54. Springer, New York.

Niedzwiedzki, G. (2010). Tetrapod trackways from early Middle Devonian period of Poland. *Nature*, 463(7277):43–48.

OGC. (2001). *Fracturing (Fracing) and Disposal of Fluids*, Information Sheet 15. Oil and Gas Commission, Canadian Centre for Energy Information (http://www.bctwa.org/Frk-BCOil&GasCom-Fracking.pdf).

Raymond, M.D. and Leffler, W.L. (2006). *Oil and Gas Production in Nontechnical Language*. PennWell Corporation, Tulsa, OK.

Sheeran, T.J. (2001). Ohio shale drilling spurs jobs hopes in Rust Belt. Associated Press, November 28.

USEPA. (2011). *Final Guidance to Protect Water Quality in Appalachian Communities from Impacts of Mountaintop Mining*. U.S. Environmental Protection Agency, Washington, DC (https://yosemite.epa.gov/opa/admpress.nsf/3881d73f4d4aaa0b85257359003f5348/1dabfc17944974d4852578d400561a13!OpenDocument).

Section II

Fracking Process and Produced Wastewater

2 Hydraulic Fracking: The Process

> The Gas Era is coming, and the landscape north and west of [New York] will inevitably be transformed as a result. When the valves start opening next year, a lot of poor farm folk may become Texas rich.
>
> **France (2008)**

SHALE GAS DRILLING DEVELOPMENT TECHNOLOGY*

When initially developed, conventional petroleum reservoirs depend on the pressure of their gas cap and oil-dissolved gas to lift the oil to the surface (i.e., *gas drive*). Water trapping the petroleum from below also exerts an upward hydraulic pressure (i.e., *water drive*). The combined pressure in petroleum reservoirs produced by the natural gas and water drives is known as the *conventional drive*. As a reservoir's production declines, lifting further petroleum to the surface, like the lifting of water, requires pumping, or *artificial lift*. In the late 1940s, drilling companies began inducing hydraulic pressure in wells to fracture the producing formation. This stimulated further production by effectively increasing the contact of a well with a formation. Advances in directional drilling technology have allowed wells to deviate from nearly vertical to extend horizontally into the reservoir formation, which further increases contact of a well with the reservoir. Directional drilling technology also enables drilling a number of wells from a single well pad, thus cutting costs while reducing environmental disturbance. Combining hydraulic fracturing with directional drilling has opened up the production of tight (less permeable) petroleum and natural gas reservoirs, particularly unconventional gas shales such as the Marcellus Shale formation.

DRILLING

From the art of divining, dowsing, and witching for water (still practiced all over the world) to the use of scientific exploration and discovery methods and innovative drilling technology, locating water and petroleum products and drilling for them have come a long way, evolving from an art to a science. Originally, drillers used cable tool rigs and percussion bits. The drill operator would raise the bit and release it to pulverize the sediment. From time to time, the driller would stop to "muck out" the pulverized rock cuttings to advance the well. Though time-consuming, this method was simple and required minimal labor. Some drillers still use this method for water wells and even some shallow gas wells. The introduction of rotary drill rigs at the beginning of the 20th century marked a big advance in drilling, particularly with development of the tricone rotary bit. (Howard Hughes, Jr., of the Hughes Tool Company, developed the modern tricone rotary bit. His father, Howard Robert Hughes, Sr., had invented the bit's ancestor, a two-cone rotary bit.) This method, as the name implies, uses a weighted rotating bit to penetrate the sediment (see Figure 2.1). Following is an explanation of the components shown in the figure:

---

* This section is adapted from Andrews, A., Ed., *Unconventional Gas Shales: Development, Technology, and Policy Issues*, CRS Report for Congress, Congressional Research Service, Washington, DC, 2009.

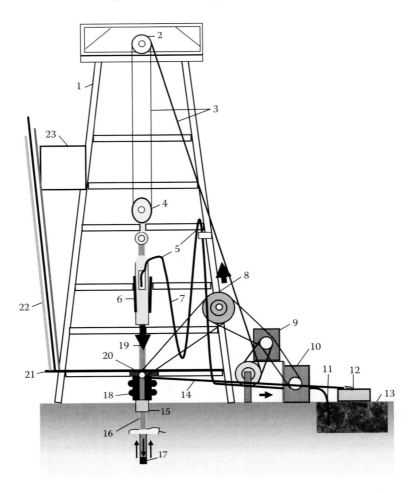

FIGURE 2.1 Simple diagram of drilling rig and its basic components (see text for details).

1. *Derrick* is the support structure for the equipment used to lower and raise the drill string into and out of the wellbore.
2. *Crown block* is the stationary end of the *block and tackle*.
3. *Drill line* is thick, stranded metal cable threaded through the two blocks (traveling and crown) to raise and lower the drill string.
4. *Traveling block* is the moving end of the block and tackle. Together, they give a significant mechanical advantage for lifting.
5. *Goose-neck* is a thick metal elbow connected to the swivel and standpipe that supports the weight of and provides a downward angle for the kelly hose to hang from.

DID YOU KNOW?

Water, as with all liquids, is compressible to a very small degree. When hydrocarbons are depleted, the reduction in pressure in the reservoir causes the water to expand slightly. Although this expansion is quite small, if the aquifer is large enough this will translate into a large increase in volume, which will push up (water-drive) on the hydrocarbons, maintaining pressure.

Hydraulic Fracking: The Process

6. *Swivel* is the top end of the kelly that allows the rotation of the drill string without twisting the block.
7. *Kelly hose* is a flexible, high-pressure hose that connects the standpipe to the kelly (or, more specifically, to the gooseneck on the swivel above the kelly) and allows free vertical movement of the kelly while facilitating the flow of drilling fluid through the system and down the drill string.
8. *Standpipe* is a thick metal pipe, situated vertically along the derrick, that facilitates the flow of drilling fluid and has attached to it and supports one end of the kelly hose.
9. *Power source motor* powers the drill line.
10. *Mud pump* is a reciprocal type of pump used to circulate drilling fluid through the system.
11. *Suction line (mud pump)* is an intake line for the mud pump to draw drilling fluid from the mud tanks.
12. *Shale shaker* separates drill cuttings from the drilling fluid before it is pumped back down the wellbore.
13. *Mud tanks*, often called *mud pits*, store drilling fluid until it is required down the wellbore.
14. *Flow line* is a large-diameter pipe that is attached to the bell nipple and extends to the shale shakers to facilitate the flow of drilling fluid back to the mud tanks.
15. *Casing head* or *well head* is a large metal flange welded or screwed onto the top of the conductor pipe (drive pipe) or the casing; it is used to bolt surface equipment such as blowout preventers or Christmas tree assemblies.
16. *Drill string* is an assembled collection of drill pipe, heavyweight drill pipe, drill collars, and any of a whole assortment of tools, connected together and run into the wellbore to facilitate drilling a well.
17. *Drill bit* is a device attached to the end of the drill string that breaks apart the rock being drilled. It contains jets through which the drilling fluid exits.
18. *Blowout preventers* are devices installed at the well head to prevent fluids and gases from unintentionally escaping from the wellbore.
19. *Kelly drive* is square-, hexagonal-, or octagonal-shaped tubing that is inserted through and is an integral part of the rotary table that moves freely vertically while the rotary table turns it.
20. *Rotary table* rotates, along with its constituent parts, the kelly and kelly bushing, the drill string, and the attached tools and bit.
21. *Pipe rack* is part of the drill floor where the stands of drill pipe stand upright. It is typically made of a metal frame structure with large wooden beams situated within it. The wood helps to protect the end of the drill pipe.
22. *Stand* is a section of two or three joints of drill pipe connected together and standing upright in the derrick.
23. *Monkey board* is the structure used to support the top end of the stands of drill pipe vertically situated in the derrick.

The key to a rotary drill's speed is the relative ease of adding new sections of drill pipe (or drill string) while the drill bit continues turning. Drilling mud (fluid) circulates down through the center of the hollow drill pipe and up through the wellbore to lift the drill cuttings to the surface. Modern drill bits are studded with industrial diamonds to make them abrasive enough to grind through any rock type. From time to time, the drill string must be removed (a process termed *tripping*) to replace dulled drill bits.

To function properly, drilling fluids must lubricate the drill bit, keep the wellbore from collapsing, and remove cuttings. The main functions of drilling mud are as follows (Schlumberger, 2012):

- Remove cuttings from well.
- Suspend and release cuttings.
- Control formation pressures.

DID YOU KNOW?

The *reciprocating pump* (or *piston pump*) is one type of positive displacement pump. This pump works just like the piston in an automobile engine—on the intake stroke, the intake valve opens, filling the cylinder with liquid. As the piston reverses direction, the intake valve is pushed closed and the discharge valve is pushed open; the liquid is pushed into the discharge pipe. With the next reversal of the piston, the discharge valve is pulled closed and the intake valve is pulled open, and the cycle then repeats. A piston pump is usually equipped with an electric motor and a gear and cam system that drives a plunger connected to the piston. Just like an automobile engine piston, the piston must have packing rings to prevent leakage and must be lubricated to reduce friction. Because the piston is in contact with the liquid being pumped, only good-grade lubricants can be used when pumping materials that will be added to drinking water. The valves must be replaced periodically as well.

- Seal permeable formations.
- Maintain wellbore stability.
- Minimize formation damage.
- Cool, lubricate, and support the bit and drilling assembly.
- Transmit hydraulic energy to tools and bit.
- Ensure adequate formation evaluation.
- Control corrosion.
- Facilitate cementing and completion.
- Minimize impact on environment.

The weight of the mud column prevents a blowout from occurring when high-pressure reservoir fluids are encountered. Drillers base the mud's composition on natural bentonite clay, a thixotropic material that is solid (viscous) when still and fluid (less viscous) when shaken, agitated, or otherwise disturbed. This essential rheological property keeps the drill cuttings suspended in the mud. The chemistry and density of the mud must be carefully monitored and adjusted as the drilling deepens (e.g., adding a barium compound increases mud density). *Mud pits*, excavated adjacent to the drill rig, provide a reservoir for mixing and holding the mud. The mud pits also serve as settling ponds for the cuttings. At the completion of drilling, the mud may be recycled at another drilling operation, but the cuttings will be disposed of into the pit. Several environmental concerns over drilling stem from the hazardous composition of the drilling mud and cuttings and from the potential for mud pits to overflow and contaminate surface water.

The most recent advance in drilling is the ability to direct the drill bit beyond the region immediately beneath the drill rig. Early directional drilling involved placing a steel wedge downhole (whipstock) that deflected the drill toward the desired target, but this approach lacked control and was time consuming. Advances such as steerable downhole drill motors that operated on the hydraulic pressure of the circulating drilling mud offered improved directional control; however, to change drilling direction the operator had to halt drill string rotation in such a position that a bend in the motor pointed in the direction of the new trajectory (referred to as the sliding mode). Rotary steerable systems introduced in the 1990s eliminated the need to slide a steerable downhole motor

DID YOU KNOW?

The oil field service company individual who is charged with maintaining a drilling fluid or completion fluid system on an oil or gas drilling rig is called the *mud engineer.*

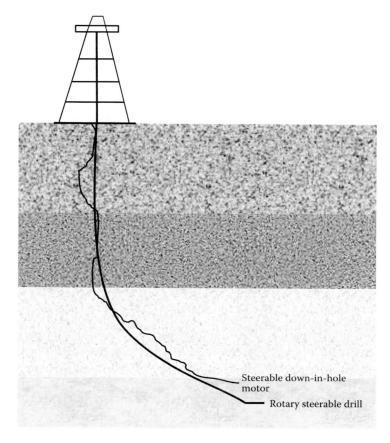

FIGURE 2.2 Directional drilling: steerable down-in-hole motor vs. rotary steerable system. (Courtesy of Schlumberger, Houston, TX.)

(Schlumberger, 2011). The newer tools drill directionally while being continuously rotated from the surface by the drilling rig. This enables a much more complex, and thus accurate, drilling trajectory. Continuous rotation also leads to higher rates of penetration and fewer incidents of the drill string sticking (see Figure 2.2).

Directional drilling offers another significant advantage in developing gas shales. In the case of thin or inclined shale formations, a long horizontal well increases the length of the wellbore in the gas-bearing formation and therefore increases the surface area for gas to flow into the well; however, the increased well surface (length) is often insufficient without some means of artificially stimulating flow. In some sandstone and carbonate formations, injecting dilute acid dissolves the natural cement that binds sand grains, thus increasing permeability. In tight formations such as shale, inducing fractures can increase flow by orders of magnitude; however, before stimulation or production can take place, the well must be completed and cased.

Well Casing Construction

Telescoping steel well casings that prevent wellbore collapse and water infiltration are commonly used in the drilling of commercial gas and oil wells and municipal water-supply wells (see Figure 2.3). The casing also conducts the produced reservoir fluids to the surface. A properly designed and cemented casing also prevents reservoir fluids (gas or oil) from infiltrating the overlying groundwater aquifers.

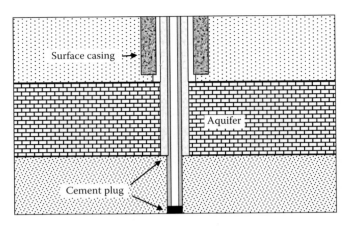

FIGURE 2.3 Theoretical well casing. (Adapted from USDOE, *Modern Shale Gas Development in the United States: A Primer*, U.S. Department of Energy, Washington, DC, 2009.)

During the first phase of drilling, termed *spudding in*, shallow casing installed underneath the platform serves to reinforce the ground surface. Drilling continues to the bottom of the water table (or the potable aquifer), at which point the drill string is tripped out (removed) in order to lower a second casing string, which is cemented in and plugged at the bottom. Drillers use special oil-well cement that expands when it sets to fill the void between the casing and the wellbore.

Surface casing and casing to the bottom of the water table prevent water from flooding the well while also protecting the groundwater from contamination by drilling fluids and reservoir fluids. (The initial drilling stages may use compressed air in place of drilling fluids to avoid contaminating the potable aquifer.) Drilling and casing then continue to the *pay zone*—the formation that produces gas or oil. The number and length of the casings, however, depend on the depth and properties of the geologic strata.

After completing the well to the target depth and cementing in the final casing, the drilling operator may hire an oil-well service company to run a *cement evaluation log*. An electrical probe lowered into the well measures the cement thickness. The cement evaluation log provides the critical confirmation that the cement will function as designed—preventing well fluids from bypassing outside the casing and infiltrating overlying formations. As mentioned earlier, additionally, state oil and gas regulatory agencies often specify the required depth of protective casings and regulate the time that is required for cement to set prior to additional drilling. These requirements are typically based on regional conditions, are established for all wildcat wells, and may be modified when field rules are designated. These requirements are instituted by state oil and gas agencies to provide protection of groundwater resources (Arkansas Oil and Gas Commission, 2012). When the casing strings have been run and cemented, there could be five or more layers or barriers between the inside of the production tubing and a water-bearing formation (salt or fresh).

Analysis of the redundant protections provided by casings and cement was presented by Michie & Associates (1988) in a series of reports and papers prepared for the American Petroleum Institute (API). These investigations evaluated the level of corrosion that occurred in Class II injection wells. Class II injection wells are used for the routine injection of water associated with oil and gas production. The research resulted in the development of a method of calculating the probability (or risk) that fluids injected into Class II injection wells could result in an impact to underground sources of drinking water (USDWs). This research began by evaluating data for oil- and gas-producing basins to determine if there were natural formation waters present that were reported to cause corrosion of well casings. The United States was divided into 50 basins, and each basin was ranked by its potential to have a casing leak resulting from such corrosion.

Detailed analysis was performed for those basins in which there was a possibility of casing corrosion. Risk probability analysis provided an upper boundary for the probability of the fracturing fluids reaching an underground source of drinking water. Based on the values calculated, a modern horizontal well completion in which 100% of the USDWs are protected by properly installed surface casings (and for geologic basins with a reasonable likelihood of corrosion), the probability that fluids injected at depth could impact a USDW would be between 2×10^{-5} (one well in 200,000) and 2×10^{-8} (one well in 200,000,000) if these wells were operated as injection wells. Other studies in the Williston Basin found that the upper-bound probability of injection water escaping the wellbore and reaching an underground source of drinking water was 7 chances in 1 million well-years where surface casings cover the drinking water aquifers (Michie and Koch, 1991). Note that these values do not account for the differences between the operation of a shale gas well and the operation of an injection well. An injection well is constantly injecting fluid under pressure and thus raises the pressure of the receiving aquifer, increasing the chance of a leak or well failure. A production well is reducing the pressure in the producing zone by giving the gas and associated fluid a way out, making it less likely that they will try to find an alternative path that could contaminate a freshwater zone. Furthermore, a producing gas well would be less likely to experience a casing leak because it is operated at a reduced pressure compared to an injection well. It would be exposed to lesser volumes of potentially corrosive water flowing through the production tubing, and it would only be exposed to the pumping of fluids into the well during fracture stimulations.

Because the API study included an analysis of wells that had been in operation for many years when the study was performed in the 1980s, it does not account for advances that have occurred in equipment and applied technologies and changes to regulations. As such, a calculation of the probability of any fluids, including hydraulic fracturing fluids, reaching a USDW from a gas well would indicate an even lower probability, perhaps by as much as two to three orders of magnitude. The API report came to another important conclusion relative to the probability of the contamination of a USDW when it stated that (Michie & Associates, 1988):

> … for injected water to reach USDW in the 19 identified basins of concern, a number of independent events must occur at the same time and go *undetected* [emphasis added]. These events include simultaneous leaks in the [production] tubing, production casing, [intermediate casing], and the surface casing coupled with the unlikely occurrence of water moving long distances up the borehole past saltwater aquifers to reach a USDW.

As indicated by the analysis conducted by the API and others, the potential for groundwater to be impacted by injection is low. It is expected that the probability for treatable groundwater to be impacted by the pumping of fluids during hydraulic fracture treatments of newly installed, deep shale gas wells when a high level of monitoring is being performed would be even less than the 2×10^{-8} estimated by the API.

In addition to the protections provided by multiple casings and cements, there are natural barriers in the rock strata that act as seals holding the gas in the target formation. Without such seals, gas and oil would naturally migrate to the surface of the Earth. A fundamental precept of oil and gas geology is that, without an effective seal, gas and oil would not accumulate in a reservoir in the first place and so could never be tapped and produced in usable quantities. These sealing strata act as barriers to vertical migration of fluids upward toward useable groundwater zones. Most shale gas wells (outside of those completed in the New Albany and the Antrim) are expected to be drilled at depths greater than 3000 feet below the land surface.

When the cement log has been run, and absent any cement voids, the well is ready for completion. A perforating tool that uses explosive shape charges punctures the casing sidewall at the pay zone. The well may then start producing under its natural reservoir pressure or, as in the case of gas shales, may require stimulation treatment. Both domestic-use gas wells and water wells are common throughout regions experiencing recent shale gas development. In the absence of regulation, domestic-use wells (gas or water) may not meet standard practices of construction. If the well head of a water supply well is improperly sealed, for example, surface water may infiltrate down along the casing exterior and contaminate the drinking water aquifer. Some domestic water wells have also produced natural gas, and some shallow gas wells have leaked into nearby building foundations. To avoid some of these problems, Pennsylvania has instituted regulations that require a minimum 2000-foot setback between a new gas well and an existing water well.

DRILLING FLUIDS AND RETENTION PITS

Drilling fluids are a necessary component of the drilling process: They circulate cuttings (rock chips created as the drill bit advances through rock, much like sawdust) to the surface to clear the borehole, lubricate and cool the drilling bit, stabilize the wellbore (preventing cave in), and control downhole fluid pressure (Schlumberger, 2008a). In order to maintain sufficient volumes of fluids onsite during drilling, operators typically use pits to store make-up water used as part of the drilling fluids. Storage pits are not used in every development situation. In the case of shale gas drilling practices, they should be adapted to facilitate development in both settings. Drilling with compressed air has become an increasingly popular alternative to drilling with fluids due to the increased cost savings from both a reduction in mud costs and the shortened drilling times as a result of air-based drilling (Singh, 1965). The air, like drilling mud, functions to lubricate, cool the bit, and remove cuttings. Air drilling is generally limited to low-pressure formations, such as the Marcellus Shale in New York (Kennedy, 2000). In rural areas, storage pits may be used to hold freshwater for drilling and hydraulic fracturing. In urban settings, due to space limitations, steel storage tanks may be used to hold drilling fluids as well as to store water and fluids for use during hydraulic fracturing. Tanks used in closed-loop drilling systems allow for the reuse of drilling fluids and the use of lesser amounts of drilling fluids (Swaco, 2006). Closed-loop drilling systems have also been used with water-based fluids in environmentally sensitive environments in combination with air-rotary drilling techniques (Oklahoma DEQ, 2008). Although closed-loop drilling has been used to address specific situations, the practice is not necessary for every well drilled. Drilling is a regulated practice managed at the state level, and although state oil and gas agencies have the ability to require operators to vary standard practices, the agencies typically do so only when it is necessary to protect the gas resources and the environment.

In rural environments, storage pits may be used to hold water. They are typically excavated containment ponds that, based on the local conditions and regulatory requirements, may be lined. Pits can also be used to store additional make-up water for drilling fluids or to store water used in the hydraulic fracturing of wells.

Water storage pits used to hold water for hydraulic fracturing purposes are typically lined to minimize the loss of water from infiltration. Water storage pits are becoming an important tool in the shale gas industry because the drilling and hydraulic fracturing of these wells often require significant volumes of water as the base fluid for both purposes (Harper, 2008).

ASPECTS OF HYDRAULIC FRACTURING

Despite the abundant natural gas content of some shales, they do not produce gas freely. Economic production depends on some means of artificially stimulating shale to liberate gas. In the late 1940s in Texas oil fields, fluids pumped down wells under pressures high enough to fracture stimulated the producing formation. Hydraulic fracturing is a formation stimulation practice used to create additional permeability in a producing formation, thus causing gas to flow more readily toward the wellbore (Jennings and Darden, 1979; Veatch et al., 1999). Hydraulic fracturing can be used to overcome natural barriers to the flow of fluids (gas or water) to the wellbore. Such barriers may include naturally low permeability common in shale formations or reduced permeability resulting from near-wellbore damage during drilling activities (Boyer et al., 2006). Hydraulic fracture stimulation treatments have been adapted to tight gas formations such as the Barnett Shale in Texas, and more

Hydraulic Fracking: The Process

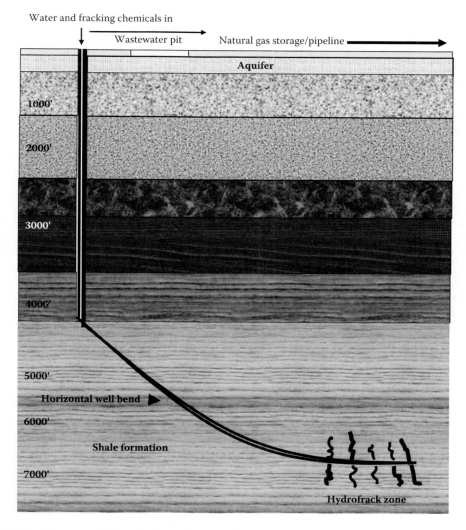

FIGURE 2.4 Diagram of hydraulic fracturing operation.

recently the Marcellus Shale. Typical fracking treatments, or frack jobs, are relatively large operations compared to some drilling operations. The oilfield service company contracted for the work may take a week to stage the job, and a convoy of trucks will be necessary to deliver the equipment and materials needed.

A company involved in developing Texas shale gas offered the following description of a frack job (Franz and Jochen, 2005):

> Shale gas wells are not hard to drill, but they are difficult to complete. In almost every case the rock around the wellbore must be hydraulically fractured before the well can produce significant amounts of gas. Fracturing involves isolating sections of the well in the producing zone, then pumping fluids and proppant (grains of sand or other material used to hold the cracks open) down the wellbore through perforations in the casing and out into the shale [see Figure 2.4]. The pumped fluid, under pressures up to 8000 psi, is enough to crack shale as much as 3000 ft in each direction from the wellbore. In the deeper high-pressure shales, operators pump slickwater (a low-viscosity water-based fluid) and proppant. Nitrogen-foamed fracturing fluids are commonly pumped on shallower shales and shales with lower reservoir pressures.

DID YOU KNOW?

Stimulations are optimized to ensure that fracture development is confined to the target formation.

As shown in Figure 2.4, hydraulic fracturing involves the pumping of fracturing fluid into a formation at a calculated, predetermined rate with enough pressure to generate fractures or cracks in the target formation. For shale gas development, fracture fluids are primarily water-based fluids mixed with additives that help the water to carry sand (or other material) proppant into the fractures. The proppant is needed to keep the fractures open when the pumping of fluid has stopped. When the fracture has initiated, additional fluids are pumped into the wellbore to continue the development of the fracture and to carry the proppant deeper into the formation. The additional fluids are needed to maintain the downhole pressure necessary to accommodate the increasing length of opened fractures in the formation. Each rock formation has inherent natural variability resulting in different fracture pressures for different formations. The process of designing hydraulic fracture treatments requires identifying properties of the target formation, including fracture pressure, and the desired length of fractures. The following discussion addresses some of the processes involved in the design of a hydraulic fracture stimulation of a shale gas formation.

FRACTURE DESIGN

Modern formation stimulation practices are sophisticated, engineered processes designed to emplace fracture networks in specific rock strata (Boyer et al., 2006). A hydraulic fracture treatment is a controlled process designed to the specific conditions of the target formation (e.g., thickness of shale, rock fracturing characteristics). Understanding the *in situ* reservoir conditions present and their dynamics is critical to successful stimulations. Hydraulic fracturing designs are continually refined to optimize fracture networking and maximize gas production. Whereas the concepts and general practices are similar, the details of a specific fracture operation can vary substantially from basin to basin and from well to well. Fracture design can incorporate many sophisticated and state-of-the-art techniques to accomplish an effective, economic, and highly successful fracture stimulation. Some of these techniques include modeling, microseismic fracture mapping, and tilt-meter analysis.

A computer model can be used to aid in candidate selection with regard to identifying wells, fields, or formations that would be good fracture candidates; such models take into consideration complex factors such as multifractured wells, non-Darcy (nonlaminar) flow, and multiphase flow. Computer modeling can also be utilized in the treatment design process through the use of refined geologic parameters to generate the final pump schedule, in the execution and analysis of the hydraulic fracture treatment (i.e., post-frack production analysis), and in simulations of hydraulic fracturing designs (Meyer & Associates, 2012). Computer models help to maximize effectiveness and to economically design a treatment event. The modeling programs allow geologists and engineers to modify the design of a hydraulic fracture treatment and evaluate the height, length, and orientation of potential fracture development (Schlumberger, 2008b). These simulators also allow designers to use the data gathered during a fracture stimulation to evaluate the success of the fracture job performed. From these data and analyses, engineers can optimize the design of future fracture stimulations.

DID YOU KNOW?

Microseismic mapping technology is rooted in the observation that when a fracture is induced into a reservoir or the bounding rock layers the *in situ* stress is disturbed, resulting in shear failure and subsequent "mini-earthquakes" (Walser, 2010).

Hydraulic Fracking: The Process

DID YOU KNOW?

Because of the length of exposed wellbore, it is usually not possible to maintain a downhole pressure sufficient to stimulate the entire length of a lateral in a single stimulation event (Overbey et al., 1988). Because of the lengths of the laterals, hydraulic fracture treatments of horizontal shale gas wells are usually performed by isolating smaller portions of the lateral.

Additional advances in hydraulic fracturing design have targeted the analysis of hydraulic fracture treatments through technologies such as microseismic fracture mapping, which is used to pinpoint fracturing and aids in well stimulation, and tilt measurements (Meyer & Associates, 2012). These technologies can be used to define the success and orientation of the fractures created, thus providing the engineers with the ability to manage the resource through the strategic placement of additional wells, taking advantage of the natural reservoir conditions and expected fracture results in new wells.

As more formation-specific data are gathered, service companies and operators can optimize fracture patterns. Operators have strong economic incentives to ensure that fractures do not propagate beyond the target formation and into adjacent rock strata (Parshall, 2008). Allowing the fractures to extend beyond the target formation would be a waste of materials, time, and money. In some cases, fracturing outside of the target formation could potentially result in the loss of the well and the associated gas resource. Fracture growth outside of the target formation can result in excess water production from bounding strata. Having to pump and handle excess water increases production costs, negatively impacting well economics. This is a particular concern in the Barnett Shale of Texas where the underlying Ellenberger Group limestones are capable of yielding significant formation water.

DESCRIPTION OF THE HYDRAULIC FRACTURING PROCESS[*]

Fracture treatments of horizontal shale gas wells are carefully controlled and monitored operations that are performed in stages. Lateral lengths in horizontal wells for shale gas development may range from 1000 feet to more than 5000 feet. Before beginning a treatment, the service company will perform a series of tests on the well to determine if it is competent to hold up to the hydraulic pressures generated by the fracture pumps. In the initial stage, hydrochloric acid (HCl) solution is pumped down the well to clean up the residue left from cementing the well casing. Each successive stage pumps discrete volumes of fluid (slickwater) and proppant down the well to open and propagate the fracture further into the formation. The treatment may last up to an hour or more, with the final stage designed to flush the well. Some wells may receive several or more treatments to produce multiple fractures at different depths or farther out into the formation in the case of horizontal wells.

The fracturing of each portion of the lateral wellbore is called a *stage*. Stages are fractured sequentially beginning with the section at the farthest end of the wellbore, moving uphole as each stage of the treatment is completed until the entire lateral well has been stimulated (Chesapeake Energy, 2008b). Horizontal wells in the various shale gas basins may be treated using two or more stages to fracture the entire perforated interval of the well. Each stage of a horizontal well fracture treatment is similar to a fracture treatment for a vertical shale gas well. For each stage of a fracture treatment, a series of different volumes of fracturing fluids, called *substages*, with specified additives and proppant concentrations, is injected sequentially. Table 2.1 presents an example of the substages of a single-stage hydraulic fracture treatment for a well completed in the Marcellus Shale (Arthur et al., 2008). This is a single-stage treatment typical of what might be performed on a vertical shale well or for each stage of a multistage horizontal well treatment. The total volume of the

---

[*] This section is based on information contained in USDOE, *Modern Shale Gas Development in the United States: A Primer*, U.S. Department of Energy, Washington, DC, 2009, pp. 55–59.

TABLE 2.1
Example of a Single Stage of a Sequenced Hydraulic Fracture Treatment

Hydraulic Fracture Treatment Substage	Volumes (gallons)	Hydraulic Fracture Treatment Substage	Volumes (gallons)
Diluted acid (15%)	5000	Prop 8	20,000
Pad	100,000	Prop 9	20,000
Prop 1	50,000	Prop 10	20,000
Prop 2	50,000	Prop 11	20,000
Prop 3	40,000	Prop 12	20,000
Prop 4	40,000	Prop 13	20,000
Prop 5	40,000	Prop 14	10,000
Prop 6	30,000	Prop 15	10,000
Prop 7	30,000	Flush	13,000

Source: Adapted from Arthur, J.D. et al., *An Overview of Modern Shale Gas Development in the United States*, ALL Consulting, Tulsa, OK, 2008.

Note: Volumes are presented in gallons (42 gallons = 1 bbl, 5000 gallons = ~120 bbl). Flush volumes are based on the total volume of open borehole; therefore, as each stage is completed, the volume of flush decreases as the volume of borehole is decreased. Total amount of proppant used is approximately 450,000 pounds.

substages in Table 2.1 is 578,000 gallons. If this were one stage of a four-stage horizontal well, the entire fracture operation would require approximately four times this amount, or 2.3 million gallons of water. Note that the actual *rate* of water usage is measured in gallons per minute (gal/min); 42 gal/min = 1 bbl/min, and 500 gal/min = ~12 bbl/min. In Table 2.1, with the exception of the 500-gal/min rate for diluted acid (15%), the other rates from pad through flush are 3000 gal/min.

Guided by state oil and gas regulatory agencies—to ensure that a well is protective of water resources and is safe for operation—operators or service companies perform a series of tests. These tests are designed to ensure that the well, well equipment, and hydraulic fracturing equipment are in proper working order and will safely withstand application of the fracture treatment pressures and pump flow rates. The tests start with the testing of well casings and cement during the drilling and well construction process. Testing continues with pressure testing of hydraulic fracturing equipment prior to the fracture treatment process (Harper, 2008). As mentioned earlier, construction requirements for wells are mandated by state oil and gas regulatory agencies to ensure that a well is protective of water resources and is safe for operation.

After the testing of equipment has been completed, the hydraulic fracture treatment process begins. The substage sequence is usually initiated with the pumping of an acid treatment. Again, this acid treatment helps to clean the near-wellbore area, which can be damaged as a result of the

DID YOU KNOW?

A single fracture treatment may consume more than 500,000 gallons of water (USDOE, 2009). Wells subject to multiple treatments consume several million gallons of water. For comparison, an Olympic-size swimming pool (164 × 82 × 6 ft deep) holds over 660,000 gallons of water, and the average daily per capita consumption of freshwater (roughly 1430 gallons per day) adds up to 522,000 gallons over one year (USGS, 2000).

Hydraulic Fracking: The Process

DID YOU KNOW?

Slickwater fracturing is a method or system of hydrofracturing that involves the addition of chemicals to water to increase the fluid flow. Fluid (friction reducer) can be pumped down the wellbore at a rate as high as 100 bbl/min to fracture the shale; otherwise, without the use of slickwater, the top speed of pumping is about 60 bbl/min.

drilling and well installation process; for example, pores and pore throats can become plugged with drilling mud or casing cement. The next sequence after the acid treatment is a slickwater pad, which is a water-based fracturing fluid mixed with a friction-reducing agent. The pad is a volume of fracturing fluid large enough to effectively fill the wellbore and the open formation area. The slickwater pad helps to facilitate the flow and placement of the proppant further into the fracture network.

After the pad is pumped, the first proppant substage combining a large volume of water with fine mesh sand is pumped. The next several substages increase the volume of fine-grained proppant while the volume of fluids pumped is decreased incrementally from 50,000 gal to 30,000 gal. This fine-grained proppant is used because the finer particle size is capable of being carried deeper in the developed fractures (Cramer, 2008). In this example, the fine proppant substages are followed by eight substages of a coarser proppant with volumes from 20,000 to 10,000 gal. After completion of the final substage of coarse proppant, the well and equipment are flushed with a volume of fresh water sufficient to remove excess proppants from the equipment and the wellbore.

Hydraulic fracture stimulations are overseen continuously by operators and service companies to evaluate and document the events of the treatment process. Every aspect of the fracture stimulation process is carefully monitored, from the well head and downhole pressures to pumping rates and density of the fracturing fluid slurry. The monitors also track the volumes of each additive and the water used and ensure that equipment is functioning properly. For a 12,000-bbl (504,000-gal) fracture treatment of a vertical shale gas well there may between 30 and 35 people onsite to monitor the entire stimulation process.

The staging of multiple fracture treatments along the length of the lateral leg of the horizontal well allows the fracturing process to be performed in a very controlled manner. By fracturing discrete intervals of the lateral wellbore, the operator is able to make changes to each portion of the completion zone to accommodate site-specific changes in the formation. These site-specific variations may include variations in shale thickness, presence or absence of natural fractures, proximity to another wellbore fracture system, and boreholes that are not centered in the formation.

Fracturing Fluids

As mentioned, the current practice of hydraulic fracturing of shale gas reservoirs is to apply a sequenced pumping event in which millions of gallons of water-based fracturing fluids mixed with proppant materials are pumped in a controlled and monitored manner into the target shale formation above fracture pressure (Harper, 2008). The fracturing fluids used for gas shale stimulations consist primarily of water but also include a variety of additives. The number of chemical additives used in a typical fracture treatment varies depending on the conditions of the specific well being

DID YOU KNOW?

The fracture is ideally represented by a vertical plane that intersects the well casing. It does not propagate in a random direction but opens perpendicular to the direction of least stress underground (which is nearly horizontal in orientation).

DID YOU KNOW?

As the term *propping* implies, the agent functions to prop or hold the fracture open. The fluid must have the proper viscosity and low friction pressure when pumped, it must break down and clean up rapidly when treatment is over, and it must provide good fluid-loss control (not dissipate). The fluid chemistry may be water, oil, or acid based, depending on the properties of the formation. Water-based fluids (slickwater) are the most widely used in shale formations because of their low cost, high performance, and ease of handling.

fractured. A typical fracture treatment will use very low concentrations of between 3 and 12 additive chemicals, depending on the characteristics of the water and shale formation being fractured. Each component serves a specific, engineered purpose (Schlumberger, 2008b). The predominant fluids currently being used for fracture treatments in the shale gas plays are water-based fracturing fluids mixed with friction-reducing additions (slickwater).

The addition of friction reducers allows fracturing fluids and proppant to be pumped to the target zone at a higher rate and reduced pressure than if water alone were used. In addition to friction reducers, other additions include biocides to prevent microorganism growth and to reduce biofouling of the fractures, oxygen scavengers and other stabilizers to prevent corrosion of metal pipes, and acids that are used to remove drilling mud damage with the near-wellbore area (Schlumberger, 2008b). These fluids function in two ways: opening the fracture and transporting the propping agent (or proppant) the length of the fracture (Economides and Nolte, 2000).

Table 2.2 lists the volumetric percentages of additives that were used for a nine-stage hydraulic fracturing treatment of a Fayetteville Shale horizontal well. The make-up of fracturing fluid varies from one geologic basin or formation to another. Evaluating the relative volumes of the components of a fracturing fluid reveals the relatively small volume of additives that are present. The additives represent less than 0.5% of the total fluid volume. Overall, the concentration of additives in most slickwater fracturing fluids is a relatively consistent 0.5% to 2%, with water making up 98% to 99.5%.

TABLE 2.2
Volumetric Composition of a Fracturing Fluid

Component	Percent (%) by Volume
Water and sand	99.51
Surfactant	0.085
KCl	0.06
Gelling agent	0.056
Scale inhibitor	0.043
pH adjusting agent	0.011
Breaker	0.01
Crosslinker	0.007
Iron control	0.004
Corrosion inhibiter	0.002
Biocide	0.001
Acid	0.123
Friction reducer	0.088

Source: Based on data (2008) from ALL Consulting for a fracture operation in the Fayetteville Shale.

Hydraulic Fracking: The Process

DID YOU KNOW?

Some fracturing fluids may include nitrogen and carbon dioxide to help foaming. Oil-based fluids find use in hydrocarbon-bearing formations susceptible to water damage, but they are expensive and difficult to use. Acid-based fluids use hydrochloric acid to dissolve the mineral matrix of carbonate formations (limestone and dolomite) and thus improve porosity; the reaction produces inert calcium chloride salt and carbon dioxide gas.

DID YOU KNOW?

Proppants hold the fracture walls apart to create conductive paths for the natural gas to reach the wellbore. Silica sands are the most commonly used proppants. Resin-coating the sand grains improves their strength.

Because the make-up of each fracturing fluid varies to meet the specific needs of each area, there is no one-size-fits-all formula for the volumes for each additive. When classifying fracturing fluids and their additives, it is important to realize that service companies that provide these additives have developed a number of compounds for different recipes with similar functional properties to be used for the same purpose in different well environments. The difference between additive formulations may be as small as a change in concentration of a specific compound. Although the hydraulic fracturing industry may have a number of compounds that can be used in a hydraulic fracturing fluid, any single fracturing job would use only a few of the available additives. In Table 2.1, for example, 12 additives are used, covering the range of possible functions that could be built into a fracturing fluid. It is not uncommon for some fracturing recipes to omit some compound categories if their properties are not required for the specific application.

Most industrial processes use chemicals, and almost any chemical can be hazardous in large enough quantities if not handled properly. Even chemicals that go into our food or drinking water can be hazardous. Drinking water treatment plants use large quantities of chlorine. When used and handled properly, it is safe for workers and nearby residents and provides clean, safe drinking water for the community. Although the risk is low, the potential exists for an unplanned release that could have serious effects on human health and the environment. By the same token, hydraulic fracturing uses a number of chemical additives that could be hazardous but are safe when properly handled according to requirements and long-standing industry practices. In addition, many of these additives are common chemicals that people regularly encounter in everyday life.

Table 2.3 provides a summary of the additives, their main compounds, the reason the additive is used in a hydraulic fracturing fluid, and some of the other common uses for these compounds. Hydrochloric acid (HCl) is the single largest liquid component used in a fracturing fluid aside from water; although the concentration of the acid may vary, a 15% HCl mix is a typical concentration. A 15% HCl mix is composed of 85% water and 15% acid; therefore, the volume of acid is diluted by 85% with water in its stock solution before it is pumped into the formation during a fracturing treatment. Once the entire stage of fracturing fluid had been injected, the total volume of acid in an example fracturing fluid from the Fayetteville Shale was 0.123%, which indicates the fluid had been diluted by a factor of 122 times before being pumped into the formation. The concentration of this acid will only continue to be diluted as it is further dispersed in additional volumes of water that may be present in the subsurface. Furthermore, if this acid comes into contact with carbonate minerals in the subsurface, it would be neutralized by chemical reaction with the carbonate minerals, producing water and carbon dioxide as a byproduct of the reaction.

TABLE 2.3
Fracturing Fluid Additives, Main Compounds, and Common Uses

Additive Type	Main Component(s)	Purpose	Common Use of Main Compound
Diluted acid (15%)	Hydrochloric acid or muriatic acid	Helps dissolve minerals and initiates cracks in the rock	Swimming pool chemical and cleaner
Biocide	Glutaraldehyde	Eliminates bacteria in the water that produce corrosive byproducts	Disinfectant; used to sterilize medical and dental equipment
Breaker	Ammonium persulfate	Allows delayed breakdown of the gel polymer chains	Bleaching agent in detergent and hair cosmetics, manufacture of household plastics
Corrosion inhibitor	N,n-dimethyl formamide	Prevents corrosion of the pipe	Pharmaceuticals, acrylic fibers, plastics
Crosslinker	Borate salts	Maintain friction between the fluid and the pipe	Laundry detergents, hand soaps, cosmetics
Friction reducer	Polyacrylamide	Minimizes friction between the fluid and the pipe	Water treatment, soil conditioner
	Mineral oil		Make-up remover, laxative, candy
Gel	Guar gum or hydroxyethyl cellulose	Thickens the water in order to suspend the sand	Cosmetics, toothpaste, sauces, baked goods, ice cream
Iron control	Citric acid	Prevents precipitation of metal oxides	Food additive, flavoring in foods and beverages; lemon juice is 7% citric acid
KCl	Potassium chloride	Creates a brine carrier fluid	Low-sodium table salt substitute
Oxygen scavenger	Ammonium bisulfite	Removes oxygen from the water to protect pipes from corrosion	Cosmetics, food and beverage processing, water treatment
pH adjusting agent	Sodium or potassium carbonate	Maintains effectiveness of other components, such as crosslinkers	Washing soda, detergents, soap, water softener, glass and ceramics
Proppant	Silica, quartz sand	Allows fractures to remain open so gas can escape	Drinking water filtration, play sand, concrete, brick mortar
Scale inhibitor	Ethylene glycol	Prevents scale deposits in pipes	Automotive antifreeze, household cleansers, deicing agent
Surfactant	Isopropanol	Used to increase viscosity of the fracture fluid	Glass cleaner, hair color, antiperspirant

Source: USDOE, *Modern Shale Gas Development in the United States: A Primer*, U.S. Department of Energy, Washington, DC, 2009.

Note: Specific compounds used in a given fracturing operation will vary depending on company preference, source water quality, and site-specific characteristics of the target formation. Compounds shown above are representative of the major compounds used in hydraulic fracturing of gas shales.

THOUGHT-PROVOKING QUESTIONS

1. Can Mother Nature correct any and all environmental damage caused by hydraulic fracturing operations? Explain.
2. How important is it to protect wildlife in a hydraulic fracturing region?
3. Do you believe that hydraulic fracturing operations damage groundwater supplies?

REFERENCES AND RECOMMENDED READING

AAPG. (1987). *Correlation of Stratigraphic Units of North America (CASUNA) Project.* American Association of Petroleum Geologists, Tulsa, OK.

Anon. (2000). Alabama lawsuit poses threat to hydraulic fracturing across U.S. *Drilling Contractor*, January/February, pp. 42–43.

Anon. (2008). *Projecting the Economic Impact of the Fayetteville Shale Play for 2008–2012.* Center for Business and Economic Research, Sam M. Walton College of Business, Fayetteville, AR.

Arkansas Oil and Gas Commission. (2012). *General Rules and Regulations.* Arkansas Oil and Gas Commission, Little Rock.

Arthur, J.D., Bohm, B., and Lane, M. (2008). *An Overview of Modern Shale Gas Development in the United States.* ALL Consulting, Tulsa, OK.

Berman, A. (2008). The Haynesville Shale sizzles while the Barnett cools. *World Oil Magazine*, 229(9):23.

Boughal, K. (2008). Unconventional plays grow in number after Barnett Shale blazed the way. *World Oil Magazine*, 229(8):77–80.

Boyer, C., Kieschnick, J., Suarez-Rivera, R., Lewis, R., and Walter, G. (2006). Producing gas from its source. *Oilfield Review*, 18(3):36–49.

Cardott, B. (2004). *Overview of Unconventional Energy Resources of Oklahoma.* Oklahoma Geological Survey, Norman, OK (www.ogs.ou.edu/fossilfuels/coalpdfs/UnconventionalPresentation.pdf).

Catacosinos, P., Harrison, W., Reynolds, R., Westjohn, D., and Wollensak. M. (2000). *Stratigraphic Nomenclature for Michigan.* Michigan Department of Environmental Quality, Geological Survey Division, and Michigan Basin Geological Survey, Kalamazoo, MI.

CBO. (1983). *Understanding Natural Gas Price Decontrol.* Congressional Budget Office, Congress of the United States, Washington, DC.

Chesapeake Energy. (2008a). Little Red River Project, paper presented to Trout Unlimited.

Chesapeake Energy. (2008b). Components of Hydraulic Fracturing, paper presented to New York Department of Environmental Conservation.

Chesapeake Energy. (2012). *Hydraulic Fracturing Facts*, http://www.hydraulicfracturing.com/Water-Usage/Pages/Information.aspxd.

Cramer, D. (2008). Stimulating unconventional reservoirs: lessons learned, successful practices, areas for improvement, in *Proceedings of SPE Unconventional Reservoirs Conference*, Keystone, CO, February 10–12. Society of Petroleum Engineers, Allen, TX.

Durham, L.S. (2008). A spike in interest and activity: Louisiana play a "company maker"? *AAPG Explorer*, 29(7):18.

France, D. (2008). The Catskills gas rush. *New York Magazine*, September 21.

Franz, Jr., J.H. and Jochen, V. (2005). *When Your Gas Reservoir Is Unconventional, So Is Our Solution*, Shale Gas White Paper 05-OF-299. Schlumberger, Houston, TX.

Gaudlip, A.W., Paugh, L.O., and Hayes, T.D. (2008). Marcellus Shale water management challenges in Pennsylvania, in *Proceedings of SPE Shale Gas Production Conference*, Fort Worth, TX, November 16–18. Society of Petroleum Engineers, Allen, TX.

Halliburton. (2008). *U.S. Shale Gas: An Unconventional Resource. Unconventional Challenges.* Halliburton, Houston, TX, 8 pp.

Harper, J. (2008). The Marcellus Shale—an old "new" gas reservoir in Pennsylvania. *Pennsylvania Geology*, 28(1):2–13.

Harrison, W. (2006). *Production History and Reservoir Characteristics of the Antrim Shale Gas Play, Michigan Basin.* Western Michigan University, Kalamazoo.

Hayden, J. and Pursell, D. (2005). *The Barnett Shale: Visitor's Guide to the Hottest Gas Play in the US.* Pickering Energy Partners, Inc., Houston, TX.

HIE. (2007). *Fayetteville Shale Power.* Hillwood International Energy, Dallas, TX.

IGS. (1986). *General Stratigraphic Column for Paleozoic Rocks in Indiana*. Indiana Geological Survey, Indiana University, Bloomington, IN (www.usi.edu/science/geology/My%20Web%20Sites/stratcolumn.pdf).

ISGS. (2011). *Mississippian Rocks in Illinois*, GeoNote 1. Illinois State Geological Survey, Champaign, IL (www.isgs.uiuc.edu/maps-data-pub/publications/geonotes/geonote1.shtml).

Jennings, Jr., A.R. and Darden, W.G. (1979). Gas well stimulation in the eastern United States, in *Proceedings of Symposium on Low Permeability Gas Reservoirs*, Denver, CO, May 20–22, 1979. Society of Petroleum Engineers, Allen, TX.

Jochen, V. (2006). *New Technology Needs to Produce Unconventional Gas*. Schlumberger, Houston, TX.

Johnston III, J., Heinrich, B., Lovelace, J., McCulluh, R., and Zimmerman, R. (2000). *Stratigraphic Charts of Louisiana*, Louisiana Geological Survey, Baton Rouge, LA.

Kennedy, J. (2000). Technology limits environmental impact of drilling. *Drilling Contractor*, July/August, pp. 31–35.

Lesley, J.P. (1892). *A Summary Description of the Geology of Pennsylvania*, Vol. II. Geological Survey of Pennsylvania, Pittsburgh.

Meyer & Associates. (2012). *Meyer Fracturing Simulators User's Guide*, http://www.mfrac.com/documentation.html.

Michie & Associates. (1988). *Oil and Gas Water Injection Well Corrosion*, prepared for the American Petroleum Institute, Washington, DC.

Michie, T.W. and Koch, C.A. (1991). Evaluation of injection-well risk management in the Williston Basin. *Journal of Petroleum Technology*, 43(6):737–741.

NaturalGas.org. (2011). *Unconventional Natural Gas Resources*, www.naturalgas.org/overview/unconvent_ng_resource.asp.

Navigant Consulting. (2008). *North American Natural Gas Supply Assessment*, prepared for American Clean Skies Foundation, Washington, DC.

NRCS. (2008). *National Water & Climate Center*. Natural Resources Conservation Service, Washington, DC (www.wcc.nrcs.usda.gov/).

Nyahay, R., Leone, J., Smith, L., Martin, J., and Jarvie, D. (2007). Update on Regional Assessment of Gas Potential in the Devonian Marcellus and Ordovician Utica Shales of New York, paper presented at the American Association of Petroleum Geologists (AAPG) Eastern Section Meeting, Lexington, KY, September 16–18, 2007 (www.searchanddiscovery.com/documents/2007/07101nyahay/).

Nyahay, R., Leone, J., Smith, L., Martin, J., and Jarvie, D. (2007). Update on Regional Assessment of Gas Potential in the Devonian Marcellus and Ordovician Utica Shales of New York, paper presented at the American Association of Petroleum Geologists (AAPG) Eastern Section Meeting, Lexington, KY, September 16–18, 2007 (www.searchanddiscovery.com/documents/2007/07101nyahay/).

Oklahoma DEQ. (2008). *Pollution Prevention Case Study for Oxy USA, Inc.* Oklahoma Department of Environmental Quality, Oklahoma City (www.deq.state.ok.us/CSDnew/P2/Casestudy/oxyusa~1.htm).

Overbey, W.K., Yost II, A.B., and Wilkins, D.A. (1988). Inducing multiple hydraulic fractures for a horizontal wellbore, in *Proceedings of SPE Annual Technical Conference and Exhibition*, Houston, TX, October 2–5. Society of Petroleum Engineers, Allen, TX.

Parshall, J. (2008). Barnett Shale showcases tight-gas development. *Journal of Petroleum Technology*, 60(9):48–55.

PLS. (2008). *Other Players Reporting Haynesville Success*. Petroleum Listing Services, Houston, TX, August 15.

Railroad Commission of Texas. (2008). *Newark, East (Barnett Shale) Field*. Oil and Gas Division, Railroad Commission of Texas, Austin.

Satterfield, J., Mantell, M., Kathol, D., Hebert, F., Patterson, K., and Lee, R. (2008). Managing Water Resource's Challenges in Select Natural Gas Shale Plays, paper presented at the Ground Water Protection Council Annual Forum, Cincinnati, OH, September 21–24, 2008.

Schlumberger. (2008a). *The Many Roles of Drilling Fluids*. Schlumberger, Houston, TX (http://www.planetseed.com/node/15300).

Schlumberger. (2008b). *PowerSTIM Service Increases Field Production*. Schlumberger, Houston, TX (http://www.slb.com/~/media/Files/dcs/case_studies/powerstim_us_escondido.ashx).

Schlumberger. (2011). *Better Turns for Rotary Steerable Drilling*. Schlumberger, Houston, TX (http://www.slb.com/resources/publications/oilfield_review/ori/ori002.aspx).

Schlumberger. (2012). *The Oil Field Glossary: Where the Oil Field Meets the Dictionary*. Schlumberger, Houston, TX (www.glossary.oilfield.slb.com/).

Singh, Jr., M.M. (1965). Mechanism of drilling wells with air as the drilling fluid, in *Proceedings of Conference on Drilling and Rock Mechanics*, Austin, TX, January 18–19. Society of Petroleum Engineers, Allen, TX.

Soeder, D.J. (1986). Porosity and permeability of Eastern Devonian gas shale. *SPE Formation Evaluation*, 1(1):116–124.

Sumi, L. (2008). *Shale Gas Focus on the Marcellus Shale*. Oil and Gas Accountability Project (OGAP)/ Earthworks, Washington, DC, 25 pp.

Swaco, M. (2006). *RECLAIM Technology: The System That Extends the Life of Oil- and Synthetic-Base Drilling Fluids While Reducing Disposal and Environmental Costs*. M-I Swaco, Houston, TX.

USDOE. (2009). *Modern Shale Gas Development in the United States: A Primer*. U.S. Department of Energy, Washington, DC.

USEIA. (2011). *Today in Energy: Haynesville Surpasses Barnett as the Nation's Leading Shale Play*. U.S. Energy Information Administration, Washington, DC (205.254.135.7/todayinenergy/detail.cfm?id=570).

USGS. (2000). *Summary of Water Use in the United States*. U.S. Geological Survey, Washington, DC (http://ga.water.usgs.gov/edu/wateruse2000.html).

USGS. (2008). *National Oil and Gas Assessment*. U.S. Geological Survey, Washington, DC (http://energy.usgs.gov/OilGas/AssessmentsData/NationalOilGasAssessment. aspx).

Veatch, Jr., R.W., Meschovitis, Z.A., and Fast, C.R. (1999). An overview of hydraulic fracturing, in Gidley, J.L. et al., Eds., *Recent Advances in Hydraulic Fracturing*. Society of Petroleum Engineers, Allen, TX.

Vulgamore, T., Clawson, T., Pope, C., Wolhart, W., Mayerhofer, M., and Waltman, C. (2007). *Applying Hydraulic Fracture Diagnostics to Optimize Stimulations in the Woodford Shale*, SPE-110029. Society of Petroleum Engineers, Allen, TX.

Walser, D. (2010). Mapping hydraulic fracturing in the Permian Basin—microseismic mapping pinpoints fracturing and aids well stimulation at drilling sites. *Upstream Pumping Solutions*, Fall (www.upstreampumping.com/article/well-completion-stimulation/mapping-hydraulic-fracturing-permian-basin-geometry-and-placemen).

West Virginia Geological and Economic Survey. (1997). *Enhancement of the Appalachian Basin Devonian Shale Resource Base in the GRI Hydrocarbon Model*, GRI Contract No. 5095-890-3478, prepared for Gas Research Institute, Arlington, VA.

Williams, P. (2008). A vast ocean of natural gas. *American Clean Skies*, Summer, pp. 44–50.

3 Fracking Water Supply

You can't frack for shale gas without water. You can't frack for shale gas without a whole bunch of water ... I'm talkin' millions of gallons, mister!

—Pennsylvania hydraulic fracturing engineer (2011)

FRACKING WATER USE

As the Pennsylvania engineer pointed out, the process of drilling and hydraulic fracturing of a horizontal shale gas well requires water—2 to 4 million gallons of water (Satterfield et al., 2008), with about 3 million gallons being most common. Note that the volume of water required may vary substantially between wells. In addition, the volume of water required per foot of wellbore appears to be decreasing as technology and methods improve over time. Table 3.1 presents data regarding estimated per-well water needs for four shale gas plays currently being developed.

Cumulatively, hydraulic fracturing activities in the United States used on average 44 billion gallons of water a year in 2011 and 2012, according to the U.S. Environmental Protection Agency's analysis of data from the FracFocus Chemical Disclosure Registry (https://fracfocus.org/). Although this represents less than 1% of total annual water use and consumption at this scale, water withdrawals could potentially impact the quantity and quality of drinking water resources at more local scales. Water for drilling and hydraulic fracturing of these wells frequently comes from surface water bodies such as rivers and lakes, but it can also come from groundwater, private water sources, municipal water, and reused produced water. Most of the producing shale gas basins contain large amounts of local water sources.

DID YOU KNOW?

For very deep wells, up to 4.5 million gallons of water may be needed to drill and fracture a shale gas well; this is equivalent to the amount of water consumed by (Chesapeake Energy, 2012):

- New York City in approximately 7 minutes
- A 1000-megawatt coal-fired power plant in 12 hours
- A golf course in 25 days
- 7.5 acres of corn in a season

DID YOU KNOW?

Water use is water withdrawn from groundwater or surface water for a specific purpose. Part or all of it may be returned to the local or urban hydrologic cycle. If no water is returned, water use equals *water consumption*. Water consumption is water that is removed from the local hydrological cycle following its use (e.g., via evaporation, transpiration, incorporation into products or crops, consumption by humans or livestock) and is therefore unavailable to other water uses (Maupin et al., 2014). In the case of hydraulic fracturing, water can be consumed by the loss of injected water to subsurface zones or via underground disposal of wastewaters, among other means.

TABLE 3.1

Estimated Water Needs for Drilling and Fracturing Wells in Select Shale Gas Plays

Shale Gas Play (gal)	Volume of Drilling Water per Well (gal)	Volume of Fracturing Water per Well (gal)	Total Volumes of Water per Well (gal)
Barnett Shale	400,000	2,300,000	2,700,000
Fayetteville Shale	60,000[a]	2,900,000	3,060,000
Haynesville Shale	1,000,000	2,700,000	3,700,000
Marcellus Shale	80,000[a]	3,800,000	3,880,000

Source: Based on data (2008) from ALL Consulting for a fracture operation in the Fayetteville Shale.

Note: These volumes are approximate and may vary substantially between wells.

[a] Drilling performed with an air mist and/or water-based or oil-based muds for deep horizontal well completions.

DID YOU KNOW?

In 2011, natural gas use in the United States accounted for 25% of all energy used. By 2035, it is estimated that 46% of all natural gas will come from fracking.

Even though water volumes needed to drill and stimulate shale gas wells are large, they occur in areas with moderate to high levels of annual precipitation. However, even in areas of high precipitation, due to growing populations, other industrial water demands, and seasonal variation in precipitation, it can be difficult to meet the needs of shale gas development and still satisfy regional needs for water.

Even though the water volumes required to drill and stimulate shale gas wells are large, they generally represent a small percentage of the total water resource used in the shale gas basins. Calculations indicate that water use will range from less than 0.1 to 0.8% by basin (Satterfield, 2008). This volume is small in terms of the overall surface water budget for an area; however, operators need this water when drilling activity is occurring (on demand), requiring that the water be procured over a relatively short period of time. Water withdrawals during periods of low stream flow could affect fish and other aquatic life, fishing and other recreational activities, municipal water supplies, and other industries such as power plants. To put shale gas water use in perspective, the consumptive use of freshwater for electrical generation in the Susquehanna River Basin alone is nearly 150 million gallons per day, whereas the projected total demand for peak Marcellus Shale activity in the same area is 8.4 million gallons per day (Gaudlip et al., 2008).

One alternative that states and operators are pursuing is to make use of seasonal changes in river flow to capture water when surface water flows are greatest. Utilizing seasonal flow differences allows planning of withdrawals to avoid potential impacts to municipal drinking water supplies or to aquatic or riparian communities. In the Fayetteville Shale play of Arkansas, one operator is constructing a 500-ac-ft impoundment to store water withdrawals from the Little Red River obtained during periods of high flow (storm events or hydroelectric power generation releases from Greer's Ferry dam upstream of the intake) when excess water is available (Chesapeake Energy, 2008a). The

DID YOU KNOW?

One acre-foot (ac-ft) of water is equivalent to the volume of water required to cover one acre with one foot of water.

DID YOU KNOW?

The average fracking well requires around 5,000,000 gal of water to operate over its lifetime.

project is limited to 1550 acre-ft of water annually. As additional mitigation, the company has constructed extra pipelines and hydrants to provide portions of this rural area with water for fire protection. Also included is monitoring of in-stream water quality as well as game and non-game fish species in the reach of river surrounding the intake. This design provides a water recovery system similar in concept to what some municipal water facilities use. It will minimize the impact on local water supplies because surface water withdrawals may still be limited to times of excess flow in the Little Red River. This project was developed with input from a local chapter of Trout Unlimited, an active conservation organization in the area, and represents an innovative environmental solution that serves both the community and the gas developer.

These water needs may challenge supplies and infrastructure in new areas of shale gas development—that is, in areas where the impact of shale gas operations is new and the potential impact is unknown to local inhabitants and governing officials. As operators look to develop new shale gas plays, a failure to communicate with local officials is not a viable option. Communication with local water planning agencies can help operators and communities to coexist and effectively manage local water resources. Understanding local water needs can help operators develop a water storage or management plan that will meet with acceptance in neighboring communities. Although the water needed for drilling an individual well may represent a small volume over a large area, the withdrawals may have a cumulative impact on watersheds over the short term. This potential impact can be avoided by working with local water resource managers to develop a plan outlining when and where withdrawals will occur (i.e., avoiding headwaters, tributaries, small surface water bodies, or other sensitive sources).

Across the United States, the vast majority of water used in hydraulic fracturing is fresh, although operators also make use of lower-quality water, including reused hydraulic fracturing wastewater. Based on available data, the median reuse of wastewater as a percentage of injected volumes is 5% nationally, with the percentage varying by location. Available data on reuse trends indicate increased reuse of wastewater over time in both Pennsylvania and West Virginia. Reuse as a percentage of injected volumes is lower in other areas, including regions with more water stress, likely because of the availability of disposal wells; for example, reused wastewater is approximately 18% of injected volumes in the Marcellus Shale in Pennsylvania's Susquehanna River Basin, whereas it is approximately 5% in the Barnett Shale in Texas.

Before a shale gas play hydraulic fracking operation is developed, not only is it a good idea to communicate with state and local government officials but it is also prudent to obtain information and data related to the urban water cycle (i.e., if the shale gas operation is near or has an impact on the surrounding area). Moreover, a study of the effect of shale gas hydraulic fracking operations on the indirect water reuse process is called for. In the planning stage, close scrutiny must be applied to water sources, outfalls, annual precipitation levels, drought histories, indirect water reuse, and (if located in or near an urban area) the urban water cycle.

DID YOU KNOW?

Reused wastewater as a percentage of injected water differs from the percentage of wastewater that is managed through reuse, as opposed to other wastewater management options. For example, in the Marcellus Shale in Pennsylvania, approximately 18% of injected water is reused produced water, while approximately 70% of wastewater or more is managed through reuse.

Today we tend to computer model this, that, and whatever, and few can doubt the worth and advantage of computer modeling. Sometimes, though, it is best to revert back to the old adage that a picture, a sketch, or simple drawing is worth a billion words. Computer models can be complicated to conceive, depict, and understand. A simple picture is … well, it is just simple, straightforward, fundamental, eye catching, and usually understandable. The simple picture referred to is a basic (*basic* being key) drawing of the local urban surface water cycle and the regional indirect surface water reuse process (Figures 3.1 and 3.2). Note that Figure 3.1 also shows the potential discharge of used but "treated" frack water to the water supply for reuse. This scenario is likely only if some type of *in situ* wastestream treatment process is available to treat the frack fluid waste with, for example, an advanced oxidation process (AOP) and reverse osmosis (RO). Such a mobile treatment option could result in 75 to 80% treated clean water for discharge into the receiving body (river or lake) and the rest could be reused on site. The basic drawings represented by Figures 3.1 and 3.2, along with data pertaining to annual rainfall levels and current usage in gallons of surface water sources, would go a long way (in understandable terminology) toward showing local officials the possible

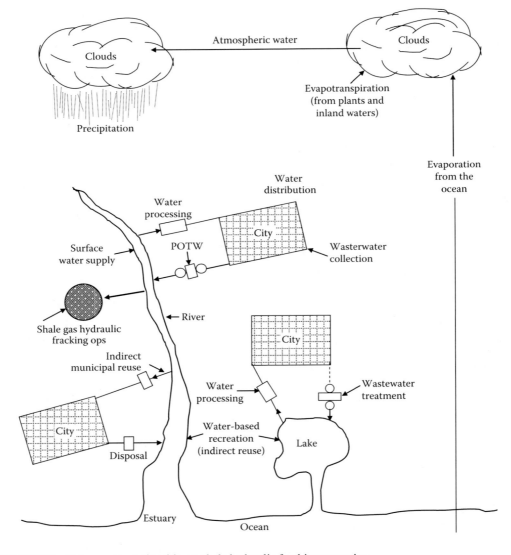

FIGURE 3.1 Urban water cycle with gas shale hydraulic fracking operation.

Fracking Water Supply

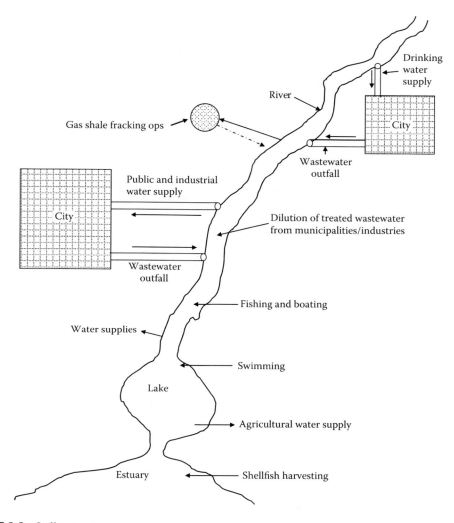

FIGURE 3.2 Indirect water reuse process.

impact of a nearby shale gas hydraulic fracking operation. When the shale gas fracking water supply is obtained from groundwater and not surface water sources, a simplified diagram along the lines of Figures 3.1 and 3.2 should be drawn showing the aquifer interface with the drilling and fracking operation in general and the water usage from the underground source in particular. Note that a more detailed discussion of frack water management is presented later in the text.

HYDRAULIC FRACTURING AND DRINKING WATER QUALITY

Cumulatively, hydraulic fracking uses billions of gallons of water each year at the national and state scales, and even in some counties. As noted earlier, hydraulic fracturing wastewater use is generally less than 1% of total annual water used and consumed at these scales. However, few counties in the United States have higher percentages. For 2011 and 2012, annual hydraulic fracturing water use was 10% or more compared to 2010 total annual water use in 6.5% of counties analyzed by the USEPA, 30% or more in 2.2% of counties, and 50% or more in 1.0% of counties. Consumption estimates followed the same general pattern. In these counties, hydraulic fracturing is a relatively large use and consumer of water (USEPA, 2015).

With regard to drinking water resources, high fracturing water use or consumption alone does not always result in impacts. Instead, impacts result from the combination of water use or consumption and water availability at local scales. To date, not one case has been reported where hydraulic fracturing water use or consumption alone caused a drinking water well or stream to run dry. This could indicate an absence of effects or a lack of documentation in the literature. Note that water availability is rarely impacted by just one use or factor alone. In Louisiana, for example, the state requested that hydraulic fracturing operations switch from using groundwater to surface water, due to concerns that groundwater withdrawals for fracturing could, in combination with other uses, adversely affect drinking water supplies.

Areas with the highest fracturing water use and lowest water availability have the greatest potential for impacts on drinking water resources. Southern and western Texas are two locations where hydraulic fracturing water use, low water availability, drought, and reliance on declining groundwater have the potential to affect the quantity of drinking water resources. Any impacts are likely to be realized locally within these areas. In a detailed case study of southern Texas, Scanlon et al. (2014) observed generally adequate water supplies for hydraulic fracturing except in specific locations. They found excessive drawdown of local groundwater in a small proportion (approximately 6% of the area) of the Eagle Ford Shale. The authors suggested that water management, particularly a shift toward brackish water use, could minimize potential future impacts to freshwater resources.

The potential for impacts to drinking water quantity due to hydraulic fracturing water use appears to be lower—but not eliminated—in other areas of the United States. Future problems could arise if hydraulic fracturing increases substantially in areas with low water availability or in times of water shortages. In detailed case studies in western Colorado and northeastern Pennsylvania, the USEPA and the author did not find current impacts but did conclude that streams could be vulnerable to water withdrawals from hydraulic fracturing. In northeast Pennsylvania, water management, such as minimum streamflow requirements, limits the potential for impacts, especially in small streams. In western North Dakota, groundwater is limited, but the industry may have sufficient supplies of surface water from the Missouri River systems. These location-specific examples emphasize the need to focus on regional and local dynamics when considering potential impacts of hydraulic fracturing water acquisition on drinking water resources.

With regard to the possible impacts on water quality of water withdrawals for hydraulic fracturing, the practice, like all other water uses for fracking operations, has the potential to alter the quality of drinking water resources. Groundwater withdrawals exceeding natural recharge rates can reduce water storage levels in aquifers, potentially mobilizing contaminants or allowing the infiltration of lower quality water from the land surface or adjacent formations. Withdrawals could also decrease groundwater discharge to streams, potentially affecting surface water quality. Areas with large amounts of sustained groundwater pumping are most likely to experience impacts, particularly drought-prone regions with limited groundwater recharge.

Surface water withdrawals also have the potential to affect water quality. Withdrawals may lower water levels and alter stream flow, potentially decreasing a stream's capacity to dilute contaminants. Case studies by the USEPA show that streams can be vulnerable to changes in water quality due to water withdrawals, particularly smaller streams and during periods of low flow. Management of the rate and timing of surface water withdrawals has been shown to help mitigate potential impacts of hydraulic fracturing withdrawals on water quality.

THOUGHT-PROVOKING QUESTIONS

1. Do you think the use of stream water for fracking is a good practice? Explain.
2. Do you think the use of groundwater for fracking is a good practice? Explain.
3. Do you think the use of any water source for fracking is a good practice? Explain.

REFERENCES AND RECOMMENDED READING

Ajani, A. and Kelkar, M. (2012). Interference Study in Shale Plays, paper presented at SPE Hydraulic Fracturing Technology Conference, The Woodlands, TX, February 6–8.

Anon. (2008). *Projecting the Economic Impact of the Fayetteville Shale Play for 2008–2012.* Center for Business and Economic Research, Sam M. Walton College of Business, Fayetteville, AR.

Arthur, J.D., Bohm, B., and Lane, M. (2008). *An Overview of Modern Shale Gas Development in the United States.* ALL Consulting, Tulsa, OK.

Bair, E.S., Freeman, D.C., and Senko, J.M. (2010). *Subsurface Gas Invasion, Bainbridge Township, Geauga County, Ohio.* Ohio Department of Natural Resources, Columbus.

Boyer, C., Kieschnick, J., Suarez-Rivera, R., Lewis, R., and Walter, G. (2006). Producing gas from its source. *Oilfield Review,* 18(3):36–49.

Cardott, B. (2004). *Overview of Unconventional Energy Resources of Oklahoma.* Oklahoma Geological Survey, Norman, OK (www.ogs.ou.edu/fossilfuels/coalpdfs/UnconventionalPresentation.pdf).

Catacosinos, P., Harrison, W., Reynolds, R., Westjohn, D., and Wollensak. M. (2000). *Stratigraphic Nomenclature for Michigan.* Michigan Department of Environmental Quality, Geological Survey Division, and Michigan Basin Geological Survey, Kalamazoo, MI.

CBO. (1983). *Understanding Natural Gas Price Decontrol.* Congressional Budget Office, Congress of the United States, Washington, DC.

Cramer, D. (2008). Stimulating unconventional reservoirs: lessons learned, successful practices, areas for improvement, in *Proceedings of SPE Unconventional Reservoirs Conference,* Keystone, CO, February 10–12. Society of Petroleum Engineers, Allen, TX.

Franz, Jr., J.H. and Jochen, V. (2005). *When Your Gas Reservoir Is Unconventional, So Is Our Solution,* Shale Gas White Paper 05-OF-299. Schlumberger, Houston, TX.

Gaudlip, A.W., Paugh, L.O., and Hayes, T.D. (2008). Marcellus Shale water management challenges in Pennsylvania, in *Proceedings of SPE Shale Gas Production Conference,* Fort Worth, TX, November 16–18. Society of Petroleum Engineers, Allen, TX.

Harrison, W. (2006). *Production History and Reservoir Characteristics of the Antrim Shale Gas Play, Michigan Basin.* Western Michigan University, Kalamazoo.

Hayden, J. and Pursell, D. (2005). *The Barnett Shale: Visitor's Guide to the Hottest Gas Play in the US.* Pickering Energy Partners, Inc., Houston, TX.

HIE. (2007). *Fayetteville Shale Power.* Hillwood International Energy, Dallas, TX.

IGS. (1986). *General Stratigraphic Column for Paleozoic Rocks in Indiana.* Indiana Geological Survey, Indiana University, Bloomington, IN (www.usi.edu/science/geology/My%20Web%20Sites/stratcolumn.pdf).

Jennings, Jr., A.R. and Darden, W.G. (1979). Gas well stimulation in the eastern United States, in *Proceedings of Symposium on Low Permeability Gas Reservoirs,* Denver, CO, May 20–22. Society of Petroleum Engineers, Allen, TX.

Jochen, V. (2006). *New Technology Needs to Produce Unconventional Gas.* Schlumberger, Houston, TX.

Johnston III, J., Heinrich, B., Lovelace, J., McCulluh, R., and Zimmerman, R. (2000). *Stratigraphic Charts of Louisiana.* Louisiana Geological Survey, Baton Rouge.

Lesley, J.P. (1892). *A Summary Description of the Geology of Pennsylvania,* Vol. II. Geological Survey of Pennsylvania, Pittsburgh.

Maupin, M.A., Kenny, J.F., Hutson, S.S., Lovelace, J.K., Barber, N.L., and Linsey, K.S. (2014). *Estimated Use of Water in the United States in 2010,* USGS Circular 1405. Reston, VA: U.S. Geological Survey.

Meyer & Associates. (2012). *Meyer Fracturing Simulators User's Guide,* http://www.mfrac.com/documentation.html.

Michie & Associates. (1988). Oil and Gas Water Injection Well Corrosion, paper prepared for the American Petroleum Institute, Washington, DC.

Michie, T.W. and Koch, C.A. (1991). Evaluation of injection-well risk management in the Williston Basin. *Journal of Petroleum Technology,* 43(6):737–741.

NRCS. (2008). *National Water & Climate Center.* Natural Resources Conservation Service, Washington, DC (www.wcc.nrcs.usda.gov/).

Overbey, W.K., Yost II, A.B., and Wilkins, D.A. (1988). Inducing multiple hydraulic fractures for a horizontal wellbore, in *Proceedings of SPE Annual Technical Conference and Exhibition,* Houston, TX, October 2–5. Society of Petroleum Engineers, Allen, TX.

Parshall, J. (2008). Barnett Shale showcases tight-gas development. *Journal of Petroleum Technology,* 60(9):48–55.

PLS. (2008). *Other Players Reporting Haynesville Success*. Petroleum Listing Services, Houston, TX, August 15.

Railroad Commission of Texas. (2008). *Newark, East (Barnett Shale) Field*. Oil and Gas Division, Railroad Commission of Texas, Austin.

Satterfield, J., Mantell, M., Kathol, D., Hebert, F., Patterson, K., and Lee, R. (2008). Managing Water Resource's Challenges in Select Natural Gas Shale Plays, paper presented at the Ground Water Protection Council Annual Forum, Cincinnati, OH, September 21–24.

Scanlon, B.R., Reedy, B.C., and Nicot, J.P. (2014). Will water scarcity in semiarid regions limit hydraulic fracturing of shale plays? *Environmental Research Letters*, 9(12).

Schlumberger. (2008a). *The Many Roles of Drilling Fluids*. Schlumberger, Houston, TX (http://www.planet-seed.com/node/15300).

Schlumberger. (2008b). *PowerSTIM Service Increases Field Production*. Schlumberger, Houston, TX (http://www.slb.com/~/media/Files/dcs/case_studies/powerstim_us_escondido.ashx).

Schlumberger. (2011). *Better Turns for Rotary Steerable Drilling*. Schlumberger, Houston, TX (http://www.slb.com/resources/publications/oilfield_review/ori/ori002.aspx).

Singh, Jr., M.M. (1965). Mechanism of drilling wells with air as the drilling fluid, in *Proceedings of Conference on Drilling and Rock Mechanics*, Austin, TX, January 18–19. Society of Petroleum Engineers, Allen, TX.

USEIA. (2011). *Today in Energy: Haynesville Surpasses Barnett as the Nation's Leading Shale Play*. U.S. Energy Information Administration, Washington, DC (205.254.135.7/todayinenergy/detail.cfm?id=570).

USEPA. (2015). *Assessment of the Potential Impacts of Hydraulic Fracturing for Oil and Gas on Drinking Water Resources*. Washington, DC: U.S. Environmental Protection Agency.

USGS. (2000). *Summary of Water Use in the United States*. U.S. Geological Survey, Washington, DC (http://ga.water.usgs.gov/edu/wateruse2000.html).

USGS. (2008). *National Oil and Gas Assessment*. U.S. Geological Survey, Washington, DC (http://energy.usgs.gov/OilGas/AssessmentsData/NationalOilGasAssessment. aspx).

Vulgamore, T., Clawson, T., Pope, C., Wolhart, W., Mayerhofer, M., and Waltman, C. (2007). *Applying Hydraulic Fracture Diagnostics to Optimize Stimulations in the Woodford Shale*, SPE-110029. Society of Petroleum Engineers, Allen, TX.

Walser, D. (2010). Mapping hydraulic fracturing in the Permian Basin—microseismic mapping pinpoints fracturing and aids well stimulation at drilling sites. *Upstream Pumping Solutions*, Fall (www.upstreampumping.com/article/well-completion-stimulation/mapping-hydraulic-fracturing-permian-basin-geometry-and-placemen).

West Virginia Geological and Economic Survey. (1997). *Enhancement of the Appalachian Basin Devonian Shale Resource Base in the GRI Hydrocarbon Model*, GRI Contract No. 5095-890-3478, prepared for Gas Research Institute, Arlington, VA.

4 Composition of Fracking Water

My sweetness is to wake in the night after days of dry heat, hearing the rain.

—Wendell Berry, novelist

A colorless, odorless, tasteless liquid, water is the only common substance that occurs naturally on the earth in all three physical states: solid, liquid, and gas. Seventy-three percent of the earth's surface is covered with it, almost 328,000,000 cubic miles. The human body is seventy percent water by weight; it is essential to the life of every living thing.

Hauser (1996)

Water, in any of its forms, also ... [has] scant respect for the laws of chemistry.

Most materials act either as acids or bases, settling on either side of a natural reactive divided. Not water. It is one of the few substances that can behave both as an acid and as a base, so that under certain conditions it is capable of reacting chemically with itself. Or with anything else.

Molecules of water are off balance and hard to satisfy. They reach out to interfere with every other molecule they meet, pushing its atoms apart, surrounding them, and putting them into solution. Water is the ultimate solvent, wetting everything, setting other elements free from the rocks, making them available for life. Nothing is safe. There isn't a container strong enough to hold it.

Watson (1988)

After sitting idle for two decades, there's steam billowing from the top of the big old steel plant in Youngstown, Ohio. This does not represent a renewal of the steel production that once created the Rust Belt. Instead, this a product of a new industry proponents say can be a game changer, not just for the depressed Youngstown–Warren area, but for the U.S. economy and the bigger energy game. It is the exploitation of [gas] shale.

Ravve (2011)

WATER CHEMISTRY

Before discussing the contents of fracking water and its specific functions, it is important to have a basic understanding of what water is. Water is a unique molecule. Although no one has seen a water molecule, we have determined that atoms in water are elaborately meshed. Moreover, although it is true that we do not know as much as we need to know about water—our growing knowledge of water is a work in progress—we have determined many things about water. A large amount of our current knowledge comes from studies of water chemistry.

Water chemistry is important because several factors about water to be treated and then distributed or returned to the environment are determined through simple chemical analysis. Probably the most important determination that the water practitioner makes about water is its hardness.

CHEMISTRY CONCEPTS AND DEFINITIONS

Chemistry, like the other sciences, has its own language; thus, to understand chemistry, it is necessary to understand the following concepts and key terms.

Concepts

Miscibility and Solubility

Substances that are *miscible* are capable of being mixed in all proportions. Simply, when two or more substances disperse themselves uniformly in all proportions when brought into contact, they are said to be completely soluble in one another, or completely miscible. The precise chemistry definition is a "homogeneous molecular dispersion of two or more substances" (Jost, 1992). Examples include the following observations:

- All gases are completely miscible.
- Water and alcohol are completely miscible.
- Water and mercury (in its liquid form) are immiscible liquids.

Between the two extremes of miscibility is a range of *solubility*; that is, various substances mix with one another up to a certain proportion. In many environmental situations, a rather small amount of a contaminant may be soluble in water in contrast to the complete miscibility of water and alcohol. The amounts are measured in parts per million (ppm).

Suspension, Sediment, Particles, and Solids

Often water carries solids or particles in *suspension*. These dispersed particles are much larger than molecules and may be comprised of millions of molecules. The particles may be suspended in flowing conditions and initially under quiescent conditions, but eventually gravity causes settling of the particles. The resultant accumulation by settling is referred to as *sediment* or *biosolids* (sludge) or *residual solids* in wastewater treatment vessels. Between this extreme of readily falling out by gravity and permanent dispersal as a solution at the molecular level are intermediate types of dispersion or suspension. Particles can be so finely milled or of such small intrinsic size as to remain in suspension almost indefinitely and in some respects similarly to solutions.

Emulsions

Emulsions represent a special case of a suspension. As you know, oil and water do not mix. Oil and other hydrocarbons derived from petroleum generally float on water with negligible solubility in water. In many instances, oils may be dispersed as fine oil droplets (an emulsion) in water and not readily separated by floating because of size or the addition of dispersal-promoting additives. Oil and, in particular, emulsions can prove detrimental to many treatment technologies and must be treated in the early steps of a multistep treatment train.

Ions

An ion is an electrically charged particle; for example, sodium chloride or table salt forms charged particles on dissolution in water. Sodium is positively charged (a cation), and chloride is negatively charged (an anion). Many salts similarly form cations and anions upon dissolution in water.

Mass Concentration

Concentration is often expressed in terms of parts per million (ppm) or mg/L. Sometimes parts per thousand (ppt) and parts per billion (ppb) are also used:

$$\text{ppm} = \text{Mass of substance} \div \text{Mass of solutions} \tag{4.1}$$

Because 1 kg of solution with water as a solvent has a volume of approximately 1 liter,

$$1 \text{ ppm} \approx 1 \text{ mg/L}$$

Composition of Fracking Water

Key Definitions

Chemistry—The science that deals with the composition and changes in composition of substances. Water is an example of this composition; it is composed of two gases: hydrogen and oxygen. Water also changes form from liquid to solid to gas but does not necessarily change composition.

Matter—Anything that has weight (mass) and occupies space. Kinds of matter include elements, compounds, and mixtures.

Solids—Substances that maintain definite size and shape. Solids in water fall into one of the following categories:

- *Dissolved solids*—Solids in a single-phase (homogeneous) solution consisting of dissolved components and water. Dissolved solids are the material in water that will pass through a glass fiber filter and remain in an evaporating dish after evaporation of the water.
- *Colloidal solids* (sols)—Solids that are uniformly dispersed in solution. They form a solid phase that is distinct from the water phase.
- *Suspended solids*—Solids in a separate phase from the solution. Some suspended solids are classified as settleable solids. Settleable solids are determined by placing a sample in a cylinder and measuring the amount of solids that have settled after a set amount of time. The size of solids increases moving from dissolved solids to suspended solids. Suspended solids are the material deposited when a quantity of water, sewage, or other liquid is filtered through a glass fiber filter.
- *Total solids*—The solids in water, sewage, or other liquids; include suspended solids (largely removable by a filter) and filterable solids (those that pass through the filter).

Liquids—Having a definite volume but not shape, liquids will fill containers to certain levels and form free level surfaces.

Gases—Having neither definite volume nor shape, gases completely fill any container in which they are placed.

Mixture—A physical, not chemical, intermingling of two of more substances. Sand and salt stirred together form a mixture.

Element—The simplest form of chemical matter. Each element has chemical and physical characteristics different from all other kinds of matter.

Compound—A substance of two or more chemical elements chemically combined. Examples include water (H_2O), which is a compound formed by hydrogen and oxygen. Carbon dioxide (CO_2) is composed of carbon and oxygen.

Molecule—The smallest particle of matter or a compound that possesses the same composition and characteristics as the rest of the substance. A molecule may consist of a single atom, two or more atoms of the same kind, or two or more atoms of different kinds.

Atom—The smallest particle of an element that can unite chemically with other elements. All the atoms of an element are the same in chemical behavior, although they may differ slightly in weight. Most atoms can combine chemically with other atoms to form molecules.

Radical—Two or more atoms that unite in a solution and behave chemically as if a single atom.

Ion—An atom or group of atoms that carries a positive or negative electric charge as a result of having lost or gained one or more electrons.

Ionization—The formation of ions by the splitting of molecules or electrolytes in solution. Water molecules are in continuous motion, even at lower temperatures. When two water molecules collide, a hydrogen ion is transferred from one molecule to the other. The water molecule that loses the hydrogen ion becomes a negatively charged hydroxide ion. The water molecule that gains the hydrogen ion becomes a positively charged hydronium ion. This process is commonly referred to as the *self-ionization of water.*

Cation—A positively charged ion.

Anion—A negatively charged ion.

Organic—Chemical substances of animal or vegetable origin having a carbon structure.

Inorganic—Chemical substances of mineral origin.

Solvent—The component of a solution that does the dissolving.

Solute—The component of a solution that is dissolved by the solvent.

Saturated solution—The physical state in which a solution will no longer dissolve more of the dissolving substance (solute).

Colloidal—Any substance in a certain state of fine division in which the particles are less than 1 micron in diameter.

Turbidity—A condition in water caused by the presence of suspended matter. Turbidity results in the scattering and absorption of light rays.

Precipitate—A solid substance that can be dissolved but is separated from the solution because of a chemical reaction or change in conditions such as pH or temperature.

WATER SOLUTIONS

A *solution* is a condition in which one or more substances are uniformly and evenly mixed or dissolved. A solution has two components, a *solvent* and a *solute*. The solvent is the component that does the dissolving. The solute is the component that is dissolved. In water solutions, water is the solvent. Water can dissolve many other substances; in fact, given enough time, not too many solids, liquids, or gases exist that water cannot dissolve. When water dissolves substances, it creates solutions with many impurities. Generally, a solution is usually transparent and not cloudy; however, a solution may have some color when the solute remains uniformly distributed throughout the solution and does not settle with time.

When molecules dissolve in water, the atoms making up the molecules come apart (dissociate) in the water. This dissociation in water is called *ionization*. When the atoms in the molecules come apart, they do so as charged atoms (both negatively and positively charged) called *ions*. The positively charged ions are called *cations* and the negatively charged ions are called *anions*. A good example of the ionization process is when calcium carbonate ionizes:

$$CaCO_3 \quad \leftrightarrow \quad Ca^{2+} \quad + \quad CO_3^{2-}$$

| Calcium carbonate | Calcium ion (cation) | Carbonate ion (anion) |

Another good example is the ionization that occurs when table salt (sodium chloride) dissolves in water:

$$NaCl \quad \leftrightarrow \quad Na^+ \quad + \quad Cl^-$$

| Sodium chloride | Sodium ion (cation) | Chloride ion (anion) |

Some of the common ions found in water and their symbols are provided below:

Hydrogen	H^+
Sodium	Na^+
Potassium	K^+
Chloride	Cl^-
Bromide	Br^-
Iodide	I^-
Bicarbonate	HCO_3^-

Composition of Fracking Water

Water dissolves polar substances better than nonpolar substances. This makes sense when we consider that water is a polar substance. Polar substances such as mineral acids, bases, and salts are easily dissolved in water, while nonpolar substances such as oils, fats, and many organic compounds do not dissolve easily in water. Water dissolves polar substances better than nonpolar substances, but only to a point; for example, only so much solute will dissolve at a given temperature. When that limit is reached, the resulting solution is saturated. When a solution becomes saturated, no more solute can be dissolved. For solids dissolved in water, if the temperature of the solution is increased, the amount of solids (solutes) required to reach saturation increases.

WATER CONSTITUENTS

Natural water can contain a number of substances (what we may call *impurities*) or constituents in water treatment operations. The concentrations of various substances in water in dissolved, colloidal, or suspended form are typically low but vary considerably. A hardness value of up to 400 ppm of calcium carbonate, for example, is sometimes tolerated in public supplies, whereas 1 ppm of dissolved iron would be unacceptable. When a particular constituent can affect the good health of the water user or the environment, it is considered a *contaminant* or *pollutant*. These contaminants, of course, are what the water operator removes from or tries to prevent from entering the water supply. In this section, we discuss some of the more common constituents of water.

Solids

Other than gases, all contaminants of water contribute to the solids content. Natural water carries many dissolved and undissolved solids. The undissolved solids are nonpolar substances and consist of relatively large particles of materials such as silt, that will not dissolve. Classified by their size and state, by their chemical characteristics, and by their size distribution, solids can be dispersed in water in both suspended and dissolved forms. The sizes of solids in water can be classified as *suspended*, *settleable*, *colloidal*, or *dissolved*. *Total solids* are the suspended and dissolved solids that remain behind when the water is removed by evaporation. Solids are also characterized as being *volatile* or *nonvolatile*. The distribution of solids is determined by computing the percentage of filterable solids by size range. Solids typically include inorganic solids such as silt and clay from riverbanks and organic matter such as plant fibers and microorganisms from natural or manmade sources.

> **Note:** Though not technically accurate from a chemical point of view because some finely suspended material can actually pass through the filter, *suspended solids* are defined as those that can be filtered out in the suspended solids laboratory test. The material that passes through the filter is defined as *dissolved solids*.

Colloidal solids are extremely fine suspended solids (particles) less than 1 micron in diameter; they are so small (though they still can make water cloudy) that they will not settle even if allowed to sit quietly for days or weeks.

Turbidity

Simply, turbidity refers to how clear the water is. The clarity of water is one of the first characteristics people notice. Turbidity in water is caused by the presence of suspended matter, which results in the scattering and absorption of light rays. The greater the amount of *total suspended solids* (TSS) in the water, the murkier it appears and the higher the measured turbidity. Thus, in plain English, turbidity is a measure of the light-transmitting properties of water. Natural water that is very clear (low turbidity) allows us to see images at considerable depths. High turbidity water, on the other hand, appears cloudy. Keep in mind that water of low turbidity is not necessarily without dissolved solids. Dissolved solids do not cause light to be scattered or absorbed; thus, the water looks clear. High turbidity causes problems for the waterworks operator, as components that cause high turbidity can cause taste and odor problems and will reduce the effectiveness of disinfection.

Color

Color in water can be caused by a number of contaminants such as iron, which changes in the presence of oxygen to yellow or red sediments. The color of water can be deceiving. In the first place, color is considered an aesthetic quality of water with no direct health impact. Second, many of the colors associated with water are not true colors but the result of colloidal suspension and are referred to as the *apparent color*. This apparent color can often be attributed to iron and to dissolved tannin extracted from decaying plant material. *True color* is the result of dissolved chemicals (most often organics) that cannot be seen. True color is distinguished from apparent color by filtering the sample.

Dissolved Oxygen

Although water molecules contain an oxygen atom, this oxygen is not what is needed by aquatic organisms living in our natural waters. A small amount of oxygen, up to about ten molecules of oxygen per million molecules of water, is actually dissolved in water. This dissolved oxygen (DO) is breathed by fish and zooplankton and is needed by them to survive. Other gases can also be dissolved in water. In addition to oxygen, carbon dioxide, hydrogen sulfide, and nitrogen are examples of gases that dissolve in water. Gases dissolved in water are important; for example, carbon dioxide is important because of the role it plays in pH and alkalinity. Carbon dioxide is released into the water by microorganisms and consumed by aquatic plants. Dissolved oxygen in water, however, is of the most importance to us here, not only because it is important to most aquatic organisms but also because dissolved oxygen is an important indicator of water quality.

Like terrestrial life, aquatic organisms need oxygen to live. As water moves past their breathing apparatus, microscopic bubbles of oxygen gas in the water—dissolved oxygen—are transferred from the water to their blood. Like any other gas diffusion process, the transfer is efficient only above certain concentrations. In other words, oxygen can be present in the water but at too low a concentration to sustain aquatic life. Oxygen also is needed by virtually all algae and macrophytes and for many chemical reactions that are important to water body functioning.

Rapidly moving water, such as in a mountain stream or large river, tends to contain a lot of dissolved oxygen, while stagnant water contains little. Bacteria in water can consume oxygen as organic matter decays; thus, excess organic material in our lakes and rivers can cause an oxygen-deficient situation to occur. Aquatic life can have a difficult time surviving in stagnant water that has a lot of rotting, organic material in it, especially in summer, when dissolved oxygen levels are at a seasonal low.

> *Note:* Solutions can become saturated with solute. This is the case with water and oxygen. As with other solutes, the amount of oxygen that can be dissolved at saturation depends on the temperature of the water. In the case of oxygen, the effect is just the opposite of other solutes. The higher the temperature, the lower the saturation level; the lower the temperature, the higher the saturation level.

Metals

Metals are elements present in chemical compounds as positive ions or in the form of cations (+ ions) in solution. Metals with a density over 5 kg/dm^3 are known as *heavy metals*. Metals are one of the constituents or impurities often carried by water. Although most of the metals are not harmful at normal levels, a few metals can cause taste and odor problems in drinking water. In addition, some metals may be toxic to humans, animals, and microorganisms. Most metals enter water as part of compounds that ionize to release the metal as positive ions. Table 4.1 lists some metals commonly found in water and their potential health hazards.

> *Note:* Metals may be found in various chemical and physical forms. These forms, or *species*, can be particles or simple organic compounds, organic complexes, or colloids. The dominating form is determined largely by the chemical composition of the water, the matrix, and in particular the pH.

TABLE 4.1
Common Metals Found in Water

Metal	Health Hazard
Barium	Circulatory system effects and increased blood pressure
Cadmium	Concentration in the liver, kidneys, pancreas, and thyroid
Copper	Nervous system damage and kidney effects; toxic to humans
Lead	Same as copper
Mercury	Central nervous system disorders
Nickel	Central nervous system disorders
Selenium	Central nervous system disorders
Silver	Gray skin
Zinc	Taste effects; not a health hazard

Organic Matter

Organic matter or compounds are those that contain the element carbon and are derived from material that was once alive (i.e., plants and animals). Organic compounds include fats, dyes, soaps, rubber products, plastics, wood, fuels, cotton, proteins, and carbohydrates. Organic compounds in water are usually large, nonpolar molecules that do not dissolve well in water. They often provide large amounts of energy to animals and microorganisms.

> *Note:* *Natural organic matter* (NOM) is used to describe the complex mixture of organic material, such as humic and hydrophilic acids, present in all drinking water sources. NOM can cause major problems in the treatment of water as it reacts with chlorine to form *disinfection byproducts* (DBPs). Many of the DBPs formed by the reaction of NOM with disinfectants are reported to be toxic and carcinogenic to humans if ingested over an extended period. The removal of NOM and subsequent reduction in DBPs are major goals in the treatment of any water source.

Inorganic Matter

Inorganic matter or compounds are carbon free, not derived from living matter, and easily dissolved in water; inorganic matter is of mineral origin. The inorganics include acids, bases, oxides, and salts. Several inorganic components are important in establishing and controlling water quality. Two important inorganic constituents in water are nitrogen and phosphorus.

Acids

Lemon juice, vinegar, and sour milk are acidic or contain acid. The common acids used in waterworks operations are hydrochloric acid (HCl), sulfuric acid (H_2SO_4), nitric acid (HNO_3), and carbonic acid (H_2CO_3). Note that in each of these acids, hydrogen (H) is one of the elements. The relative strengths of acids in water (listed in descending order of strength) are shown in Table 4.2.

> *Note:* An acid is a substance that produces hydrogen ions (H^+) when dissolved in water. Hydrogen ions are hydrogen atoms stripped of their electrons. A single hydrogen ion is nothing more than the nucleus of a hydrogen atom.

> *Note:* Acids and bases become solvated and loosely bond to water molecules.

Bases

A base is a substance that produces hydroxide ions (OH^-) when dissolved in water. Lye, or common soap, contains bases. The bases used in waterworks operations are calcium hydroxide, $Ca(OH)_2$; sodium hydroxide, NaOH; and potassium hydroxide, KOH. Note that the hydroxyl group (OH) is

TABLE 4.2
Relative Strengths of Acids in Water

Acid	Formula
Perchloric acid	$HClO_4$
Sulfuric acid	H_2SO_4
Hydrochloric acid	HCl
Nitric acid	HNO_3
Phosphoric acid	H_3PO_4
Nitrous acid	HNO_2
Hydrofluoric acid	HF
Acetic acid	CH_3COOH
Carbonic acid	H_2CO_3
Hydrocyanic acid	HCN
Boric acid	H_3BO_3

found in all bases. In addition, note that bases contain metallic substances, such as sodium (Na), calcium (Ca), magnesium (Mg), and potassium (K). These bases contain the elements that produce the alkalinity in water.

Salts

When acids and bases chemically interact, they neutralize each other. The compound (other than water) that forms from the neutralization of acids and bases is called a *salt*. Salts constitute, by far, the largest group of inorganic compounds. A common salt used in waterworks operations, copper sulfate, is utilized to kill algae in water. This copper sulfate that is added intentionally should not be confused with the naturally occurring sulfates found in drinking water. With regard to produced wastewater, salinity is its general attribute; further, it is a combination of inorganic and organic compounds (Fakhru'l-Razo et al., 2009). The properties of produced wastewater vary depending on the geographic location of the field, the geological host formation, and the type of hydrocarbon product being produced (Veil et al., 2004). Salinity or salt concentration, described as total dissolved solids (TDS), can vary in conventional oil and gas well produced waters from 1000 to 400,000 mg/L (USGS, 2002). Variations in TDS are related to geologic variations between basins, well location in a well field, and the resource produced.

pH

pH is a measure of the hydrogen ion (H^+) concentration. Solutions range from very acidic (having a high concentration of H^+ ions) to very basic (having a high concentration of OH^- ions). The pH scale ranges from 0 to 14, with 7 being the neutral value. The pH of water is important to the chemical reactions that take place within water, and pH values that are too high or low can inhibit the growth of microorganisms. High pH values are considered basic, and low pH values are considered acidic. Stated another way, low pH values indicate a high H^+ concentration, and high pH values indicate a low H^+ concentration. Because of this inverse logarithmic relationship, each pH unit represents a tenfold difference in H^+ concentration. Natural water varies in pH depending on its source. Pure water has a neutral pH, with equal H^+ and OH^-. Adding an acid to water causes additional positive ions to be released, so the H^+ ion concentration goes up and the pH value goes down:

$$HCl \leftrightarrow H^+ + Cl^-$$

To control water coagulation and corrosion, the waterworks operator must test for the hydrogen ion concentration of the water to determine the pH of the water. In a coagulation test, as more alum

Composition of Fracking Water

(acid) is added, the pH value lowers. If more lime (alkali) is added, the pH value raises. This relationship should be remembered—if a good floc is formed, the pH should then be determined and maintained at that pH value until the raw water changes.

Pollution can change the pH of water, which in turn can harm animals and plants living in the water. Water coming out of an abandoned coal mine, for example, can have a pH of 2, which is very acidic and would definitely affect any fish crazy enough to try to live in that water. By using the logarithm scale, this mine-drainage water would be 100,000 times more acidic than neutral water— so stay out of abandoned mines.

Note: Seawater is slightly more basic (the pH value is higher) than most natural freshwater. Neutral water (such as distilled water) has a pH of 7, which is in the middle of being acidic and alkaline. Seawater happens to be slightly alkaline (basic), with a pH of about 8. Most natural water has a pH range of 6 to 8, although acid rain can have a pH as low as 4.

Alkalinity

Alkalinity is defined as the capacity of water to accept protons; it can also be defined as a measure of the ability of water to neutralize an acid. Bicarbonates, carbonates, and hydrogen ions cause alkalinity and create hydrogen compounds in a raw or treated water supply. Bicarbonates are the major components because of carbon dioxide action on basic materials of soil; borates, silicates, and phosphates may be minor components. The alkalinity of raw water may also contain salts formed from organic acids such as humic acids.

Alkalinity in water acts as a buffer that tends to stabilize and prevent fluctuations in pH. In fact, alkalinity is closely related to pH, but the two must not be confused. Total alkalinity is a measure of the amount of alkaline materials in the water. The alkaline materials act as buffers to changes in the pH. If the alkalinity is too low (below 80 ppm), the pH can fluctuate rapidly because of insufficient buffer. High alkalinity (above 200 ppm) results in the water being too buffered. Thus, having significant alkalinity in water is usually beneficial, because it tends to prevent quick changes in pH that interfere with the effectiveness of common water treatment processes. Low alkalinity also contributes to the corrosive tendencies of water.

Note: When alkalinity is below 80 mg/L, it is considered to be low.

Water Temperature

Water temperature is important not only to fishermen but also to industries and even fish and algae. A lot of water is used for cooling purposes in power plants that generate electricity. These plants need to cool the water to begin with and then generally release warmer water back to the environment. The temperature of the released water can affect downstream habitats. Temperature can also affect the ability of water to hold oxygen as well as the ability of organisms to resist certain pollutants.

Specific Conductance

Specific conductance is a measure of the ability of water to conduct an electrical current. It is highly dependent on the amount of dissolved solids (such as salt) in the water. Pure water, such as distilled water, will have a very low specific conductance, and seawater will have a high specific conductance. Rainwater often dissolves airborne gases and airborne dust while it is in the air and thus often has a higher specific conductance than distilled water. Specific conductance is an important water quality measurement because it gives a good idea of the amount of dissolved material in the water. When electrical wires are attached to a battery and light bulb and the wires are put into a beaker of distilled water, the light will not light. But, the bulb does light up when the beaker contains saline (saltwater). In saline water, the salt has dissolved and released free electrons, so the water will conduct an electric current.

TABLE 4.3
Water Hardness

Water Hardness Classification	mg/L CaCo$_3$
Soft	0–75
Moderately hard	75–150
Hard	150–300
Very hard	Over 300

Hardness

Hardness may be considered a physical or chemical characteristic or parameter of water. It represents the total concentration of calcium and magnesium ions, reported as calcium carbonate. Simply, the amount of dissolved calcium and magnesium in water determines its hardness. Hardness causes soaps and detergents to be less effective and contributes to scale formation in pipes and boilers. Hardness is not considered a health hazard; however, water that contains hardness must often be softened by lime precipitation or ion exchange. Hardwater can even shorten the life of fabrics and clothes. Low hardness contributes to the corrosive tendencies of water. Hardness and alkalinity often occur together, because some compounds can contribute both alkalinity and hardness ions. Hardness is generally classified as shown in Table 4.3.

CHEMICAL MIXING

Technological advances have permitted the industry to drill deeper in various directions, tapping into gas reserves with greater facility and profitability. These advances have allowed the mining of vast, newly discovered gas deposits; however, the new technology depends heavily on the use of undisclosed types and amounts of toxic chemicals (Colborn et al., 2010). Hydraulic fracturing fluids (mixtures) are developed from many of these undisclosed types and amounts of toxic chemicals to perform specific functions, such as creating and extending fractures, transporting proppant, and placing proppant in the fractures. The fluid generally consists of three parts: (1) the base fluid, which is the largest constituent by volume (typically 90%) and is usually water (see Figure 4.1); (2) the additives, which can be a single chemical or a mixture of chemicals, including around 750 compounds ranging from chemical additives (0.5% to 2%) found in food and common household cleaners to known carcinogens; and (3) the proppant (9.5%). Additives are chosen to serve a specific purpose (e.g., adjust pH, increase viscosity, limit bacterial growth). Chemicals generally comprise a small percentage (typically 0.5 to 2%) of the overall injected volume. Because over 1 million gallons of fluids are typically injected per well, thousands of gallons of chemicals can be potentially stored onsite and used during hydraulic fracturing activities. Onsite storage, mixing, and pumping of chemicals and hydraulic fracturing fluids have the potential to result in accidental releases, such as spills or leaks. Potential impacts to drinking water resources from spills of hydraulic fracturing fluids and chemicals depend on the characteristics of the spills, and the fate, transport, and the toxicity of the chemicals spills.

Types of Fracturing Fluids and Additives[*]

Service companies have developed a number of different oil- and water-based fluids and treatments to more efficiently induce and maintain permeable and productive fractures. The composition of these fluids varies significantly, from simple water and sand to complex polymeric substances with

---

[*] This section is adapted from USEPA, *The Central Appalachian Coal Basin—Attachment 6: Evaluation of Impacts to Underground Sources of Drinking Water by Hydraulic Fracturing of Coalbed Methane Reserves*, EPA 816-R-04-003, U.S. Environmental Protection Agency, Washington, DC, 2004; USEPA, *Hydraulic Fracturing Study Plan*, U.S. Environmental Protection Agency, Washington, DC, 2011.

Composition of Fracking Water

FIGURE 4.1 Water is combined with sand and a cocktail of chemicals to help in the fracking process.

a multitude of additives. Each type of fracturing fluid has unique characteristics, and each possesses its own positive and negative performance traits. For ideal performance, fracturing fluids should possess the following four qualities (Powell et al., 1999):

- Be viscous enough to create a fracture of adequate width.
- Measure fluid travel distance to extend fracture length.
- Be able to transport large amounts of proppant into the fracture.
- Require minimal gelling agent to allow for easier degradation or "breaking" and reduced cost.

The main fluid categories are

- Gelled fluids, including linear or cross-linked gels
- Foamed gels
- Plain water and potassium chloride (KCl) water
- Acids
- Combination treatments (any combination of two or more of the aforementioned fluids)

Gelled Fluids

Water alone is not always adequate for fracturing certain formations because its low viscosity limits its ability to transport proppant. In response to this problem, the industry developed linear and cross-linked fluids, which are higher viscosity fracturing fluids. Water gellants or thickeners are used to create these gelled fluids. Gellant selection is based on formation characteristics such as pressure, temperature, permeability, porosity, and zone thickness. These gelled fluids are described in more detail below.

Linear Gels

A substantial number of fracturing treatments are completed using thickened, water-based linear gels. The gelling agents used in these fracturing fluids are typically guar gum, guar derivatives such as hydroxypropylguar (HPG), and carboxymethylhydroxypropylguar (CMHPG), or cellulose derivatives such as carboxymethylguar or hydroxethylcelluslose (HEC). In general, these products are biodegradable. Guar is a polymeric substance derived from the ground endosperm of seeds of the guar plant (Ely, 1994). Guar gum, also called *guaran*, on its own, is nontoxic and, in fact, is a food-grade product commonly used to increase the viscosity and elasticity of foods such as ice cream, baked goods, pastry fillings, dairy products, meat, and condiments, in addition to being used in dry soups, instant oatmeal, sweet desserts, and frozen food and animal feed. Its industrial applications include uses in textiles, paper, explosives, pharmaceuticals, mining, oil and gas drilling, and other products.

To formulate a viscous fracturing gel, guar powder or concentrate is dissolved in a carrier fluid such as water or diesel fuel. Increased viscosity improves the ability of the fracturing fluid to transport proppant and decreases the need for more turbulent flow. Concentrations of guar gelling agents within fracturing fluids have decreased over the past several years, as it was determined that reduced concentrations provide better and more complete fractures (Powell et al., 1999); this decreased use may make the food industry happier.

Diesel fuel has frequently been used in lieu of water to dissolve the guar powder because its carrying capacity per unit volume is much higher (Haliburton, 2002). Diesel is a common solvent additive, especially in liquid gel concentrates, that is used by many service companies for continuous delivery of gelling agents in fracturing treatments (Penny and Conway, 1996). Diesel does not enhance the efficiency of the fracturing fluid—it is merely a component of the delivery system, and using diesel instead of water minimizes the number of transport vehicles required to carry the liquid gel to the site (Haliburton, 2002). Based on typical practice and observation, the percentage of diesel fuel in the slurried thickener can range between 30% and almost 100%. Diesel fuel is a petroleum distillate that may contain known carcinogens. One such component of diesel fuel is benzene, which, according to literature sources, can make up anywhere between 0.003 and 0.1% by weight of diesel fuel (Clark and Brown, 1977; Morrison & Associates, 2001). Slurried diesel and gel are diluted with water prior to injection into the subsurface. The dilution is approximately 4 to 10 gal of concentrated liquid gel (guar slurried in diesel) per 1000 gal of make-up water to produce an adequate polymer slurry (CIS, 2001; USEPA, 2004).

Cross-Linked Gels

The development of cross-linked gels in 1968 was one of the major advances in fracturing fluid technology (Ely, 1994). When cross-linking agents are added to linear gels, the result is a complex, high-viscosity fracturing fluid that provides higher proppant transport performance than do linear gels (Ely, 1994; Messina, 2001; USEPA, 2004). Cross-linking reduces the need for fluid thickener and extends the viscous life of the fluid indefinitely. The fracturing fluid remains viscous until a breaking agent is introduced to break the cross-linker and eventually the polymer. Although cross-linkers make the fluid more expensive, they can considerably improve hydraulic fracturing performance. Cross-linked gels are typically metal ion–cross-linked guar (Ely, 1994). Service companies have used metal ions such as chromium, aluminum, titanium, and other metal ions to achieve cross-linking, and low-residue (cleaner) forms of cross-linked gels, such as cross-linked hydroxypropylguar, have been developed (Ely, 1994). Cross-linked gels may contain boric acid, sodium tetraborate decahydrate, ethylene glycol, and monoethylamine. These constituents are hazardous in their undiluted form and can cause kidney, liver, heart, blood, and brain damage through prolonged or repeated exposure. According to a Bureau of Land Management environmental impact statement, cross-linkers may contain hazardous constituents such as ammonium chloride, potassium hydroxide, zirconium nitrate, and zirconium sulfate (USDOI, 1998).

Foamed Gels

Foam fracturing technology uses foam bubbles to transport and place proppant into fractures. The most widely used foam fracturing fluids employ nitrogen or carbon dioxide as their base gas. Incorporating inert gases with foaming agents and water reduces the amount of fracturing liquid required. Foamed gels use fracturing fluids with higher proppant concentrations to achieve highly effective fracturing. The gas bubbles in the foam fill voids that would otherwise be filled by fracturing fluid. The high concentrations of proppant allow for an approximately 75% reduction in the overall amount of fluid that would be necessary using a conventional linear or cross-linked gel (Ely, 1994; USEPA, 2004). Foaming agents can be used in conjunction with gelled fluids to achieve an extremely effective fracturing fluid. Foam emulsions experience high leakoff; therefore, typical protocol involves the addition of fluid-loss agents, such as fine sands (Ely, 1994; USEPA, 2004). Foaming agents suspend air, nitrogen, or carbon dioxide within the aqueous phase of a fracturing treatment. The gas/liquid ratio determines if a fluid will be true foam or simply a gas-energized liquid (Ely, 1994). Carbon dioxide can be injected as a liquid, whereas nitrogen must be injected as a gas to prevent freezing (USEPA, 2004). Foaming agents can contain diethanolamine and alcohols such as isopropanol, ethanol, and 2-butoxyethanol. They can also contain such hazardous substances as glycol ethers (USDOI, 1998). One type of foaming agent can cause negative liver and kidney effects. The final concentration is typically 3 gal of foamer per 1000 gal of gel (USEPA, 2004).

Water and Potassium Chloride Water Treatments

Many shale gas service companies use groundwater pumped directly from the formation or treated water for their fracturing jobs. In some well stimulations, proppants are not needed to prop fractures open, so simple water or slightly thickened water can be a cost-effective substitute for an expensive polymer of foam-based fracturing fluid with proppant (Ely, 1994). Hydraulic fracturing performance is not exceptional with plain water, but, in some cases, the production rates achieved are adequate. Plain water has a lower viscosity than gelled water, which reduces proppant transport capacity.

Acids

Acids are used in limestone formations that overlay or are interbedded within shale gas formations to dissolve the rock and create a conduit through which formation water and shale gas can travel. Typically, the acidic stimulation fluid is hydrochloric acid or a combination of hydrochloric and acetic or formic acid. For acid fracturing to be successful, thousands of gallons of acid must be pumped far into the formation to etch the face of the fracture; some of the cellulose derivatives used as gelling agents in water and water/methanol fluids can be used in acidic fluids to increase treatment distance (Ely, 1994). Note that acids may also be used as a component of breaker fluids. In addition, acid can be used to clean perforations of the cement surrounding the well casing prior to fracturing fluid injection. The cement is perforated at the zone of injection to ease fracturing fluid flow into the formation (Halliburton, 2002; USEPA, 2004). Acids, such as formic and hydrochloric acids, are corrosive and can be extremely hazardous in concentrated form. Acids are substantially diluted with water-based or water- and gas-based fluids prior to injection into the subsurface. The injected concentration is typically 1000 times weaker than the concentrated versions (USEPA, 2004).

Use of Chemical Additives

Chemicals are used throughout fracking operations to reach gas shale and release natural gas. Several fluid additives have been developed to enhance the efficiency and increase the success of fracturing fluid treatments. Chemicals are used not only in drilling for shale gas but also in fracking fluids for the purposes listed in Table 4.4.

TABLE 4.4
Fracking Fluid Additions and Function

Addition	Function
Acids	Acids, typically hydrochloric acid, are pumped into the formation to dissolve some of the rock material (minerals and clays) to clean out pores and to allow gas and fluid to flow more readily into the well.
Biocides	Because the watery fracking fluid used to fracture rocks gets hotter when pumped into the ground at high pressure and high speed, bacteria and mold multiply. When the bacteria grow, they secrete enzymes that break down the gelling agent, which causes a black slime or ooze to form in the lines and reduces viscosity. Reduced viscosity translates into poor proppant placement and poor fracturing performance. Biocides work to reduce the formation of bacteria and mold.
Breakers	Water-based gels are used in the gas industry to viscosify fluids used in the fracking of production wells, where they serve to increase the force applied to the rock and to improve the transport of proppants used to maintain the fracture after formation. The various types of breakers include time-release breakers and temperature-dependent breakers. Most breakers are typically acids, oxidizers, or enzymes (Messina, 2001) According to a Bureau of Land Management environmental impact statement, breakers may contain hazardous constituents, including ammonium persulfate, ammonium sulfate, copper compounds, ethylene glycol, and glycol ethers (USDOI, 1998). After fracturing, the gel must be degraded to a low viscosity with enzymes or gel breakers (Barati et al., 2011).
Clay stabilizers	These additives prevent clay swelling in the shale rocks and minimize migration of clay fines.
Corrosion inhibitors	These additives reduce the potential for rusting in pipes and casings. Corrosion inhibitors are required in acid fluid mixtures because acids will corrode steel tubing, well casings, tools, and tanks. The solvent acetone is a common additive in corrosion inhibitors (Penny and Conway, 1996). Corrosion inhibitors are quite hazardous in their undiluted form. These products are diluted to a concentration of 1 gal per 1000 gal of make-up water and acid mixture (USEPA, 2004). Acids and acid corrosion inhibitors are used in very small quantities in fracturing operations (500 to 2000 gal per treatment).
Cross-linkers	Cross-linkers are used to thicken fluids, often with metallic salts, to increase viscosity and improve proppant transport.

(continued)

Defoamers	These additives are used to reduce foaming after it is no longer needed in order to lower surface tension and allow trapped gas to escape.
Foamers	Foamers are used to increase carrying capacity while transporting proppants and decreasing the overall volume of fluid required.
Fluid-loss additives	These additives restrict leakoff of the fracturing fluid into the exposed rock at the fracture face. Because the additives prevent excessive leakoff, fracturing fluid effectiveness and integrity are maintained. Fluid-loss additives of the past and present include bridging materials such as 100-mesh sand, 100-mesh soluble resin, and silica flour, or plastering materials such as starch blends, talc silica flour, and clay (Ely, 1994).
Friction reducers	These additives make water slick, minimize the friction created under high pressure, and increase the rate and efficiency of moving the fracking fluid. Friction reducers are typically latex polymers or copolymers of acrylamides. They are added to slickwater treatments (water with solvent) at concentrations of 0.25 to 2.0 lb per 1000 gal (Ely, 1994). Some examples of friction reducers are oil-soluble anionic liquid, cationic polyacrilate liquid, and cationic friction reducer (Messina, 2001).
Gellants	Gellants are used to increase viscosity and suspend sand during proppant transport.
pH control	The addition of buffers maintains pH and ensures maximum effectiveness of various additions.
Proppants	Proppants, usually composed of sand and occasionally glass beads, prop or hold fissures open, allowing gas to flow out of the cracked formation. An ideal proppant should produce maximum permeability in a fracture. Fracture permeability is a function of proppant grain roundness, proppant purity, and crush strength. Larger proppant volumes allow for wider fractures, which facilitate more rapid flowback to the production well. Over a period of 30 minutes, 4500 to 15,000 gal of fracturing fluid will typically transport and place approximately 11,000 to 25,000 lb of proppant into the fracture (Powell et al., 1999).
Scale control	Scale control prevents the buildup of mineral scale that can block fluid and gas passage through the pipes.
Surfactants	These additives are used to decrease liquid surface tension and improve fluid passage through pipes in either direction.

DID YOU KNOW?

Similar to plain water, another fracturing fluid uses water with potassium chloride (KCl), which is harmless if ingested at lower concentrations, in addition to small quantities of gelling agents, polymers, and surfactants (Ely, 1994).

DID YOU KNOW?

As a result of the growing use of hydraulic fracturing, natural gas production in the United States reached 21,577 billion cubic feet in 2010, a level not achieved since a period of high natural gas production between 1970 and 1974 (USEIA, 2012).

Hydraulic fracturing creates access to more natural gas supplies, but the process requires the use of large quantities of water and fracturing fluids, which are injected underground at high volumes and pressure. Oil and gas service companies design fracturing fluids to create fractures and transport sand or other granular substances to properly open the fractures. The composition of these fluids varies by formation, ranging from a simple mixture of water and sand to more complex mixtures with a multitude of chemical additives. Fracking companies may use these chemical additives (see Table 4.4) to thicken or thin the fluids, improve the flow of the fluid, or kill bacteria that can reduce fracturing performance (USEPA, 2004). Some of these chemicals, if not disposed of safely or if allowed to leach into the drinking water supply, could damage the environment or pose a risk to human health. During hydraulic fracturing, fluids containing chemicals are injected deep underground, where their migration is not entirely predictable. Well failures (such as those due to the use of insufficient well casing) could lead to the release of these fluids at shallower depths, closer to drinking water supplies. Although some fracturing fluids are removed from the well at the end of the fracturing process, a substantial amount remains underground (Veil, 2010).

Although most underground injections of chemicals are subject to the protections of the Safe Drinking Water Act (SDWA), Congress in 2005 modified the law to exclude "the underground injection of fluids or propping agents (other than diesel fuels) pursuant to hydraulic fracturing operations related to oil, gas, or geothermal production activities" from the Act's protection (42 USC §300h(d)). Unless oil and gas service companies use diesel in the hydraulic fracturing process, the permanent underground injection of chemicals used for hydraulic fracturing is not regulated by the USEPA.

DID YOU KNOW?

Many have dubbed 42 USC §300h(d) the *Halliburton loophole* because of Halliburton's ties to then Vice President Cheney and its role as one of the largest providers of hydraulic fracturing services.

DID YOU KNOW?

The U.S Energy Information Administration (USEIA) projects that the United States possesses 2552 trillion cubic feet of potential natural gas resources, enough to supply the country for approximately 110 years. Natural gas from shale resources accounts for 827 trillion cubic feet of this total, which is more than double what the USEIA estimated in 2010 (USEIA, 2012).

Composition of Fracking Water

DID YOU KNOW?

Wyoming enacted relatively strong disclosure regulations, requiring disclosure on a well-by-well basis and, for each stage of the well stimulation program, the chemical additives, compounds, and concentrations or rates proposed to be mixed and injected (WCWR 055-000-003 Sec. 45). Similar regulations are in effect in Arkansas (Arkansas Oil and Gas Commission Rule B-19). In Wyoming, much of this information, after an initial period of review, is available to the public. Other states, however, do not insist on such robust disclosure. West Virginia, for example, has no disclosure requirements for hydraulic fracturing and expressly exempts fluids used during fracking from the disclosure requirements applicable to the underground injection of fluids for purposes of waste storage.

Concerns also have been raised about the ultimate outcome of chemicals that are recovered and disposed of as wastewater. This wastewater is stored in tanks or pits at the well site, where spills are possible (Urbina, 2011; USEPA, 2012b). For final disposition, well operators must recycle the fluids for use in future fracturing jobs, inject the fluids into underground storage wells (which, unlike the fracturing process itself, are subject to the Safe Drinking Water Act), discharge them to nearby surface water, or transport them to wastewater treatment facilities (Veil, 2010).

Any risk or impact to the environment and human health posed by fracking fluids depends in large part on their contents. Federal law, however, contains no public disclosure requirements for oil and gas producers or service companies involved in hydraulic fracturing, and state disclosure requirements vary greatly. Although the industry has recently announced that it will soon create a public database of fluid components, reporting to this database is strictly voluntary, disclosure will not include the chemical identity of products labels as proprietary, and there is no way to determine if companies are accurately reporting information for all wells.

The absence of a minimum national baseline for disclosure of fluids injected during the hydraulic fracturing process and the exemption of most hydraulic fracturing injections from regulation under the Safe Drinking Water Act has left an information void concerning the contents, chemical concentrations, and volumes of fluids that go into the ground during fracturing operations and return to the surface in the form of wastewater. As a result, regulators and the public are unable to effectively assess any impact that the use of these fluids may have on the environment or public health.

NATURALLY OCCURRING RADIOACTIVE MATERIAL[*]

Before presenting a detailed discussion of the major chemical constituents that make up hydraulic fracking fluids currently in use it is important to briefly discuss naturally occurring radioactive material (NORM) that could be involved in the fracking process. Some soils and geologic formations contain low levels of radioactive material. This naturally occurring radioactive material emits low levels of radiation to which everyone is exposed on a daily basis. Radiation from natural sources is also referred to as *background radiation*. Other sources of background radiation include radiation from space and sources that occur naturally in the human body. This background radiation accounts for about 50% of the total exposures for Americans. Most of this background exposure is from radon gas encountered in homes (35% of the total exposure). The average person in the United States is exposed to about 360 millirem (mrem) of radiation from natural sources each year (a mrem, or 1/1000 of a rem, is a measure of radiation exposure) (RRC, 2012). The other 50% of exposures for Americans comes primarily from medical sources. Consumer products and industrial and occupational sources contribute less than 3% of the total exposure (NCRP, 2009).

---

[*] This section is adapted from USDOE, *Modern Shale Gas Development in the United States: A Primer*, U.S. Department of Energy, Washington, DC, 2009.

In addition to the background radiation normally found at the surface of the Earth, NORM can also be brought to the surface in the natural gas production process. When NORM is associated with oil and natural gas production, it begins as small amounts of uranium and thorium within the rock. These elements, along with some of their decay elements, notably radium-226 and radium-228 (USGS, 1999), can be brought to the surface in drill cuttings and produced water. Radon-222, a gaseous decay element of radium, can come to the surface along with the shale gas.

When NORM is brought to the surface, it remains in the rock pieces of the drill cuttings, remains in solution with produced water, or, under certain conditions, precipitates out in scales or sludge. The radiation from this NORM is weak and cannot penetrate dense materials such as the steel used in pipes and tanks (Smith et al., 1996). The principal concern for NORM in the oil and gas industry is that, over time, it can become concentrated in field production equipment (API, 2004) and as sludge or sediment inside tanks and process vessels that have an extended history of contact with formation water (BSEEC, 2012). Because the general public does not come into contact with oilfield equipment for extended periods, there is little exposure risk from oilfield NORM. Studies have shown that exposure risks for workers and the public are low for conventional oil and gas operations (BSEEC, 2012; Smith et al., 1996).

If measured NORM levels exceed state regulatory levels or U.S. Occupational Safety and Health Administration (OSHA) exposure dose risks (29 CFR 1910.1096), the material is taken to licensed facilities for proper disposal. In all cases, OSHA requires employers to evaluate radiation hazards, post caution signs, and provide personal protection equipment for workers when radiation doses could exceed 5 mrem in 1 hour or 100 mrem in any 5 consecutive days. In addition to these federal worker protections, states have regulations that require operators to protect the safety and health of both workers and the public.

Currently, no existing federal regulations specifically address the handling and disposal of NORM wastes (but the USEPA does have drinking water standards for NORM). Instead, states producing oil and gas are responsible for promulgating and administering regulations to control the reuse and disposal of NORM-contaminated equipment, produced water, and oilfield wastes. Although regulations vary by state, generally, if NORM concentrations are less than regulatory standards, operators are allowed to dispose of the material by methods approved for standard oilfield waste. Conversely, if NORM concentrations are above regulatory limits, then the material must be disposed of at a licensed facility. These regulations, standards, and practices ensure that oil and gas operations present negligible risk to the general public with respect to potential NORM exposure. They also present negligible risk to workers when proper controls are implemented (Smith et al., 1996).

FRACKING FLUIDS AND THEIR CONSTITUENTS

In 2011, the U.S. Congressional Committee on Energy and Commerce published a report, *Chemicals Used in Hydraulic Fracturing*, which lauded hydraulic fracturing as a new technological device in the ongoing pursuit of oil and natural gas products. Moreover, the report pointed out that hydraulic fracturing has opened access to vast domestic reserves of natural gas that could provide an important stepping stone to a clean energy future. Yet, the Committee also observed that questions about the safety of hydraulic fracturing persist and are compounded by the secrecy surround the chemicals used in fracking fluids. The report indicated that, between 2005 and 2009, the 14 leading hydraulic fracturing companies in the United States (Basic Energy Services, BJ Services, Calfrac Well Services, Complete Production Services, Frac Tech Services, Halliburton, Key Energy Services, RPC, Sanjel Corporation, Schlumberger, Superior Well Services, Trican Well Service, Universal Well Services, and Weatherford) used over 2500 hydraulic fracturing products containing 750 compounds. More that 650 of these products contained chemicals that are known or possible human carcinogens, are regulated under the Safe Drinking Water Act for their risks to human health, or are listed as hazardous air pollutants under the Clean Air Act. Overall, these companies

Composition of Fracking Water

DID YOU KNOW?

Each hydraulic fracturing product is a mixture of chemicals or other components designed to achieve a certain performance goal, such as increasing the viscosity of water. Some oil and gas service companies create their own products, but most purchase these products from chemical vendors. The service companies then mix these products together at the well site to formulate the hydraulic fracturing fluids that they pump underground.

used 780 million gal of hydraulic fracturing products in their fluids in this period of time. This volume does not include water that the companies added to the fluids at well sites before injection. Hydraulic fracturing products are comprised of a wide range of chemicals. Some are seemingly harmless, such as sodium chloride (salt), gelatin, walnut hulls, instant coffee, and citric acid. Others, though, could pose severe risks to human health or the environment.

Some of the components are surprising. One company told the Congressional Committee that it used instant coffee as one of the components of a fluid designed to inhibit acid corrosion. Two companies reported using walnut hulls as part of a breaker (a product used to degrade the fracturing fluid viscosity, which helps to enhance post-fracturing fluid recovery). Another company reported using carbohydrates as a breaker. One company used tallow soap (soap made from beef, sheep, or other animals) to reduce the loss of fracturing fluid into the exposed rock.

COMMONLY USED CHEMICAL COMPONENTS

The most widely used chemical in hydraulic fracturing from 2005 to 2009, as measured by the number of products containing the chemical, was methanol. Methanol is a hazardous air pollutant and a candidate for regulation under the Safe Drinking Water Act. It was a component in 342 hydraulic fracturing products. Some of the other most widely used chemicals include isopropyl alcohol, which was used in 274 products, and ethylene glycol, which was used in 119 products. Crystalline silica (silicon dioxide) appeared in 207 products, generally proppants used to hold open fractures. Table 4.5 provides a list of the most commonly used compounds in hydraulic fracturing fluids.

Hydraulic fracturing companies used 2-butoxyethanol (2-BE) as a foaming agent or surfactant in 126 products. According to USEPA scientists, 2-BE is easily absorbed and rapidly distributed in humans following inhalation, ingestion, or dermal exposure. Studies have shown that exposure to 2-BE can cause hemolysis (destruction of red blood cells) and damage to the spleen, liver, and bone marrow (USEPA, 2010a). The hydraulic fracturing companies injected 21.9 million gallons of products containing 2-BE between 2005 and 2009. The highest volume of products containing 2-BE

TABLE 4.5
Chemical Components Appearing Most Often in Hydraulic Fracturing Products (2005–2009)

Chemical Component	No. of Products Containing Chemical
Methanol (methyl alcohol)	342
Isopropanol (isopropyl alcohol, propan-2-ol)	274
Crystalline silica (quartz) (SiO_2)	207
Ethylene glycol monobutyl ether (2-butoxyethanol)	126
Ethylene glycol (1,2-ethanediol)	119
Hydrotreated light petroleum distillates	89
Sodium hydroxide (caustic soda)	80

TABLE 4.6
States with the Highest Volume of Hydraulic Fracturing Fluids Containing 2-Butoxyethanol (2005–2009)

State	Fluid Volume (gal)
Texas	12,031,734
Oklahoma	2,186,613
New Mexico	1,871,501
Colorado	1,147,614
Louisiana	890,068
Pennsylvania	747,416
West Virginia	464,231
Utah	382,874
Montana	362,497
Arkansas	348,959

was found in Texas, which accounted for more than half of the volume used. The USEPA recently found this chemical in drinking water wells tested in Pavillion, Wyoming (USEPA, 2010b). Table 4.6 shows the use of 2-BE by state.

TOXIC CHEMICALS

The oil and gas service companies used hydraulic fracturing products containing 29 chemicals that are (1) known or possible human carcinogens, (2) regulated under the Safe Drinking Water Act for their risks to human health, or (3) listed as hazardous air pollutants (HAPs) under the Clean Air Act. These 29 chemicals were components of 652 different products used in hydraulic fracturing. Table 4.7 lists these toxic chemicals and their frequency of use.

Carcinogens

Between 2005 and 2009, the hydraulic fracturing companies used 95 products containing 13 different carcinogens. These included naphthalene (a possible human carcinogen), benzene (a known human carcinogen), and acrylamide (a probable human carcinogen). Overall, these companies injected 102 million gal of fracturing products containing at least one carcinogen. The companies used the highest volume of fluids containing one or more carcinogens in Texas, Colorado, and Oklahoma. Table 4.8 shows the use of these chemicals by state.

> **DID YOU KNOW?**
> Diesel contains benzene, toluene, ethylbenzene, and xylene (USEPA, 2004).

> **DID YOU KNOW?**
> Here, a chemical is considered a carcinogen if it is on one of two lists: (1) substances identified by the National Toxicology Program as "known to be human carcinogens" or as "reasonably anticipated to be human carcinogens"; and (2) substances identified by the International Agency for Research on Cancer, part of the World Health Organization, as "carcinogenic" or "probably carcinogenic" to humans (IARC, 2011; NTP, 2005).

TABLE 4.7
Chemicals Components of Concern: Carcinogens, SDWA-Regulated Chemicals, and Hazardous Air Pollutants

Chemical Component	Chemical Category	No. of Products
Methanol (methyl alcohol)	HAP	342
Ethylene glycol (1,2-ethanediol)	HAP	119
Diesel	Carcinogen, SDWA, HAP	51
Naphthalene	Carcinogen, HAP	44
Xylene	SDWA, HAP	44
Hydrogen chloride (hydrochloric acid)	HAP	42
Toluene	SDWA, HAP	29
Ethylbenzene	SDWA, HAP	28
Diethanolamine (2,2-iminodiethanol)	HAP	14
Formaldehyde	Carcinogen, HAP	12
Sulfuric acid	Carcinogen	9
Thiourea	Carcinogen	9
Benzyl chloride	Carcinogen, HAP	8
Cumene	HAP	6
Nitrilotriacetic acid	Carcinogen	6
Dimethyl formamide	HAP	5
Phenol	HAP	5
Benzene	Carcinogen, SDWA, HAP	3
Di(2-ethylhxyl)phthalate	Carcinogen, SDWA, HAP	3
Acrylamide	Carcinogen, SDWA, HAP	2
Hydrogen fluoride (hydrofluoric acid)	HAP	2
Phthalic anhydride	HAP	2
Acetaldehyde	Carcinogen, HAP	1
Acetophenone	HAP	1
Copper	SDWA	1
Ethylene oxide	Carcinogen, HAP	1
Lead	Carcinogen, SDWA, HAP	1
Propylene oxide	Carcinogen, HAP	1
p-Xylene	HAP	1
Number of products containing a component of concern		652

Note: HAP, hazardous air pollutant; SDWA, Safe Drinking Water Act.

TABLE 4.8
States with at Least 100,000 gal of Hydraulic Fracturing Fluids Containing a Carcinogen (2005–2009)

State	Fluid Volume (gal)
Texas	3,877,273
Colorado	1,544,388
Oklahoma	1,098,746
Louisiana	777,945
Wyoming	759,898
North Dakota	557,519
New Mexico	511,186
Montana	394,873
Utah	382,338

TABLE 4.9

States with at Least 100,000 gal of Hydraulic Fracturing Fluids Containing a SDWA-Regulated Chemical (2005–2009)

State	Final Volume (gal)
Texas	9,474,631
New Mexico	1,157,721
Colorado	375,817
Oklahoma	202,562
Mississippi	108,809
North Dakota	100,479

Safe Drinking Water Act Chemicals

Under the Safe Drinking Water Act, the USEPA regulates 53 chemicals that may have an adverse effect on human health and are known to or are likely to occur in public drinking water systems at levels of public health concern. Between 2005 and 2009, the hydraulic fracturing companies used 67 products containing at least one of eight SDWA-regulated chemicals. Overall, these companies injected 11.7 million gal of fracturing products containing at least one chemical regulated under SDWA. Most of these chemicals were injected in Texas. Table 4.9 shows the use of these chemicals by state.

The vast majority of these SDWA-regulated chemicals were the BTEX compounds—benzene, toluene, ethylbenzene, and xylene. The BTEX compounds appeared in 60 hydraulic fracturing products used between 2005 and 2009 and were used in 11.4 million gal of hydraulic fracturing fluids. The Department of Health and Human Services, the International Agency for Research on Cancer, and the USEPA have determined that benzene is a human carcinogen (ATSDR, 2007). Chronic exposure to toluene, ethylbenzene, or xylene can also damage the central nervous system, liver, and kidneys (USEPA, 2012b).

In addition, it is common to use diesel in hydraulic fracturing fluids; for example, the hydraulic fracturing companies injected more than 32 million gal of diesel fuel or hydraulic fracturing fluids containing diesel fuel in wells in 19 states. The USEPA has stated that the "use of diesel fuel in fracturing fluids poses the greatest threat" to underground sources of drinking water (USEPA, 2004). Diesel fuel contains toxic constituents, including BTEX compounds; thus, the use of diesel fuel should be avoided. According to the company Halliburton, "Diesel does not enhance the efficiency of the fracturing fluid; it is merely a component of the delivery system." According to the USEPA, it is technologically feasible to replace diesel with nontoxic delivery systems, such as plain water (Earthworks, 2016). The USEPA has created candidate contaminant lists of contaminants that are currently not subject to national primary drinking water regulations but are known or anticipated to occur in public water systems and may require regulation under the Safe Drinking Water Act in the future (USEPA, 2012a). These lists include, among others, pesticides, disinfection byproducts, chemicals used in commerce, waterborne pathogens, pharmaceuticals, and biological toxins. Nine listed chemicals were pertinent to or used in hydraulic fracking between 2005 and 2009: 1-butanol, acetaldehyde, benzyl chloride, ethylene glycol, ethylene oxide, formaldehyde, methanol, *n*-methyl-2-pyrrolidone, and propylene oxide.

TRADE SECRETS AND PROPRIETARY CHEMICALS

Many chemical components of hydraulic fracturing fluids used by fracking service companies are listed on the Material Safety Data Sheets (MSDSs) as being "proprietary" or "trade secret." OSHA's 29 CFR 1910.1200 (Hazard Communication Standard) section (i) states:

Composition of Fracking Water

(i)(1) The chemical manufacturer, importer, or employer may withhold the specific chemical identity, including the chemical name and other specific identification of a hazardous chemical, from the Material Safety Data Sheet (MSDS), provided that:

(i)(1)(i) The claim that the information withheld is a trade secret can be supported;

(i)(1)(ii) Information contained in the Material Safety Data Sheet concerning the properties and effects of the hazardous chemical is disclosed;

(i)(1)(iii) The Material Safety Data Sheet indicates that the specific chemical identity is being withheld as a trade secret; and,

(i)(1)(iv) The specific chemical identity is made available to health professionals, employees, and designated representatives in accordance with the applicable provisions of this paragraph.

Information on chemicals used during oil and gas development can also be obtained from Tier II reports and from websites such as FracFocus or state agency sites; however, Colborn et al. (2010) of the Endocrine Disruption Exchange have enumerated several problems with the information in MSDS and Tier II reports. MSDSs and Tier II reports are fraught with gaps in information about the formulation of the products. OSHA provides only general guidelines for the format and content of MSDSs, and manufacturers of the products are left to determine what information is revealed on their MSDSs. The forms are not submitted to OSHA for review unless they are part of an inspection under the Hazard Communication Standard. Some MSDSs report little or no information about the chemical composition of a product. Those MSDSs that do may report only a fraction of the total composition, sometimes less than 0.1%. Some MSDSs provide only a general description of the content, such as "plasticizer" or "polymer," while others describe the ingredients as "proprietary" or just a chemical class. Under the existing regulatory system, all of the above identifiers are permissible; consequently, it is not surprising that a study by the U.S. General Accounting Office revealed that MSDSs could easily be inaccurate and incomplete. Tier II reports can be similarly uninformative, as reporting requirements vary from state to state, country to country, and company to company. Some Tier II forms include only a functional category name (e.g., "weight materials" or "biocides") with no product name. The percent of the total composition of the product is rarely reported on these forms.

THOUGHT-PROVOKING QUESTIONS

1. Should chemical manufacturers be allowed to claim "trade secret" instead of revealing constituents of chemicals used for fracking?
2. We do not know what we do not know about the damage caused by fracking fluids. Does this statement make sense? Is it realistic? Explain.
3. There is a lot of secrecy surrounding the chemicals used in fracking fluids. Do you have any concerns about that? Why?

REFERENCES AND RECOMMENDED READING

API. (2004). *Naturally Occurring Radioactive Material in North American Oilfields.* American Petroleum Institute, Washington, DC.

ATSDR. (2007). *Public Health Statement for Benzene.* Agency for Toxic Substances and Disease Registry, Atlanta, GA.

ATSDR. (2011). *Medical Management Guidelines for Hydrogen Fluoride.* Agency for Toxic Substances and Disease Registry, Atlanta, GA (www.atsdr.cdc.gov/mmg/mmg.asp?id=1142&tid=250).

Barati, R., Jonson, S.J., McCool, S., Green, D.W., Willhite, G.P., and Liang, J.T. (2011). Fracturing fluid cleanup by controlled release of enzymes for polyelectrolyte complex nanoparticles. *Journal of Applied Polymer Science*, 121:1292–1298.

BSEEC. (2012). *Environment*. Barnett Shale Energy Education Council, Fort Worth, TX (www.Bseec.org/index.php/content/facts/environment).

CIS. (2001). *Hydraulic Fracturing Site Visit Notes: Western Interior Coal Region, State of Kansas*. Consolidated Industrial Services, Signal Hill, CA.

Clark, R.C. and Brown, D.W. (1977) Petroleum: properties and analysis in biotic and abiotic systems, in Malins, D.C., Ed., *Effects of Petroleum on Arctic and Subartic Environments and Organisms*. Vol. 1. *Nature and Fate of Petroleum*. Academic Press, New York, pp. 1–89.

Colborn, T., Kwiatkowski, C., Schultz, K., and Bachran, M. (2010). Natural gas operations from a public health perspective. *International Journal of Human and Ecological Risk Assessment*, 17(5):1039–1056.

Earthworks. (2016). *Hydraulic Fracturing 101*. Earthworks, Washington, DC (https://www.earthworksaction.org/issues/detail/hydraulic_fracturing_101#.WDXeg5LAoTk).

Ely, J.W. (1994). *Stimulation Engineering Handbook*. PennWell, Tulsa, OK.

Ely, J.W., Zbitowski, R.I., and Zuber, M.D. (1990). How to develop a coalbed methane prospect: a case study of an exploratory five-spot well pattern in the Warrior basin, Alabama, in *Proceedings of 1990 SPE Annual Technical Conference and Exhibition*, New Orleans, LA, September 23–26, SPE Paper 206666. Society of Plastics Engineers, Newtown, CT, pp. 487–496.

Fakhru'l-Razo. A., Pendashteh, A., Abdullah, L.C. et al. (2009). Review of technologies for oil and gas produced water treatment. *Journal of Hazardous Materials*, 170(2–3):530–555.

Halliburton. (2002). Personal communication with Halliburton staff, fracturing fluid experts Joe Sandy, Pat Finley, and Steve Almond.

Hauser, B.A. (1996). *Practical Manual of Wastewater Chemistry*. Lewis Publishers, Boca Raton, FL.

IARC. (2011). *Agents Classified by the IARC Monographs*, Vols. 1–104. International Agency for Research on Cancer, Lyon, France.

Jost, N.J. (1992). Surface and ground water pollution control technology, in Knowles, P.-C., Ed., *Fundamentals of Environmental Science and Technology*. Government Institutes, Rockville, MD.

Messina. (2001). *Fracturing Chemicals*. Messina, Dallas, TX (www.messinachemicals.com/).

Morrison, R. & Associates. (2001). Does diesel #2 fuel oil contain benzene? *Environmental Tool Box*, Fall.

NCRP. (2009). *Ionizing Radiation Exposure of the Populations of the United States*, Report No. 160. National Council on Radiation Protection and Measurements, Bethesda, MD.

Nikiforuk, A. (2011). Truth comes out on "fracking" toxins. *The Tyee*, April 20 (http://thetyee.ca/Opinion/2011/04/20/FrackingToxins/).

NTP. (2005). *The Report on Carcinogens*, 11th ed. Public Health Service, National Toxicology Program, Washington, DC.

Penny, G.S. and Conway, M.W. (1996). *Coordinated Studies in Support of Hydraulic Fracturing of Coalbed Methane*, Final Report #GRI-95/0283, prepared by STIM-Lab. Gas Research Institute, Chicago, IL.

Powell, R.J., McCabe, M.A., Salbaugh, B.F., Terracina, J.M., Yaritz, J.G., and Ferrer, D. (1999). Applications of a new, efficient hydraulic fracturing fluid system. *SPE Production and Facilities*, 14(2):139–143.

Ravve, R. (2011). Shale oil in America: economy fix or dangerous fantasy? *Foxnews.com*, www.foxnews.com/us/2011/12/27/shale-oil-in-america-economy-fix-or-dangerous-fantasy/.

Rohrlich, J. (2011). Food industry blames fracking for guar gum bubble. *Minyanville*, www.minyanville.com/dailyfeed/2011/09/02/food-industry-blames -fracking-for/.

RRC. (2012). *NORM—Naturally Occurring Radioactive Material*. Railroad Commission of Texas, Austin (www.rrc.state.tx.us/environmental/publications/norm/index.php).

Smith, K.P., Blunt, D.L., Williams, G.P., and Tebes, C.L. (1996). *Radiological Dose Assessment Related to Management of Naturally Occurring Radioactive Materials Generated by the Petroleum Industry*. Environmental Assessment Division, Argonne National Laboratory, Argonne, IL.

Urbina, I. (2011). Regulation lax as gas wells' tainted water hits rivers. *The New York Times*, February 26.

USDOI. (1998). *Glenwood Springs Resource Area: Oil & Gas Leasing & Development—Draft Supplemental Environmental Impact Statement*. Bureau of Land Management, Colorado State Office, Lakewood.

USEIA. (2011). *What Is Shale Gas and Why Is It Important?* U.S. Energy Information Administration, Washington, DC (http://www.eia.gov/energy_in_brief/article/about_shale_gas.cfm).

USEIA. (2012). *Natural Gas: U.S. Dry Natural Gas Production*. U.S. Energy Information Administration, Washington, DC (www.eia.gov/dnav/ng/hist/n9070us1A.htm).

USEIA. (2016). *Natural Gas*. U.S. Energy Information Administration, Washington, DC (http://www.eia.gov/dnav/ng/hist/n9070us1A.htm).

USEPA. (2004). *The Central Appalachian Coal Basin—Attachment 6: Evaluation of Impacts to Underground Sources of Drinking Water by Hydraulic Fracturing of Coalbed Methane Reserves*, EPA 816-R-04-003. U.S. Environmental Protection Agency, Washington, DC, 26 pp.

USEPA. (2010a). *Toxicological Review of Ethylene Glycol Monobutyl Ether.* U.S. Environmental Protection Agency, Washington, DC (www.epa.gov/iris/toxreviews/0500tr.pdf).

USEPA. (2010b). *Pavillion, Wyoming, Groundwater Investigation: January 2010 Sampling Results and Site Update.* U.S. Environmental Protection Agency, Washington, DC.

USEPA. (2011). *Hydraulic Fracturing Study Plan.* U.S. Environmental Protection Agency, Washington, DC.

USEPA. (2012a). *Water: Contaminant Candidate List 3.* U.S. Environmental Protection Agency, Washington, DC (http://water.epa.gov/scitech/drinkingwater/dws/ccl/ccl3.cfm).

USEPA. (2012b). *Basic Information about Toluene in Drinking Water* (http://water.epa.gov/drink/contaminants/basicinformation/toluene.cfm); *Basic Information about Ethylbenzene in Drinking Water* (http://water.epa.gov/drink/contaminants/basicinformation/ethylbenzene.cfm); *Basic Information about Xylenes in Drinking Water* (http://water.epa.gov/drink/contaminants/basicinformation/xylenes.cfm); *Basic Information about Lead in Drinking Water* (http://water.epa.gov/drink/contaminants/basicinformation/lead.cfm). U.S. Environmental Protection Agency, Washington, DC.

USGS. (1999). *Naturally Occurring Radioactive Materials (NORM) in Produced Water and Oil-Field Equipment—An Issue for the Energy Industry.* U.S. Geological Survey, Washington, DC (pubs.usgs.gov/fs/fs-0142-99/fs-0142-99.pdf).

USGS. (2002). *Produced Waters.* U.S. Geological Survey, Washington, DC (http://energy.usgs.gov/EnvironmentalAspects/EnvironmentalAspectsofEnergyProductionandUse/ProducedWaters.aspx#3822111-news).

Veil, J.A. (2010). *Water Management Technologies Used by Marcellus Shale Gas Producers,* ANL/EVS/R-10/3, prepared by the Environmental Science Division, Argonne National Laboratory, for the U.S. Department of Energy, Office of Fossil Energy, National Energy Technology Laboratory, Washington, DC.

Veil, J.A, Pruder, M.G., Elcock, D., and Redweik, Jr., R.J. (2004). *A White Paper Describing Produced Water from Production of Crude Oil, Natural Gas, and Coal Bed Methane.* U.S. Department of Energy, National Energy Technology Laboratory, Washington, DC.

Watson, L. (1988). *The Water Planet: A Celebration of the Wonder of Water.* Crown Publishers, New York.

Section III

Fracking Water Management Choices

5 Fracking Wastewater Treatment

The use of the term "unconventional" to describe a gas resource is open to interpretation; as technology advances and discrete reservoirs become limited, the reserves considered "unconventional" a few decades ago are more economically viewed as conventional by modern standards (Santoro et al., 2011). In this book, the term "conventional" refers specifically to discrete reservoirs of associated or unassociated natural gas and "unconventional" refers to tight-gas formations.

The six main categories of unconventional natural gas are deep natural gas, tight natural gas, shale gas, coalbed methane, geopressurized zones, and methane hydrates:

- *Deep natural gas*—Natural gas that exists in deposits very far underground, beyond conventional drilling depths (typically, 15,000 feet or deeper).
- *Tight natural gas*—Gas that is stuck in a very tight formation underground, trapped in unusually impermeable hard rock or in a sandstone or limestone formation that is unusually nonporous (tight sand).
- *Shale gas*—Certain shale basins contain natural gas, usually when two thick, black shale deposits sandwich a thinner area of shale.
- *Coalbed methane*—Many coal seams contain natural gas, either within the seam itself or the surrounding rock.
- *Geopressurized zones*—Areas formed by layers of clay that are deposited and compacted very quickly on the top of more porous, absorbent material such as sand or silt. Water and natural gas trapped within this clay are squeezed out by the rapid compression of the clay and enter the more porous sand or silt deposits.
- *Methane hydrates*—Formations made up of a lattice of frozen water, which forms a sort of cage around molecules of methane.

FRACKING WATER TREATMENT OPTIONS

This chapter catalogs advanced water treatment technologies currently used in hydraulic fracturing to treat oil and gas produced wastewater. Each technology has positives and negatives with respect to chemical requirements, energy requirements, footprint, cost, and removal capability. General information is included on a number of categories of applied technologies, including a brief technology description, applicable contaminants removed, removal mechanisms, and qualitative notes on advantages and disadvantages. In fracking operations, depending on location, water may be the most valuable and elusive operational substance used. Oil and gas extraction creates substantial quantities of produced wastewaters with varying levels of contamination that must be disposed of through treatment or injection. Produced wastewater coexists naturally with oil and gas deposits and is brought to the surface during well production. Produced wastewater is extracted at an average rate of 2.4 billion gallons per day (gpd), and over 80% of production occurs in the western United States (Clark and Veil, 2009). The produced wastewater from flowback is a valuable resource that might be treatable and reusable. If this wastewater is not disposed of by injection (98% of produced wastewater is disposed in this manner), various produced wastewater treatment technologies offer opportunities for the oil and gas industry to reuse flowback water from fracking applications. This is important because wells from horizontal drilling require more water than traditional vertical wells. Moreover, water reuse by the oil and gas industry can offset freshwater requirements and reduce demand on regional water systems. Produced wastewater represents the largest wastestream associated with oil and gas production, with estimates of almost 2.7 million acre feet per year.

Typically, 98% of produced wastewater is reinjected for disposal due to its salinity or for other reasons such as cost and the unavailability of other options, but in water-stressed areas, this water can be treated and managed for uses such as the following:

- Well drilling
- Emergency drought supply
- Livestock water
- Irrigation water
- Surface water augmentation
- Drinking water applications

Most produced wastewater requires treatment to make it suitable for recycling or beneficial use. Produced water varies widely in quantity and quality, depending on the method of extraction, type of oil and gas reservoir, geographical locations, and the geochemistry of the producing formation; therefore, many different types of technologies exist to treat produced wastewater. Furthermore, the industry uses these technologies in both upstream and downstream applications. Also, advanced water treatment technology can be used for

- Treating alternative water sources for fracking
- Internal industry reuse onsite at decentralized facilities
- Beneficial use of produced wastewater and flowback water for alternative applications offsite

PRODUCED WASTEWATER MANAGEMENT COSTS

Costs for water management in the oil and gas industry are highly variable. Cost calculations for water sourcing, transportation, storage, treatment, and disposal are not commonly presented on a whole-cost basis; thus, comparisons between management costs can be difficult to quantify. General ranges of costs are presented as a basis of understanding the relative cost of water management.

TRANSPORTATION

Water transportation is required to move water onsite for well development and offsite for treatment or disposal. Trucking costs may range from $0.50 to $8.00 per barrel depending on the state and transportation distance.

WATER SOURCING

Water required for well development may be purchased for use from local landowners or municipalities for $0.24 to $1.75 per barrel (Boschee, 2012).

DISPOSAL

Disposal is commonly managed through injection wells with costs for underground injection ranging from $0.07 to $1.60 per barrel of produced wastewater. Options such as impoundments or evaporation ponds are not always available due to permitting restrictions.

TREATMENT

Treatment costs both onsite and offsite vary considerably based on technology, water quality, and end use. Estimates depend on the site location and type of project and range from $0.20 to $8.50 per barrel.

PRODUCED WASTEWATER QUALITY AND CONSTITUENTS

This section lists and describes the quality of and constituents typically contained in produced wastewater. Generally, produced wastewater requires treatment to make it suitable for recycling or beneficial use due to the naturally occurring constituents and chemical additives in the water. Water quality varies widely in quantity and quality—depending on the method of extraction, type of oil and gas reservoir, geographical location, and the geochemistry of the producing formation (Guerra et al., 2011). Following are some constituents that commonly occur in produced wastewater.

SALINITY

Salinity in flowback and produced wastewater originates from water associated with the producing formation. Salinity in produced wastewater commonly consists of sodium and chloride. Calcium, magnesium, potassium, and sulfate may also exist due to mineral ion exchange. Salinity levels vary greatly nationally and even in one location over the lifetime of a well. Produced wastewater ranges from fresh (<700 mg/L) to highly saline (>200,000 mg/L), depending on the location and type of hydrocarbon produced. Produced wastewater with lower levels of salinity is generally diluted by fresh groundwater recharge and exhibits meteoric water composition. Ionic mineral exchange and residual high-salinity water from paleogeologic attributes influence both the concentration and composition of high-salinity waters.

SUSPENDED SOLIDS

Suspended solids accumulate in flowback and produced wastewater as residual particles from the fracturing process, naturally occurring granular material from the formation, and aggregate biological or chemical compounds. Concentrations are highly visible, based on the hydrocarbon produced and well location.

OIL AND GREASE, HYDROCARBONS, AND NATURAL ORGANIC MATTER

Organic contaminants exist naturally in the formation and are expected to be present in flowback and produced wastewater. Oil and gas water separators are not 100% efficient at separating these compounds from produced wastewater. Furthermore, these constituents may be difficult to remove in gravity processes as they are suspended and generally lighter than water. The concentration and type of organic contaminant will vary by well type and location.

DISSOLVED GAS AND VOLATILE COMPOUNDS

Naturally occurring dissolved gases and volatile compounds exist in flowback and produced wastewater. Dissolved gas may be present due to the hydrocarbons produced (e.g., methane) or due to other saturated gases in the formation, such as carbon dioxide or hydrogen sulfide. Additionally, volatile compounds, such as benzene, toluene, ethylbenzene, and xylene, may also exist in produced water. Care should be taken when dealing with certain volatile constituents as they may be hazardous to human health.

IRON AND MANGANESE

The concentrations and occurrence of iron and manganese vary by location, but these compounds are generally present in flowback and produced wastewater, as they are naturally present in the surrounding geology. Water existing naturally in the formation is generally anoxic; therefore, iron and manganese occur in their reduced forms. When iron and manganese are exposed to oxygen, they form oxides that precipitate on equipment and in pipelines. Furthermore, iron sulfides are detrimental to the hydraulic fracturing process and must be removed if water is to be recycled.

BARIUM AND STRONTIUM

Barium and strontium exist naturally in subsurface geology. Dissolution and weathering of these minerals may result in their presence in groundwater. These compounds are of particular concern, as they form sparingly soluble salts that precipitate in certain treatment processes and cause decreased efficiency or damage. These constituents occur at much higher concentrations in produced wastewater than in groundwater resources.

BORON AND BROMIDE

Both boron and bromide are present in seawater and may occur in flowback and produced water. Both of these compounds are mentioned as they may be difficult to remove or detrimental to downstream users. Boron may not be readily removed in reverse osmosis, for example, without raising the pH or adding ion-exchange treatment to polish the reverse osmosis permeate. Bromate is not always measured, and trace amounts may cause the formation of disinfection byproducts when combined with ozone as a disinfectant.

TRACE METALS

Mineral dissolution leads to the presence of various trace metals in flowback and produced wastewater. Certain constituents, such as arsenic and chromium, are detrimental to certain end uses. Care should be taken, as these constituents may be harmful—even at trace levels. Trace metals are not commonly measured in produced wastewater, so subsequent analyses may be necessary to identify occurrence and concentration.

RADIONUCLIDES

Radionuclides occur naturally in subsurface formations and generally depend on subsurface geology; therefore, certain areas nationally are predisposed to high concentrations of radioactive compounds. Care must be taken in areas with high concentrations, as process equipment and solids bound for landfills may accumulate radioactive material, making them hazardous.

WELL ADDITIVES AND FRACTURING CHEMICALS

Chemicals are added during well development and well production to maintain well operations or to improve fracturing conditions. A wide range of chemicals can reduce scaling, improve cross-linking, or act as biocides to removal microbial growth. Chemical disclosure is generally an industry practice, and databases with chemical registries may be used to identify compounds used by the industry.

THOUGHT-PROVOKING QUESTIONS

1. Should chemical manufacturers be allowed to claim "trade secret" instead of revealing constituents of chemicals used for fracking?
2. We do not know what we do not know about the damage caused by fracking fluids. Does this statement make sense? Explain.
3. How should fracturing fluids be treated and reused or disposed of?
4. Should fracturing fluids be treated in conventional wastewater treatment plants? Why or why not?

REFERENCES AND RECOMMENDED READING

API. (2004). *Naturally Occurring Radioactive Material in North American Oilfields.* American Petroleum Institute, Washington, DC.

ATSDR. (2007). *Public Health Statement for Benzene.* Agency for Toxic Substances and Disease Registry, Atlanta, GA.

ATSDR. (2011). *Medical Management Guidelines for Hydrogen Fluoride.* Agency for Toxic Substances and Disease Registry, Atlanta, GA (www.atsdr.cdc.gov/mmg/mmg.asp?id=1142&tid=250).

Barati, R., Jonson, S.J., McCool, S., Green, D.W., Willhite, G.P., and Liang, J.T. (2011). Fracturing fluid cleanup by controlled release of enzymes for polyelectrolyte complex nanoparticles. *Journal of Applied Polymer Science*, 121:1292–1298.

Boschee, P. (2012). Handling produced water from hydraulic fracturing. *Oil and Gas Facilities*, 1(1):23–26.

CIS. (2001). *Hydraulic Fracturing Site Visit Notes, Western Interior Coal Region, State of Kansas.* Consolidated Industrial Services, Signal Hills, CA.

Colborn, T, Kwiatkowski, C., Schultz, K., and Bachran, M. (2011). Nature gas operations from a public health perspective. *Human and Ecological Risk Assessment: An International Journal*, 17(5).

Earthworks. (2016). *Hydraulic Fracturing 101.* Earthworks, Washington, DC (https://www.earthworksaction.org/issues/detail/hydraulic_fracturing_101#.WDXeg5LAoTk).

Ely, J.W., Zbitowski, R.I., and Zuber, M.D. (1990). How to develop a coalbed methane prospect: a case study of an exploratory five-spot well pattern in the Warrior basin, Alabama, in *Proceedings of 1990 SPE Annual Technical Conference and Exhibition, New Orleans, LA, September 23–26*, SPE Paper 206666. Society of Plastics Engineers, Newtown, CT, pp. 487–496.

Ely, J.W. (1994). *Stimulation Engineering Handbook.* PennWell, Tulsa, OK.

Guerra, K., Dahm, K., and Dundorf, S. (2011). *Oil and Gas Produced Water Management and Beneficial Use in the Western United States*, Bureau of Reclamation DWPR Report #157. Bureau of Reclamation, Washington, DC.

Halliburton. (2002). Personal communication with Halliburton staff, fracturing fluid experts Joe Sandy, Pat Finley, and Steve Almond.

IARC. (2011). *Agents Classified by the IARC Monographs*, Vols. 1–104. International Agency for Research on Cancer, Lyon, France.

NCRP. (2009). *Ionizing Radiation Exposure of the Populations of the United States*, Report No. 160. National Council on Radiation Protection and Measurements, Bethesda, MD.

NTP. (2005). *The Report on Carcinogens*, 11th ed. Public Health Service, National Toxicology Program, Washington, DC.

Penny, G.S. and Conway, M.W. (1996). *Coordinated Studies in Support of Hydraulic Fracturing of Coalbed Methane*, Final Report #GRI-95/0283, prepared by STIM-Lab. Gas Research Institute, Chicago, IL.

Powell, R.J., McCabe, M.A., Salbaugh, B.F., Terracina, J.M., Yaritz, J.G., and Ferrer, D. (1997). Applications of a new, efficient hydraulic fracturing fluid system. *SPE Production and Facilities*, 14(2):139–143.

Rohrlich, J. (2011). Food industry blames fracking for guar bum bubble. *Minyanville*, www.minyanville.com/dailyfeed/2011/09/02/food-industry-blames-fracking-for/.

RRC. (2012). *NORM—Naturally Occurring Radioactive Material.* Railroad Commission of Texas, Austin (www.rrc.state.tx.us/environmental/publications/norm/index.php).

Smith, K.P., Blunt, D.L., Williams, G.P., and Tebes, C.L. (1996). *Radiological Dose Assessment Related to Management of Naturally Occurring Radioactive Materials Generated by the Petroleum Industry.* Environmental Assessment Division, Argonne National Laboratory, Argonne, IL.

Urbina, I. (2011). Regulation lax as gas wells' tainted water hits rivers. *The New York Times*, February 26.

USDOI. (1998). *Glenwood Springs Resource Area: Oil & Gas Leasing & Development—Draft Supplemental Environmental Impact Statement.* Bureau of Land Management, Colorado State Office, Lakewood.

USEIA. (2011). *What Is Shale Gas and Why Is It Important?* U.S. Energy Information Administration, Washington, DC (http://www.eia.gov/energy_in_brief/article/about_shale_gas.cfm).

USEIA. (2012). *Natural Gas: U.S. Dry Natural Gas Production.* U.S. Energy Information Administration, Washington, DC (www.eia.gov/dnav/ng/hist/n9070us1A.htm).

USEIA. (2016). *Natural Gas.* U.S. Energy Information Administration, Washington, DC (http://www.eia.gov/dnav/ng/hist/n9070us1A.htm).

USEPA. (2004). *Evaluation of Impacts to Underground Sources of Drinking Water by Hydraulic Fracturing of Coalbed Methane Reserves*, EPA 816-R-04-003. U.S. Environmental Protection Agency, Washington, DC.

USEPA. (2010a). *Toxicological Review of Ethylene Glycol Monobutyl Ether.* U.S. Environmental Protection Agency, Washington, DC (www.epa.gov/iris/toxreviews/0500tr.pdf).

USEPA. (2010b). *Pavillion, Wyoming, Groundwater Investigation: January 2010 Sampling Results and Site Update*. U.S. Environmental Protection Agency, Washington, DC.

USEPA. (2012a). *Water: Contaminant Candidate List 3*. U.S. Environmental Protection Agency, Washington, DC (http://water.epa.gov/scitech/drinkingwater/dws/ccl/ccl3.cfm).

USEPA. (2012b). *Basic Information about Toluene in Drinking Water* (http://water.epa.gov/drink/contaminants/basicinformation/toluene.cfm); *Basic Information about Ethylbenzene in Drinking Water* (http://water.epa.gov/drink/contaminants/basicinformation/ethylbenzene.cfm); *Basic Information about Xylenes in Drinking Water* (http://water.epa.gov/drink/contaminants/basicinformation/xylenes.cfm); *Basic Information about Lead in Drinking Water* (http://water.epa.gov/drink/contaminants/basicinformation/lead.cfm). U.S. Environmental Protection Agency, Washington, DC.

USGS. (1999). *Naturally Occurring Radioactive Materials (NORM) in Produced Water and Oil-Field Equipment—An Issue for the Energy Industry*. U.S. Geological Survey, Washington, DC (pubs.usgs.gov/fs/fs-0142-99/fs-0142-99.pdf).

Veil, J.A. (2010). *Water Management Technologies Used by Marcellus Shale Gas Producers*, ANL/EVS/R-10/3, prepared by the Environmental Science Division, Argonne National Laboratory, for the U.S. Department of Energy, Office of Fossil Energy, National Energy Technology Laboratory, Washington, DC.

6 Produced Wastewater Treatment Technology

We hunger for energy, but we also thirst for water. The country is flush with natural gas as a result of new drilling techniques that have enabled energy companies to tap vast supplies that were out of reach not so long ago. The country's natural gas surplus has been growing even as the country burns record amounts.

Fahey (2012)

The *rule of capture* generally permits a landowner to drain or "capture" oil and natural gas from a neighboring property without liability or recourse.

TREATMENT OF PRODUCED WASTEWATER

Because produced wastewater may contain many different types of contaminants and the concentration of contaminants varies significantly, numerous types of treatment technologies have been proposed and used to treat produced wastewater. Most often, an effective produced wastewater treatment system, like a conventional wastewater treatment system, will consist of a train of many different types of individual unit processes used in series to remove a wide suite of contaminants that may not be removed with a single process. Organic and particulate removal (often classified as pretreatment), desalination, and disinfection are the major classifications of produced wastewater treatment technologies.

This chapter presents a qualitative comparison of produced wastewater technologies to provide an assessment of the benefits and limitations of each technology for produced water applications. The criteria used to compare the technologies are robustness, reliability, mobility, flexibility, modularity, cost, chemical and energy demand, and brine or residual disposal requirements. Based on research, communication with industry experts, and site visits, the following criteria were deemed most important for produced wastewater applications (Guerra et al., 2011):

- *Robustness* refers to the ability of the equipment to withstand harsh environmental conditions, have high mechanical strength, and to represent a technology in which the failure of an individual component does not significantly affect the overall performance of the technology.
- *Reliability* means that the technology will require minimal down time, can produce consistent water quality, and is not prone to failure.
- *Mobility* measures the ease with which the equipment can be moved from one site to another.
- *Flexibility* is the measure of the capability of the technology to accommodate a wide range of feed water qualities and to handle an upset in water quality without failure or reduced product water quality.
- *Modularity* refers to the ability to implement the technology as a unit process in a train of treatment technologies and the ease with which the system can be modified to handle changing water volumes.

Several physical separation technologies can be used to remove oil and grease and other organics from produced water. These technologies, which include advanced separators, hydrocyclones, filters, and centrifuges, are primarily deployed at offshore facilities where produced water is treated prior to ocean discharge. Oil and grease occur in at least three forms:

83

- *Free oil* (large droplets) is readily removable by gravity separation methods.
- *Dispersed oil* (small droplets) is somewhat difficult to remove.
- *Dissolved oil* (hydrocarbons and other similar materials dissolved in the water stream) is very challenging to eliminate.

Oil and water separation at onshore sites generally involves some form of oil/water separator or free water knockout vessel (for separation of the free oil). In offshore settings, oil/water separators and skim piles are deployed to remove oil droplets greater than 100 μm in diameter. More physical separation steps are added to remove any remaining free oil and some dispersed oil. Additional treatment iterations may be required to achieve compliance with all applicable discharge limits.

ORGANIC, PARTICULATE, AND MICROBIAL INACTIVATION/REMOVAL TECHNOLOGIES

Organic chemical and particulate removal is most often required as a pretreatment step when desalination technologies must be employed to treat produced wastewater. The technologies considered in this section include the following: biological aerated filter, hydrocyclone, dissolved air flotation, adsorption, media filtration, oxidation, granular activated carbon, ultraviolet disinfection, and ceramic and polymeric micro- and ultrafiltration. Table 6.1 presents a helpful guide for selecting treatment equipment based on the size of the particles that must be removed. The technologies discussed are either emerging technologies or an established technology and have been used for treatment of produced wastewater.

ADVANCED SEPARATORS

Separators rely on the difference in specific gravity between oil droplets and produced water. The lighter oil rises at a rate dependent on the droplet diameter and the fluid viscosity (Stokes' law). Smaller diameter droplets rise more slowly. If insufficient retention time is provided, the water exits the separator before the small droplets have risen through the water to collect as a separate oil layer. Likewise, inclined plate separators show better performance. Advanced separators contain additional internal structures that shorten the path followed by the oil droplets before they are collected. This gives smaller oil droplets the opportunity to reach a surface before the produced water overflows and exits the separator.

TABLE 6.1
Particle Size Removal Capabilities

Technology	Removal Capacity by Particle Size (μm)
API gravity separator	150
Corrugated plate separator	40
Induced gas flotation without chemical addition	25
Induced gas flotation with chemical addition	3–5
Hydrocyclone	10–15
Mesh coalescer	5
Media filter	5
Centrifuge	2
Membrane filter	0.01

Source: Frankiewicz, T., Understanding the Fundamentals of Water Treatment, the Dirty Dozen—12 Common Causes of Poor Water Quality, paper presented at 11th Produced Water Seminar, Houston, TX, January 17–19, 2001.

Produced Wastewater Treatment Technology

Biological Aerated Filters

Biological aerated filters (BAFs) are a class of technologies that include fixed-film and attached growth processes (see Figure 6.1), roughing filters, intermittent filters, packed-bed media filters, and conventional trickling filters (see Figure 6.2). A BAF consists of permeable media, such as rocks, gravel, or plastic media. The water to be treated flows downward over the media, which, over time, generates a microbial film on the surface of the media. The media facilitates biochemical oxidation/ removal of organic constituents. This is an aerobic process, and aerobic conditions are maintained by pumps and fans in the system. The thickness of the microbial layer continues to increase as the filter is used. Eventually, the microbial layer becomes thick enough that part of the slime layer becomes anaerobic and the microbial layer begins to slough off in the filter effluent (Spellman, 2014; USEPA, 1991). Media should have high a surface area per unit volume and be durable and inexpensive. The type of media used often is determined based on what materials are available at the site. Media can be field stone or gravel, and each stone should be between 1 and 4 inches in diameter to generate a pore space that does not prohibit flow through the filter and will not clog when sloughing occurs (Spellman, 2014; USEPA, 1980).

Biological aerated filters can remove oil, suspended solids, ammonia, nitrogen, chemical oxygen demand, biological oxygen demand, iron, manganese, heavy metals, organics, trace organics, and hydrogen sulfide. Iron and manganese removal in BAFs is mainly due to chemical oxidation rather than a biological process. Because BAFs do not remove dissolved constituents, however, high concentrations of salts can decrease the effectiveness of this technology due to salt toxicity effects. At

---

DID YOU KNOW?

The *rotating biological contactor* (RBC) is a biological treatment system (see Figure 6.1) and is a variation of the *attached-growth* idea provided by the trickling filter. Still relying on microorganisms that grow on the surface of the media, the RBC is a *fixed-film* biological treatment device, but the basic biological process is similar to that occurring in the trickling filter. An RBC consists of a series of closely spaced (mounted side by side), circular, plastic (synthetic) disks, that are typically about 3.5 m in diameter and attached to a rotating horizontal shaft (see Figure 6.1). Approximately 40% of each disk is submerged in a tank containing the wastewater to be treated. As the RBC rotates, the attached biomass film (zoogleal slime) that grows on the surface of the disk moves into and out of the wastewater. While submerged in the wastewater, the microorganisms absorb organics; when they are rotated out of the wastewater, they are supplied with needed oxygen for aerobic decomposition. As the zoogleal slime reenters the wastewater, excess solids and waste products are stripped off the media as sloughings. These sloughings are transported with the wastewater flow to a settling tank for removal (Spellman, 2014).

DID YOU KNOW?

Trickling filters have been used to treat wastewater since the 1890s. It was found that if settled wastewater was passed over rock surfaces, slime grew on the rocks and the water became cleaner. Today, we still use this principle, but in many installations instead of rocks we use plastic media. In most wastewater treatment systems, the trickling filter follows primary treatment and includes a secondary settling tank or clarifier as shown in Figure 6.2. Trickling filters are widely used for the treatment of domestic and industrial wastes. The process is a fixed-film biological treatment method designed to remove biochemical oxygen demand and suspended solids (Spellman, 2014).

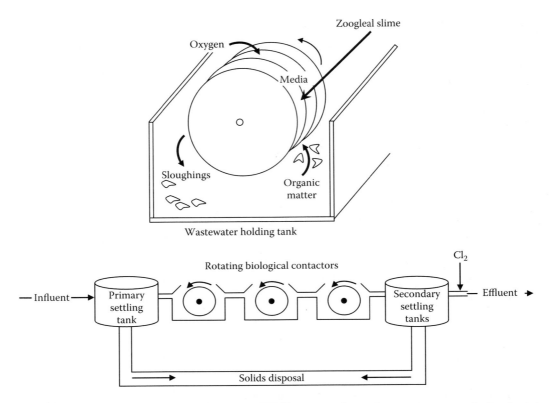

FIGURE 6.1 (Top) Rotating biological contactor (RBC) cross-section and treatment system; (bottom) rotating biological contactor treatment system.

chloride levels below 6600 mg/L, there is no diminished contaminant removal with BAFs; at 20,000-mg/L chloride levels, there will be a reduction in slime growth and BOD removal (Ludzack and Noran, 1965). This technology can be used to treat water with much greater organic contaminant concentrations than typically found in coalbed methane (CBM) produced wastewater. BAF is a well-established technology that has been used for produced wastewater treatment for many years (Doran, 1997; Su et al., 2007). Because of this technology's ability to remove oil and grease, it has been primarily used for oilfield produced wastewater treatment (Su et al., 2007). Informal versions of BAFs require minimal equipment and can be made by flowing water over rock beds. These types of BAFs also have been used in CBM produced water treatment for iron removal and suspended solids removal.

Biological aerated filtration is most effective on waters with chloride levels below 6600 mg/L (Ludzack and Noran, 1965), oil concentrations less than 60 mg/L, chemical oxygen demand less than 400 mg/L, and biological oxygen demand less than 50 mg/L. The maximum feed water constituent concentration for which this technology can be employed depends on desired removal and target water quality requirements. Removal capability of BAFs is dependent on the hydraulic loading rate on the filter and the raw water quality. The following are approximate removal capabilities of this technology (Ball, 1994; Su et al., 2007; USEPA, 1991):

- 60 to 90% nitrification
- 50 to 70% total nitrogen
- 70 to 80% oil
- 30 to 60% COD
- 85 to 95% BOD
- 75 to 85% suspended solids

Produced Wastewater Treatment Technology

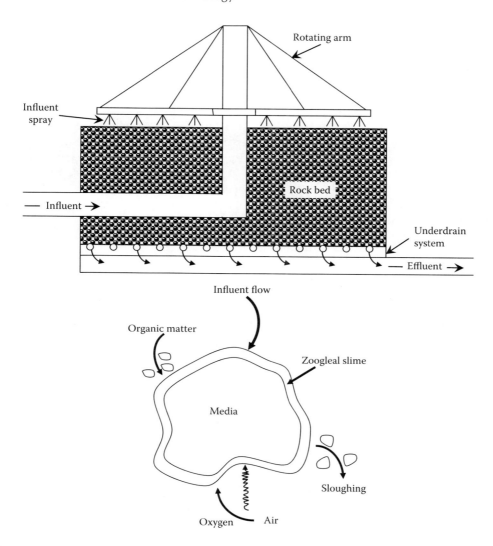

FIGURE 6.2 (Top) Cross-section of a trickling filter; (bottom) filter media showing the biological activities that take place on the surface.

There is nearly 100% water recovery from this process. The residuals generated are from the settling of the microbial layer that sloughs off the media. The residuals generation, which is highly dependent on the water quality, is approximately 0.4 to 0.7 pounds of dry solids per 1000 gallons of water treated (for conventional wastewater treatment) (Ball, 1994). Solids disposal is required for the sludge that accumulates in the sedimentation basins, and solids disposal can account for up to 40% of the total cost of the technology.

This technology has a long expected lifespan; however, BAFs require upstream and downstream sedimentation. For this reason, they have a large footprint and are not very mobile or modular. Very little monitoring is required, and occasional emptying of sedimentation ponds is required; the use of this technology does not require skilled operators. BAF can easily accommodate highly varying water quantity and quality. There is little down time or need for maintenance. Electricity is required for pumps and for fans for aeration and circulating water. The majority of the overall cost of this technology is capital, and operations and maintenance (O&M) costs are very low.

DID YOU KNOW?

Biological *nitrification* is the first basic step of *biological nitrification–denitrification*. In nitrification, the secondary effluent is introduced into another aeration tank, trickling filter, or biodisc. Because most of the carbonaceous biological oxygen demand has already been removed, the microorganisms that drive this advanced step are the nitrifying bacteria *Nitrosomonas* and *Nitrobacter*. In nitrification, the ammonia nitrogen is converted to nitrate nitrogen, producing a *nitrified effluent*. At this point, the nitrogen has not actually been removed, only converted to a form that is not toxic to aquatic life and that does not cause an additional oxygen demand. The nitrification process can be limited (performance affected) by alkalinity (requires 7.3 parts alkalinity to 1.0 part ammonia nitrogen); pH; dissolved oxygen availability; toxicity (ammonia or other toxic materials); and process mean cell residence time (sludge retention time). As a general rule, biological nitrification is more effective and achieves higher levels of removal during the warmer times of the year.

Primary sedimentation should be employed upstream of BAFs to allow the full bed of the filter to be used for removing nonsettling, colloidal, and dissolved particles if the water requires a large degree of contaminant removal. Sedimentation also should follow BAFs to remove the microbial layer that sloughs off of the filter. In addition to pumps and fans for aeration, other equipment such as distribution nozzles may be required. The estimated energy demand for BAFs is 1 to 4 kilowatt-hours per day (kWh/day). No chemicals are necessary for this treatment process (USEPA, 1980). The criteria ratings (high, moderate, low, none) for BAF applications in treating produced wastewater are as follows (Guerra et al., 2011):

- *Robustness*—high
- *Reliability*—high
- *Mobility*—low
- *Flexibility*—high
- *Modularity*—low

Hydrocyclone

Hydrocyclones have been used for surface treatment of produced wastewater for several decades. By the mid-1990s, over 300 hydrocyclones were deployed at offshore platforms (Hashimi et al., 1994). Hydrocyclones are referred to as liquid–liquid deoilers or enhanced gravity separators and are further classified as static or dynamic hydrocyclones. Hydrocyclones are used to separate hydrocarbon solids from liquids based on the density of the materials to be separated. Hydrocyclones normally have a cylindrical section at the top where the liquid is fed tangentially and a conical base (see Figure 6.3). The angle of the conical section determines the performance and separating capability of the hydrocyclone. Hydrocyclones can be made from metal, plastic, or ceramic and have no moving parts. The hydrocyclone has two exits—one at the bottom called the *underflow* or *reject* for the more dense fraction, and one at the top called the *overflow* or *product* for the less dense fraction of the original stream (Wagner et al., 1986).

Hydrocarbons can be used to separate liquids and solids or liquids of different densities (Stokes' law). Hydrocyclones can be used to remove particulates and oil from produced wastewater. Depending on the model of hydrocyclone employed, they can remove particles in the range of 5 to 15 μm (NETL, 2016a). Hydrocyclones will not remove soluble oil and grease components (Hayes and Arthur, 2004). Hydrocyclones have been used extensively to treat produced wastewater and are marketed by numerous companies for produced wastewater (NETL, 2016b; Sinker, 2007). Hydrocyclones were used to treat fracturing brine in the Barnett Shale play (Burnett, 2005); in

Produced Wastewater Treatment Technology 89

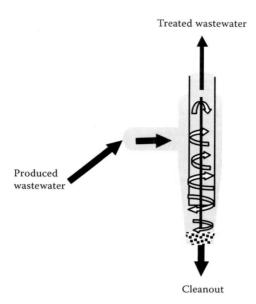

FIGURE 6.3 Hydrocyclone.

this research study, hydrocyclones were used in combination with organoclays as a pretreatment to reverse osmosis. Hydrocyclones can be used to treat water with high solids and organic chemical concentrations and can reduce oil and grease concentrations to 10 ppm. High product water recovery is possible with this technology. The waste generated from a hydrocyclone is a slurry of concentrated solids. This is the only residual that requires disposal.

Hydrocyclones do not require any pre- or post-treatment. The hydrocyclone itself does not require any chemicals or energy; however, a forwarding pump may be necessary to deliver water to the hydrocyclone or to recover pressure lost through the hydrocyclone. The hydrocyclone is the only piece of equipment necessary. There are no energy requirements unless the plant setup requires a forwarding pump to deliver water to the hydrocyclone. Depending on the size and configuration of the hydrocyclone, a large pressure drop can occur across the hydrocyclone.

Hydrocyclones have a long operational life due to the fact that they have no moving parts; however, they may suffer from abrasion when treating water with high particulate concentrations. Solid material can block the inlet, and scale formation can occur that requires cleaning; however, cleaning is typically minimal. The criteria ratings (high, moderate, low, none) for hydrocyclone applications in treating produced wastewater are as follows (Guerra et al., 2011):

- *Robustness*—high
- *Reliability*—high
- *Mobility*—low
- *Flexibility*—low
- *Modularity*—none

Dissolved Air/Gas Flotation

Flotation is a process in which fine gas bubbles are used to separate small, suspended particles that are difficult to separate by settling or sedimentation (see Figure 6.4). Gas is injected into the water to be treated; particulates and oil droplets suspended in the water are attached to the air bubbles, and they both rise to the surface. As a result, foam develops on the surface, which is commonly removed

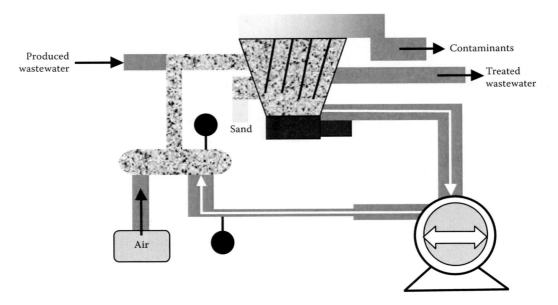

FIGURE 6.4 Dissolved air flotation.

by skimming. The dissolved gas can be air, nitrogen, or another type of inert gas. Dissolved air/gas flotation also can be used to remove volatile organics and oil and grease. Dissolved air flotation units have been widely used for the treatment of produced wastewater (Cakmakce et al., 2008; Casaday, 1993; Hayes, 2004).

Gas flotation technology is divided into dissolved gas flotation (DGF) and induced gas flotation (IGF). The two technologies differ by the method used to generated gas bubbles and the resultant bubble sizes. In DGF units, gas (usually air) is fed into the flotation chamber, which is filled with a fully saturated solution. Inside the chamber, the gas is released by applying a vacuum or by creating a rapid pressure drop. IGF technology uses mechanical shear or propellers to create bubbles that are introduced into the bottom of the flotation chamber (NETL, 2016b). Coagulation can be used as a pretreatment to flotation.

The efficiency of the flotation process depends on the density differences of the liquid and contaminants to be removed. It also depends on the oil droplet size and temperature. Minimizing gas bubble size and achieving an even gas bubble distribution are critical to removal efficiency (Casaday, 1993). Flotation works well in cold temperatures (but is not ideal for high-temperature feed streams) and can be used for waters with both high and low total organic carbon (TOC) concentrations. It is excellent for removing natural organic matter (NOM) and can be used to treat water containing TOC, oil and grease, and particulates < 7% solids (Burke, 1997).

Dissolved air flotation (DAF) can remove particles as small as 25 μm. If coagulation is added as pretreatment, DAF can remove contaminants 3 to 5 μm in size (NETL, 2016a). In one reported study, flotation achieved an oil removal of 93% (ALL Consulting, 2003). Flotation cannot remove soluble oil constituents from water. Product wastewater is nearly 100% with this technology.

Because flotation involves dissolving a gas into the water stream, flotation works best at low temperatures. If high temperatures are present, a higher pressure is required to dissolve the gas in the water. Energy is required to pressurize the system to dissolve gas in the feed stream. Coagulant chemical may be added to enhance the removal of target contaminants. Chemical coagulant and pumping costs are the major components of O&M costs for flotation. Treatment costs are estimated to be $0.60 per cubic meter (Cakmakce et al., 2008). Solids disposal will be required for the sludge generated from flotation. The criteria ratings (high, moderate, low, none) for flotation applications in treating produced wastewater are as follows (Guerra et al., 2011):

Produced Wastewater Treatment Technology 91

- *Robustness*—high
- *Reliability*—high
- *Mobility*—low
- *Flexibility*—low
- *Modularity*—low

ADSORPTION AND MEDIA FILTRATION

Adsorption and media filtration can be accomplished using a variety of materials, including zeolites, organoclays, activated alumina, and activated carbon. Chemicals are not required for normal operation of adsorptive processes, but chemicals may be used to regenerate media when all active sites are occupied. The media must be backwashed periodically to remove large particulates trapped between the voids in the media. Typically, these processes can be gravity fed and do not require an energy supply, except during backwash. Adsorbents are capable of removing iron; manganese; total organic carbon; benzene, toluene, ethylbenzene, and xylene (BTEX) compounds; heavy metals; and oil from produced water. Adsorption is generally utilized as a unit process in a treatment train rather than as a stand-alone process. The adsorbent can be easily overloaded with large concentrations of organics, so this process is best used as a polishing step rather than as a primary treatment process (NETL, 2016b). Adsorption is capable of removing over 80% of heavy metals (Spellman, 2014) and can accomplish nearly 100% product wastewater recovery. The rate of media usage represents one of the main operational costs for adsorptive processes, as the media may require frequent replacement or regeneration depending on media type and feed water quality. When all active sites of the adsorptive material have been consumed, the material must be either regenerated or disposed of. Regenerating the materials will result in a liquid waste for disposal. Solid waste disposal is necessary when the material must be replaced entirely. A loss of pressure will be incurred across the filter; however, depending on the plant configuration, this may not require any additional pumps. Pumps will be necessary to backwash the filters. Adsorption is best used as a polishing step to avoid rapid usage of adsorbent material.

Filtration can be accomplished using a variety of different types of media: walnut shell, sand, and anthracite, among others. Filtration is a widely used technology for produced wastewater, especially walnut shell filters for removing oil and grease. Many vendors market filtration technologies specifically for produced wastewater. Filtration does not remove dissolved ions, and the performance of filters is not affected by high salt concentrations; therefore, filtration can be used for all total dissolved solids (TDS) bins regardless of salt type. Filtration can be used to remove TOC as well as oil and grease from produced wastewater, with greater than 90% oil and grease removal. Removal efficiencies can be improved by employing coagulation upstream of the filter. Nearly 100% water recovery is achieved with filtration; some filtrate may be used for backwashes. Minimal energy is required for these processes. Energy is required for backwashing the filter. Coagulant may be added to the feed water to increase particle size and enhance separation, and chemicals may be required for media regeneration. A loss of pressure will be incurred across the filter; however, depending on the plant configuration, this may not require any additional pumps. Pumps will be necessary to backwash the filters. Solid waste disposal is required for spent media or the waste produced during regeneration of the media.

The criteria ratings (high, moderate, low, none) for adsorption and media filtration applications in treating produced wastewater are as follows (Guerra et al., 2011):

- *Robustness*—high
- *Reliability*—high
- *Mobility*—high
- *Flexibility*—high
- *Modularity*—high

OXIDATION

Chemical oxidation treatment can be used to remove iron, manganese, sulfur, color, tastes, odor, and synthetic organic chemicals. Chemical oxidation relies on oxidation/reduction reactions, which consist of two half-reactions: the oxidation reaction in which a substance loses or donates electrons, and a reduction reaction in which a substance accepts or gains electrons. Oxidation and reduction reactions will always occur together, because free electrons cannot exist in solution and electrons must be conserved (AWWA, 2005). Oxidants commonly used in water treatment applications include chlorine, chlorine dioxide, permanganate, oxygen, and ozone. The appropriate oxidant for a given application depends on many factors including raw water quality, specific contaminants present in the water, and local chemical and power costs (AWWA, 2005). Chemical oxidation is well established and reliable and requires minimal equipment (USDOI, 2010). Oxidation can be employed to remove organics and some inorganic compounds (i.e., iron and manganese) from produced wastewater. The removal or oxidation rate can be controlled by the applied chemical dose and contact time between the oxidants and water, with 100% feed water recovery.

Chemical metering is required. Energy usage usually accounts for approximately 18% of the total O&M for oxidation processes coupled with high chemical costs. Critical components of the oxidation process are the chemical metering pumps. Chemical metering equipment can have a life expectancy of 10 years or greater. Periodic calibration and maintenance of chemical meter pumps are required. Capital costs can be near to $0.01 per gallon per day (gpd), and O&M costs can be approximately $0.05 per kilogallon (kgal) (>$0.01 per bbl). No waste is generated from oxidation processes.

No pretreatment is required for oxidation. Solid separation post-treatment might be required to remove oxidized particles. Chemical metering pumps are required for dosing. Some equipment may be required to generate the oxidant onsite, and chemical costs may be high. Solids disposal will be required for the sludge generated from flotation. The criteria ratings (high, moderate, low, none) for oxidation applications in treating produced wastewater are as follows (Guerra et al., 2011):

- *Robustness*—high
- *Reliability*—high
- *Mobility*—high
- *Flexibility*—high
- *Modularity*—high

SETTLING PONDS OR BASINS

Settling can be achieved using a pond or a basin. The primary goals of produced wastewater ponds focus on simplicity and flexibility of operation, protection of the water environment, and protection of public health. Moreover, ponds are relatively easy to build and manage, they accommodate large fluctuations in flow, and they can also provide treatment that approaches conventional systems (producing a highly purified effluent) at much lower cost. It is the cost (the economics) that drives many managers to decide on the pond option of treatment. The actual degree of treatment provided in a pond depends on the type and number of ponds used. Ponds can be used as the sole type of treatment, or they can be used in conjunction with other forms of produced wastewater treatment—that is, other treatment processes followed by a pond or a pond followed by other treatment processes. Ponds can be classified based on their location in the system, by the type of wastes they receive, and by the main biological process occurring in the pond.

In the produced wastewater pond, particulates are removed by gravity settling. Settling ponds require a large footprint and environmental mitigation to protect wildlife. The volume of the settling basin required depends on the hydraulic residence time required for the desired level of contaminant removal. Settling pounds most likely will be used in combination with other treatment unit

> **DID YOU KNOW?**
>
> Acre-feet (ac-ft) is a unit that can cause confusion, especially for those not familiar with pond or lagoon operations. One acre-foot is the volume of a box with a 1-acre top and 1 foot of depth—but the top does not have to be an even number of acres in size to use acre-feet.

processes. There are no chemical requirements, but chemicals can be used to enhance sedimentation. Infrastructure requirements include liners. Settling ponds are used to remove large particulates from water sources. The degree of particle removal and size of particles removed depends on the water detention time in the pond. The criteria ratings (high, moderate, low, none) for settling ponds and basins in treating produced wastewater are as follows (Guerra et al., 2011):

- *Robustness*—high
- *Reliability*—high
- *Mobility*—none
- *Flexibility*—moderate
- *Modularity*—none

AIR STRIPPING

Air stripping primarily is used for removing volatile organic chemicals (VOCs), oxidizing contaminants such as iron and manganese, improving taste, or removing odor. Air stripping is a U.S. Environmental Protection Agency (USEPA) best available technology (BAT) for some VOCs, including benzene, toluene, xylene, tri/tetrachloroethylene, trihalomethanes, and vinyl chloride, among others.

Air stripping is the process of transferring a contaminant from the liquid phase to the gas phase. In the air stripping process, air and water are contacted in a packed column designed to maximize the contact surface area between the water and air. Air stripping performance depends on such factors as the following:

- Characteristics of the volatile material (e.g., partial pressure, Henry's constant, gas-transfer resistance) (AWWA, 2005)
- Water and ambient air temperature
- Turbulence in gaseous and liquid phases
- Area-to-volume ratio
- Exposure time

Appropriate design of the packed column is necessary to ensure the desired level of contaminant removal based on the process operating temperature and the Henry's constant of the target contaminant. Scaling can occur when calcium exceeds 40 mg/L, iron exceeds 0.3 mg/L, magnesium exceeds 10 mg/L, and manganese exceeds 0.05 mg/L. Biological fouling also may occur depending on the feed water quality (USACE, 2001).

Spray aerators dissipate water at a vertical or inclined angle, breaking the water into small drops. Multiple-tray aerators use uniquely designed trays to increase the surface area for aeration. Cascade and cone aerators allow water to flow in a downward direction over a series of baffles or pans. The two main types of pressure aerators are one that sprays water on top of a tank that is constantly supplied with compressed air and one that injects compressed air directly into a pressurized piping, adding fine air bubbles into the flowing water. Diffusion-type aerators are similar to pressure aerators but are designed to allow air bubbles to diffuse upward through the tank of water to help produce turbulence and mixing. Mechanical aerators use a motor-driven impeller to achieve air mixing; occasionally, it also is used in combination with an air injection device (AWWA, 2005).

Air stripping is a proven and widely used technology that can serve as a low-profile addition to a treatment process, with a high contaminant removable efficiency (>99%) (USACE, 2001). Air stripping systems must be properly designed to provide the proper air and water balance to prevent flooding or excess air flow (USACE, 2001); scaling and biological fouling may impact the performance of the air stripper. The criteria ratings (high, moderate, low, none) for air stripping systems in treating produced wastewater are as follows (Guerra et al., 2011):

- *Robustness*—high
- *Reliability*—high
- *Mobility*—moderately
- *Flexibility*—none
- *Modularity*—none

SURFACTANT-MODIFIED ZEOLITE/VAPOR-PHASE BIOREACTOR

Zeolites are naturally occurring hydrated aluminosilicates with a large surface area. Because of the natural shape and size of zeolites, they are suitable for flow through applications or media for fluidized beds. Treatment of zeolites with cationic surfactants changes the surface chemistry of the zeolites, allowing them to absorb nonpolar organic solutes. These materials also are capable of cation or anion exchange; however, their usefulness as a desalination technology is questionable. A substantial amount of research has been conducted on the use of surfactant modified zeolites for organic chemical removal from produced wastewater. This research suggests that this process is a promising produced wastewater treatment technology (Altare et al., 2007; Bowman, 2005; Ranck et al., 2005). The criteria ratings (high, moderate, low, none) for surfactant modified zeolite vapor-phase bioreactors in treating produced wastewater are as follows (Guerra et al., 2011):

- *Robustness*—none
- *Reliability*—none
- *Mobility*—none
- *Flexibility*—none
- *Modularity*—high

CONSTRUCTED WETLANDS

One of the major benefits of constructed wetlands is that these systems have low construction and operating costs. The estimated cost of constructed wetlands treatment is $0.01 to $0.02 per barrel; however, these systems are not efficient and the treatment rate is slow compared to other technologies. The average lifespan of a constructed wetland is approximately 20 years (Shutes, 2001). A study was conducted to look at a hybrid reverse osmosis (RO) constructed wetland system for produced wastewater treatment. This study showed that, although RO removes the majority of organic and inorganic constituents, it does not sufficiently remove dissolved organic compounds. These compounds were removed by the constructed wetland. Toxicity tests were conducted with a few different types of bacteria. All species of bacteria experienced a greater survival rate with water that had been passed through the constructed wetland. Many water quality parameters were increased following the wetland construction, such as total dissolved solids and calcium. The sodium absorption ratio (SAR) was decreased. Boron removal was not addressed in this study, and mention was made that, for irrigation, the levels of boron in the treated water may be too high (Murry-Gulde, 2003). The criteria ratings (high, moderate, low, none) for constructed wetlands in treating produced wastewater are as follows (Guerra et al., 2011):

- *Robustness*—high
- *Reliability*—high
- *Mobility*—none
- *Flexibility*—moderate
- *Modularity*—none

GRANULAR ACTIVATED CARBON

Granular activated carbon (GAC) can be used to remove the following contaminants from produced water: mercury, cadmium, natural organic matter, BTEX compounds, and synthetic organic chemicals—specifically benzo(*a*)pyrene, di(2-ethylhexyl)adipate, di(2-ethylhexyl)phthalate, hexachloro-benzene, dioxin, and radionuclides. For source water with a large amount of bacteria, pretreatment in the form of filtration and disinfection prior to carbon treatment may be required. Filtration prior to the use of GAC also may be required when dealing with high total suspended solids (TSS) waters. GAC has an extremely large amount of adsorption surface area, generally around 73 acres/lb (650 m^2/g) to 112 acres/lb (1000 m^2/g) (AWWA, 2005). GAC is made of tiny clusters of carbon atoms stacked upon one another and is produced by heating the carbon source (coal, lignite, wood, nut-shells, or peat) in the absence of air, which produces a high-carbon-content material. The adsorption isotherm for carbon and the source water will determine the total contaminant removal capacity.

The physical removal of a contaminant by adsorption onto the carbon surface is achieved in the mass transfer zone. *Breakthrough* is defined as the point at which the concentration of a contaminant in the effluent adsorption unit exceeds the treatment required (AWWA, 2005). The breakthrough time is important to note so that treatment goals are not exceeded and backwashing rates can be optimized. Backwashing a GAC system follows the same general procedures as for conventional granular gravity filter systems. The GAC typically will expand up to 75 to 100% in volume, but the expansion may be only as much as 50% (AWWA, 2005). Empty bed contact time is the volume of the empty bed divided by the volumetric flow rate of water through the carbon. A typical bed depth can contain up to 50% freeboard excess capacity beyond the designed capacity to allow for bed expansion during backwashing. Surface loading rates, or the volume of water that is passing through a given area, typically range from 2 to 6 gallons per minute per square foot (5 to 15 meters per hour) (AWWA, 2005).

Biological growth can be desirable (to a point) within GAC, which results in what is known as biologically active carbon (BAC). BAC can be beneficial by removing assimilable organic carbon (AOC) and other biodegradable compounds. If BAC is to be utilized, the GAC filters typically are preceded by ozonation that breaks down the organic carbon into a more assimilable form. This process can enhance the overall contaminant removal of the GAC process. However, the biological growth needs to be controlled with frequent backwashing (once every 5 days). The use of chlorine prior to the beds will not prevent growth, will produce disinfection byproducts (DBPs) that take up

DID YOU KNOW?

In conventional wastewater treatment, granular activated carbon (GAC) adsorption can be used following filtration to remove additional natural organic matter (NOM). For most applications, empty bed contact times in excess of 20 minutes are required, with regeneration frequencies on the order of 2 to 3 months. These long control times and frequent regeneration requirements make GAC an expensive treatment option. In cases where prechlorination is practiced, the chlorine rapidly degrades the GAC. Addition of a disinfectant to the GAC bed can result in specific reactions in which previously absorbed compounds leach into the treated water (Spellman, 2014).

more GAC adsorption sites, and will make the carbon more brittle. To prevent biological growth in the distribution system disinfection is recommended after the GAC filters; additionally, this practice helps to achieve the highest removal of AOC within the system. Uncontrolled biological growth can lead to odor problems and the growth of undesirable organisms.

Note that in GAC operation, regular reactivation or replacement of the carbon media is required. If the GAC plant is large enough, regeneration can be done onsite, but it is typically performed offsite. If the carbon exhaustion rate is larger than 910 kg/day, onsite regeneration may not be effective (AWWA, 2005). Reactivation frequency is dependent on contaminant type, concentration, rate of water usage, and type of carbon used (Guerra et al., 2011). Around the time of startup and breakthrough, careful monitoring and testing are necessary to be sure that contaminant removal is being achieved. Flushing is required if the carbon filter is not used for several days, and regular backwashing may be required to prevent bacterial growth (AWWA, 2005). The criteria ratings (high, moderate, low, none) for GAC systems in treating produced wastewater are as follows (Guerra et al., 2011):

- *Robustness*—high
- *Reliability*—high
- *Mobility*—high
- *Flexibility*—high
- *Modularity*—high

ULTRAVIOLET DISINFECTION

Although ultraviolet (UV) disinfection was recognized as a method for achieving disinfection in the late 19th century, its application virtually disappeared with the evolution of chlorination technologies. However, in recent years, there has been a resurgence in its use in the water and wastewater fields, largely as a consequence of concern for the discharge of toxic chlorine residual. Even more recently, UV has gained attention because of the tough new regulations on chlorine use imposed by both the Occupational Safety and Health Administration (OSHA) and USEPA. Because of this relatively recent increased regulatory pressure, many facilities are actively engaged in substituting various disinfection alternatives for chlorine. Moreover, many improvements have been made to UV technology that make this an attractive disinfection alternative. Ultraviolet light has very good germicidal qualities and is very effective in destroying microorganisms. It is used in hospitals, biological testing facilities, and many other similar settings. In wastewater treatment, the plant effluent is exposed to ultraviolet light of a specified wavelength and intensity for a specified contact period. The effectiveness of the process is dependent on

- UV light intensity
- Contact time
- Wastewater quality (turbidity)

For any one treatment plant, disinfection success is directly related to the concentration of colloidal and particulate constituents in the wastewater.

The Achilles' heel of UV for disinfecting wastewater is turbidity. If the wastewater quality is poor (i.e., opaque), the ultraviolet light will be unable to penetrate the solids, and the effectiveness of the process decreases dramatically. For this reason, many states limit the use of UV disinfection to facilities that can reasonably be expected to produce an effluent containing ≤ 30 mg/L of BOD_5 and total suspended solids.

The main components of a UV disinfection system are mercury arc lamps, a reactor, and ballasts. The source of UV radiation is either a low-pressure or a medium-pressure mercury arc lamp with low or high intensities, and replacement UV lamps must be readily available. The best lamps are those with a stated operating life of at least 7500 hours and those that do not produce significant amounts

Produced Wastewater Treatment Technology

of ozone or hydrogen peroxide. The lamps must also meet technical specifications for intensity, output, and arc length. If the UV light tubes are submerged in the wastestream, they must be protected inside quartz tubes, which not only protect the lights but also make cleaning and replacement easier.

Contact tanks must be used with UV disinfection. They must be designed with the banks of UV lights in a horizontal position, either parallel or perpendicular to the flow, or with the banks of lights placed in a vertical position perpendicular to the flow.

Note: The contact tank must provide, at a minimum, a 10-second exposure time.

Turbidity has been an ongoing problem with using UV in wastewater treatment; however, if turbidity is its Achilles' heel, then the need for increased maintenance (as compared to other disinfection alternatives) is the toe of the same foot. UV maintenance requires that the tubes be cleaned on a regular basis or as needed. In addition, periodic acid washing is also required to remove chemical buildup. Routine monitoring is required. Monitoring to check on bulb burnout, buildup of solids on quartz tubes, and UV light intensity is necessary.

Note: UV light is extremely hazardous to the eyes. Never enter an area where UV lights are in operation without proper eye protection. Never look directly into the ultraviolet light.

Advantages
- UV disinfection is effective at inactivating most viruses, spores, and cysts.
- UV disinfection is a physical process rather than a chemical disinfectant; it eliminates the need to generate, handle, transport, or store toxic, hazardous, or corrosive chemicals.
- There is no residual effect that can be harmful to humans or aquatic life.
- UV disinfection is user friendly for operators.
- UV disinfection has a shorter contact time when compared with other disinfectants (approximately 20 to 30 seconds with low-pressure lamps).
- UV disinfection equipment requires less space than other methods.

Disadvantages
- Low dosages may not effectively inactivate some viruses, spores, and cysts.
- Organisms can sometimes repair and reverse the destructive effects of UV through a repair mechanism known as photoreactivation or, in the absence of light, as dark repairs.
- A preventive maintenance program is necessary to control fouling of tubes.
- Turbidity and total suspended solids (TSS) in the wastewater can render UV disinfection ineffective. UV disinfection with low-pressure lamps is not as effective for secondary effluent with TSS levels above 30 mg/L.
- UV disinfection is not as cost competitive when chlorination/dechlorination is used and fire codes are met.

Applicability
When choosing a UV disinfection system, three critical areas must be considered. The first is primarily determined by the manufacturer and the second by design and by operations and maintenance (O&M); the third has to be controlled at the treatment facility. Choosing a UV disinfection system depends on the three critical factors listed below:

- *Hydraulic properties of the reactor*—Ideally, a UV disinfection system should have a uniform flow with enough axial motion (radial mixing) to maximize exposure to UV radiation. The path that an organism takes in the reactor determines the amount of UV radiation it will be exposed to before inactivation. A reactor must be designed to eliminate short-circuiting and/or dead zones, which can result in inefficient use of power and reduced contact time.

- *Intensity of the UV radiation*—Factors affecting the intensity are the age of the lamps, lamp fouling, and the configuration and placement of lamps in the reactor.
- *Wastewater characteristics*—These include the flow rate, suspended and colloidal solids, initial bacterial density, and other physical and chemical parameters. Both the concentration of TSS and the concentration of particle-associated microorganisms determine how much UV radiation ultimately reaches the target organisms. The higher these concentrations, the lower the UV radiation absorbed by the organisms. UV disinfection can be used in plants of various sizes that provide secondary or advanced levels of treatment.

The criteria ratings (high, moderate, low, none) for UV systems in treating produced wastewater are as follows (Guerra et al., 2011):

- *Robustness*—low
- *Reliability*—high
- *Mobility*—high
- *Flexibility*—high
- *Modularity*—high

MICROFILTRATION/ULTRAFILTRATION

A filter system can be employed as an alternative filtering system to reduce turbidity and remove *Giardia* and *Cryptosporidium*. A standard filter is made of synthetic media contained in a plastic or metal housing. These systems are normally installed in a series of three or four filters. Each filter contains media successively smaller than in the previous filter. The media sizes typically range from 50 to 5 μm or less. Microfiltration (MF) has the largest pore size (0.1 to 3 μm); ultrafiltration (UF) pore sizes range from 0.01 to 0.1 μm. Ultrafiltration is not fundamentally different from microfiltration. Both techniques separate based on size exclusion or particulate capture. In terms of pore size, MF fills the gap between ultrafiltration and granular media filtration. In terms of characteristic particle size, the MF range covers the lower portion of the conventional clays and the upper half of the range for humic acids. UF membranes are defined by the molecular weight cutoff of the membrane used. The filter arrangement is dependent on the quality of the water, the capability of the filter, and the quantity of water needed. Ultrafiltration is applied in cross-flow or dead-end mode. The USEPA and state agencies have established criteria for the selection and use of filters. Generally, MF and UF filter systems are regulated in the same manner as other filtration systems.

Because of new regulatory requirements and the need to provide more efficient removal of pathogenic protozoans (e.g., *Giardia, Cryptosporidium*) from water supplies, *membrane filtration systems* are finding increased application in water treatment systems. A membrane is a thin film separating two different phases of a material that acts as a selective barrier to matter transported by some driving force. Simply, a membrane can be regarded as a sieve with very small pores. Membrane filtration processes are typically pressure, electrically, vacuum, or thermally driven. The types of drinking water membrane filtration systems include microfiltration, ultrafiltration, nanofiltration, and reverse osmosis. A typical membrane filtration process has one input and two outputs. Membrane performance is largely a function of the properties of the materials to be separated and can vary throughout operation.

Common Filter Problems

Two common types of filter problems occur: those caused by filter runs that are too long (infrequent backwash) and those caused by inefficient backwash (cleaning). Running a filter too long can cause *breakthrough* (the pushing of debris removed from the water through the media and into the effluent) and *air binding* (the trapping of air and other dissolved gases in the filter media). Air binding occurs when the rate at which water exits the bottom of the filter exceeds the rate at which the water penetrates the top of the filter. When this happens, a void and partial vacuum occur inside the

Produced Wastewater Treatment Technology

filter media. The vacuum causes gases to escape from the water and fill the void. When the filter is backwashed, the release of these gases may cause a violent upheaval in the media and destroy the layering of the media bed, gravel, or underdrain. Two solutions to the problems are to (1) check the filtration rates to ensure that they are within the design specifications, and (2) remove the top 1 inch of media and replace with new media. This keeps the top of the media from collecting the floc and sealing the entrance into the filter media.

Another common filtration problem associated with poor backwashing practices is the formation of *mud balls* that get trapped in the filter media. In severe cases, mud balls can completely clog a filter. Poor agitation of the surface of the filter can form a crust on top of the filter; the crust later cracks under the water pressure, causing uneven distribution of water through the filter media. Filter cracking can be corrected by removing the top 1 inch of the filter media, increasing the backwash rate, or checking the effectiveness of the surface wash (if installed). Backwashing at too high a rate can cause the filter media to wash out of the filter over the effluent troughs and may damage the filter underdrain system. Two possible solutions are to (1) check the backwash rate to be sure that it meets the design criteria, and (2) check the surface wash (if installed) for proper operation. The criteria ratings (high, moderate, low, none) for microfiltration/ultrafiltration systems in treating produced wastewater are as follows (Guerra et al., 2011):

- *Robustness*—high
- *Reliability*—moderate
- *Mobility*—high
- *Flexibility*—moderate
- *Modularity*—high

THOUGHT-PROVOKING QUESTIONS

1. Is technology the answer to mitigating the environmental problems created by hydraulic fracturing?
2. Are contractors' claims of "trade secrets" regarding the chemicals added to fracking water covering up the creation of environmental problems?
3. Can fracking water pass the "yuck" test; that is, can it be made drinkable again? Will people want to drink it? Does it really matter if anyone wants to drink treated fracking water?

REFERENCES AND RECOMMENDED READING

ALL Consulting. (2003). *Handbook on Coal Bed Ethane Produced Water: Management and Beneficial Use Alternatives.* ALL Consulting, Tulsa, OK.

Altare, C.R., Bowman, R.S. et al. (2007). Regeneration and long-term stability of surfactant-modified zeolite for removal of volatile organic compounds from produced water. *Microporous and Mesoporous Materials*, 105(3):305–316.

AWWA. (2005). *Water Treatment Plant Design*, 4th ed. McGraw-Hill, New York.

Ball, H.L. (1994). Nitrogen reduction in an onsite trickling filter/upflow filter wastewater treatment system, in *Proceedings of the Seventh International Symposium on Individual and Small Community Sewage Systems*, Atlanta, GA, December 11–13.

Boschee, P. (2012). Handling produced water from hydraulic fracturing. *Oil and Gas Facilities*, 1(1):23–26.

Bowman, R.S. (2005). Applications of surfactant-modified zeolites to environmental remediation. *Microporous and Mesoporous Materials*, 61(1–3):43–56.

Burke, D. (1997). *Application of the Anoxic Gas Flotation Process.* Environmental Energy Company, Olympia, WA.

Burnett, D. and Barrufet, M. (2005). *Membranes Used in Produced Brine Desalination*, DOE/PERF Project Review. U.S. Department of Energy, Washington, DC.

Cakmakce, M., Kayaalp, N. et al. (2008). Desalination of produced water from oil production fields by membrane processes. *Desalination*, 222(1–3):176–186.

Casady, A.L. (1993). *Advances in Flotation Unit Design for Produced Water Treatment*, SPE #25472. Society of Petroleum Engineers, Richardson, TX.

Dahm, K. and Chapman, M. (2014). *Produced Water Treatment Primer: Case Studies of Treatment Applications*. U.S. Department of the Interior, Bureau of Reclamation, Washington, DC.

Doran, G. (1997). *Developing a Cost Effective Environmental Solution for Produced Water and Creating a New Water Resource*, Report No. DE-FC22-95MT95008. U.S. Department of Energy, Washington, DC.

Fahey, J. (2012). Natural gas price plunge aids families, businesses. *Associated Press*, January 15.

Faucher, M. and Sellman, E. (1998). Produced Water Deoiling Using Disc Stack Centrifuges, paper presented at the API Produced Water Management Technical Forum and Exhibition, Lafayette, LA, November 17–18.

Frankiewicz, T. (2001). Understanding the Fundamentals of Water Treatment, the Dirty Dozen—12 Common Causes of Poor Water Quality, paper presented at 11th Produced Water Seminar, Houston, TX, January 17–19.

Guerra, K., Dahm, K., and Durden, S. (2011). *Oil and Gas Produced Water Management and Beneficial Use in the Western United States*, Science and Technology Program Report No. 157. U.S. Department of Interior, Bureau of Reclamation, Washington, DC.

Hashmi, K.A., Hamza, H.A., and Thew, M.T. (1994). Liquid–liquid hydrocyclone for deoiling produced waters in heavy oil recovery, in *Proceedings of the International Petroleum Environmental Conference*, Houston, TX, March 2–4.

Hayes, T. and Arthur, D. (2004). Overview of Emerging Produced Water Treatment Technologies, paper presented at 11th Annual International Petroleum Conference, Albuquerque, NM, October 12–15.

Ludzack, F.J. and Noran, D.K. (1965). *Tolerance of High Salinities by Conventional Wastewater Treatment Processes*. Water Environment Federation, Arlington, VA.

Murry-Gulde, C. (2003). Performance of a hybrid reverse-osmosis-constructed wetland treatment system for brackish oil field produced water. *Water Research*, 37(3):705–713.

NETL. (2016a). *Produced Water Management Technology Descriptions*. National Energy Technology Laboratory, Washington, DC (www.netl.doe.gov/research/coal/crosscutting/pwmis/tech-desc).

NETL. (2016b). *Fact Sheet—Industrial Use*. National Energy Technology Laboratory, Washington, DC (www.netl.doe.gov/research/coal/crosscutting/pwmis/tech-desc/induse).

NETL. (2016c). *Fact Sheet—Physical Separation*. National Energy Technology Laboratory, Washington, DC (www.netl.doe.gov/research/coal/crosscutting/pwmis/tech-desc/physep).

Ranck, J.M., Bowman, R.S. et al. (2005). *Removal of BTEX from Produced Water Using Surfactant-Modified Zeolite*. National Ground Water Association, Westerville, OH.

Shutes, F.B.E. (2001). Artificial wetlands and water quality improvement. *Environmental International*, 26(2001):441–447.

Sinker, A. (2007). Produced Water Treatment Using Hydrocyclones: Theory and Practical Applications, paper presented at 14th Annual International Petroleum Environmental Conference, Houston, TX, November 6–9.

Spellman, F.R. (2014). *Handbook of Water and Wastewater Treatment Plant Operations*, 3rd ed. CRC Press, Boca Raton, FL.

Su, D., Wang, J. et al. (2007). Kinetic performance of oil-field produced water treatment by biological aerated filter. *Chinese Journal of Chemical Engineering*, 15(4):591–594.

Tyrie, C.C. (1998). The Technology and Economics of the Various Filters That Are Used in Oil Field Produced Water Clean Up, paper presented at API Produced Water Management Technical Forum and Exhibition, Lafayette, LA, November 17–18.

USACE. (2001). *Engineering and Design: Air Stripping*, Design Guide No. 1110-1-3. U.S. Army Corps of Engineers, Washington, DC.

USDOI. (2010). *Water Treatment Primer for Communities in Need*, Desalination Series Report No. 68. U.S. Department of the Interior, Bureau of Reclamation, Denver, CO.

USEPA. (1980). *Design Manual: Onsite Wastewater Treatment and Disposal Systems*, EPA 625/1-80-012. U.S. Environmental Protection Agency, Washington, DC.

USEPA. (1991). *Assessment of Single-Stage Trickling Filter Nitrification*, EPA 430/09-91-005. U.S. Environmental Protection Agency, Washington, DC.

Veil, J.A., Puder, M.G., Elcock, D., and Redweik, Jr., R.J. (2004). *A White Paper Describing Produced Water from Production of Crude Oil, Natural Gas, and Coal Bed Methane*. U.S. Department of Energy, Washington, DC.

Wagner, K. et al. (1986). *Remedial Action Technology for Waste Disposal Sites*. William Andrew, Inc., Norwich, NY.

7 Desalination

There is more evidence to prove that saltness [of water] is due to the admixture of some substance, besides that which we have adduced. Make a vessel of wax and put it in the sea, fastening its mouth in such a way to prevent any water getting in. Then the water that percolates through the wax sides of the vessel is sweet, the earthly stuff, the admixture of which makes the water salt, being separated off as it were by a filter [i.e., desalination by osmosis].

—**Aristotle, Greek philosopher (384–322 BC)**

INTRODUCTION

Environmental effects associated with the production of oil and gas have had significant impacts, including the degradation of soils, groundwater, surface water, and the ecosystems they support by releases of suspended and dissolved hydrocarbons and coproduced saline water. Produced wastewater salts are less likely than hydrocarbons to be adsorbed by mineral phases in the soil and sediment and are not subject to degradation by biological processes. Sodium is a major dissolved constituent in most produced wastewaters, and it causes substantial degradation of soils through the alteration of clays and soil textures and subsequent erosion. Produced wastewater salts seem to have the most wide-ranging effects on soils, water quality, and ecosystems. Trace elements, including boron, lithium, bromine, fluorine, and radium, also occur in elevated concentrations in some produced wastewaters. Many trace elements are phytotoxic; they are adsorbed and may remain in soils after the saline water has been flushed away. Radium-bearing scale and sludge found in oilfield equipment and discarded on soils pose additional hazards to human health and ecosystems.

DESALINATION TECHNOLOGIES*

In order to utilize coproduced wastewater for beneficial use, it is necessary to lower the total dissolved solids (TDS) concentrations of ions that are too high. To accomplish this, desalination technologies are necessary; these technologies fall into the categories of membrane, thermal, and alternative technologies. Combined together with hybrid technologies, desalination technologies are often employed to reduce the energy cost of the process or to enhance the product water recovery. The membrane processes discussed in this chapter include reverse osmosis (RO), nanofiltration (NF), electrodialysis (ED), and forward osmosis. Also considered are the hybrid membrane processes of two-pass nanofiltration, dual RO with chemical precipitation, dual RO with HERO™, dual RO with seeded slurry precipitation, high-efficiency electrodialysis, and electrodeionization. The thermal desalination technologies included in this discussion are membrane distillation, multistage flash distillation, multiple-effect distillation, and vapor compression.

The technologies presented in this chapter are emerging and established technologies. Many of the technologies have previously been employed for treatment of produced wastewater. A discussion of the total dissolved solids range of applicability of these technologies and their salt rejection and product wastewater recoveries is also included. Also addressed are specific sodium, organic,

---

* Based on Otton, J.K., *Environmental Aspects of Produced-Water Salt Releases in Onshore and Coastal Petroleum-Producing Areas of the Conterminous U.S.—A Bibliography*, U.S. Geological Survey, Reston, VA, 2008.

and heavy metal rejection capabilities. A particular emphasis is placed on reverse osmosis because it is widely deployed in practice. To ensure that the essential and practical details regarding RO are easily understandable, the basic concepts, units of expression, and pertinent nomenclature are also presented. A few of the terms and concepts that follow were presented earlier but are repeated here to ensure understanding.

MISCIBILITY AND SOLUBILITY

Substances that are *miscible* are capable of being mixed in all proportions. Simply, when two or more substances disperse themselves uniformly in all proportions when brought into contact, they are said to be completely soluble in one another, or completely miscible. The precise chemistry definition is a "homogeneous molecular dispersion of two or more substances" (Jost, 1992). Examples include the following observations:

- All gases are completely miscible.
- Water and alcohol are completely miscible.
- Water and mercury (in its liquid form) are immiscible liquids.

Between the two extremes of miscibility is a range of *solubility*; that is, various substances mix with one another up to a certain proportion. In many environmental situations, a rather small amount of a contaminant may be soluble in water in contrast to the complete miscibility of water and alcohol. The amounts are measured in parts per million (ppm).

SUSPENSION, SEDIMENT, PARTICLES, AND SOLIDS

Often water carries solids or particles in *suspension*. These dispersed particles are much larger than molecules and may be comprised of millions of molecules. The particles may be suspended in flowing conditions and initially under quiescent conditions, but eventually gravity causes settling of the particles. The resultant accumulation by settling is referred to as *sediment* or *biosolids* (sludge) or *residual solids* in wastewater treatment vessels. Between this extreme of readily falling out by gravity and permanent dispersal as a solution at the molecular level are intermediate types of dispersion or suspension. Particles can be so finely milled or of such small intrinsic size as to remain in suspension almost indefinitely and in some respects similarly to solutions.

EMULSION

Emulsions represent a special case of a suspension. As you know, oil and water do not mix. Oil and other hydrocarbons derived from petroleum generally float on water with negligible solubility in water. In many instances, oils may be dispersed as fine oil droplets (an emulsion) in water and not readily separated by floating because of size or the addition of dispersal-promoting additives. Oil and, in particular, emulsions can prove detrimental to many treatment technologies and must be treated in the early steps of a multi-step treatment train.

ION

An ion is an electrically charged particle; for example, sodium chloride or table salt forms charged particles when dissolved in water. Sodium is positively charged (a cation), and chloride is negatively charged (an anion). Many salts similarly form cations and anions when dissolved in water.

Desalination

Mass Concentration

Concentration is often expressed in terms of parts per million (ppm) or mg/L. Sometimes parts per thousand (ppt) and parts per billion (ppb) are also used. These are known as units of expression. One ppm is analogous to a full shot glass of swimming pool water as compared to the entire contents of a standard swimming pool full of water. One ppb is analogous to one drop of water from an eye dropper added to a standard swimming pool full of water.

$$\text{Parts per million (ppm)} = \text{Mass of substance} \div \text{Mass of solutions} \qquad (7.1)$$

Because 1 kg of a solution with water as the solvent has a volume of approximately 1 liter,

$$1 \text{ ppm} \approx 1 \text{ mg/L}$$

Permeate

The portion of the feed stream that passes through a reverse osmosis membrane is the permeate.

Concentrate, Reject, Retentate, Brine, or Residual Stream

The concentrate, reject, retentate, brine, or residual stream is the membrane output stream that contains water that has not passed through the membrane barrier and concentrated feed water constituents that are rejected by the membrane.

Tonicity

Tonicity is a measure of the effective osmotic pressure gradient (as defined by the water potential of the two solutions) of two solutions separated by a semipermeable membrane. It is important to point out that, unlike osmotic pressure, tonicity is only influenced by solutes that cannot cross this semipermeable membrane, as only these exert an effective osmotic pressure. Solutes able to freely cross do not affect tonicity because they will always be in equal concentrations on both sides of the membrane. There are three classifications of tonicity that one solution can have relative to another (Sperelakis, 2012):

- *Hypertonic* refers to a greater concentration. In biology, a hypertonic solution is one with a higher concentration of solutes outside the cell than inside the cell; the cell will lose water by osmosis.
- *Hypotonic* refers to a lesser concentration. In biology, a hypotonic solution has a lower concentration of solutes outside the cell than inside the cell; the cell will gain water through osmosis.
- *Isotonic* refers to a solution in which the solute and solvent are equally distributed. In biology, a cell normally wants to remain in an isotonic solution, where the concentration of the liquid inside it equals the concentration of liquid outside it; there will be no net movement of water across the cell membrane.

Osmosis

Osmosis is the naturally occurring transport of water through a membrane from a solution of low salt content to a solution of high salt content in order to equalize salt concentrations.

Osmotic Pressure

Osmotic pressure is a measurement of the potential energy difference between solutions on either side of a semipermeable membrane due to osmosis. Osmotic pressure is a colligative property, meaning that the property depends on the concentration of the solute, but not on its identity.

Osmotic Gradient

The osmotic gradient is the difference in concentration between two solutions on either side of a semipermeable membrane. It is used to indicate the difference in percentages of the concentration of a specific particle dissolved in a solution. Usually, the osmotic gradient is used when comparing solutions that have a semipermeable membrane between them, allowing water to diffuse between the two solutions, toward the hypertonic solution. Eventually, the force of the column of water on the hypertonic side of the semipermeable membrane will equal the force of diffusion on the hypotonic side, creating equilibrium. When equilibrium is reached, water continues to flow, but it flows both ways in equal amounts as well as force, thus stabilizing the solution.

Membrane

A membrane is thin layer of material capable of separating materials as a function of their chemical or physical properties when a driving force is applied.

Semipermeable Membrane

A semipermeable membrane is a membrane permeable only by certain molecules or ions.

Reverse Osmosis System Flow Rating

Although the influent and reject flows are usually not indicated, the product flow rate is used to derive an RO system flow rating. A 600-gpm RO system, for example, yields 600 gpm of permeate.

Recovery or Conversion Rate

The recovery or conversion rate is the ratio of the permeate flow to the feed flow, which is fixed by the designer and is generally expressed as a percentage. It is used to describe what volume percentage of influent water is recovered. Exceeding the design recovery can result in accelerated and increased fouling and scaling of the membranes.

$$\% \text{ Recovery} = (\text{Recovery flow/Feed flow}) \times 100 \tag{7.2}$$

Concentration Factor

The concentration factor is the ratio of solute concentration in the concentrate stream to the solute concentration in the feed system. The concentration factor is related to recovery in that at 40% recovery, for example, the concentrate would be 2/5 that of the influent water.

Rejection

The term *rejection* is used to describe what percentage of an influent species a membrane retains. For example, 97% rejection of salt means that the membrane will retain 97% of the influent salt. It also means that 3% of influent salt will pass through the membrane into the permeate; this is known as *salt passage*. Equation 7.3 is used to calculate the rejection of a given species:

Desalination

TABLE 7.1

Estimated Reverse Osmosis Rejection Percentages of Selected Impurities for Thin-Film Composite Membranes

Impurity	Rejection Percentage	Impurity	Rejection Percentage
Aluminum	97–98%	Lead	96–98%
Ammonium	85–95%	Magnesium	96–98%
Arsenic	95–96%	Manganese	96–98%
Bacteria	99+%	Mercury	96–98%
Bicarbonate	95–96%	Nickel	97–99%
Boron	50–70%	Nitrate	93–96%
Bromide	93–96%	Phosphate	99+%
Cadmium	96–98%	Radioactivity	95–98%
Calcium	96–98%	Radium	97%
Chloride	94–95%	Selenium	97%
Chromium	96–98%	Silica	85–90%
Copper	97–99%	Silicate	95–97%
Cyanide	90–95%	Silver	95–97%
Detergents	97%	Sodium	92.98%
Fluoride	94–96%	Sulfate	99+%
Herbicides	97%	Sulfite	96–98%
Insecticides	97%	Virus	99+%
Iron	98–99%	Zinc	98–99%

Source: Adapted from Pure Water Products, *Reverse Osmosis Rejection Percentages*, Pure Water Products, LLC, Denton, TX, 2014.

$$\% \text{ Rejection} = [(C_i - C_p)/C_i] \times 100 \tag{7.3}$$

where

C_i = Influent concentration of a specific component.

C_p = Permeate concentration of a specific component.

The RO system uses a semipermeable membrane to reject a wide variety of impurities. Table 7.1 is a partial list of the general rejection ability of the most commonly used thin-film composite (TFC) RO membranes. Note that these percentages are averaged based on experience and are generally accepted within the industry. They are not a guarantee of performance. Actual rejection can vary according to the chemistry of the water, temperature, pressure, pH, and other factors (Pure Water Products, 2014).

FLUX

The word *flux* comes from the Latin *fluxus* ("flow") or *fluere* ("to flow") (Weekley, 1967). This term was first introduced into differential calculus as *fluxion* by Sir Isaac Newton. With regard to RO systems, flux is the rate of water flow (volumetric flow rate) across a unit surface area (membrane); it is expressed as gallons of water per square foot of membrane area per day (gfd) or liters per square meter per hour (LMH). In general, flux is proportional to the density of flow; it varies by how the boundary faces the direction of flow and is proportional within the area of the boundary.

Specific Flux (Permeability)

Specific flux, or permeability, refers to the membrane flux normalized for temperature and pressure, expressed as gallons per square foot per day per pound per square inch (gfd/psi) or liters per square meter per hour per bar (LMH/bar). Specific flux is sometimes discussed when comparing the performance of one type of membrane with another. In comparing membranes, the higher the specific flux the lower the driving pressure required to operate the RO system (Kucera, 2010).

Concentration Polarization

Similar to the flow of water through a pipe (see Figure 7.1A,B), concentration polarization is the phenomenon of increased solute (e.g., salt) concentration relative to the bulk solution that occurs in a thin boundary layer at the membrane surface on the feed side (Figure 2.1C). Let's look first at Figure 7.1A, which shows that flow may be *laminar* (streamline), and then look at Figure 7.1B, where the flow may be *turbulent*. Laminar flow occurs at extremely low velocities. The water moves in straight parallel lines, called *streamlines* or *laminae*, which slide upon each other as they travel rather than mixing up. Normal pipe flow is turbulent flow, which occurs because of friction encountered on the inside of the pipe. The outside layers of flow are thrown into the inner layers, and the result is that all of the layers mix and are moving in different directions and at different velocities, although the direction of flow is forward. Figure 7.1C shows the hydraulic boundary layer formed by fluid flow through a pipe. Concentration polarization has a negative effect on the performance of an RO membrane; specifically, it reduces the throughput of the membrane (Kucera, 2010). Flow may be steady or unsteady. For our purposes, we consider steady-state flow only; that is, most of the hydraulic calculations in this text assume steady-state flow.

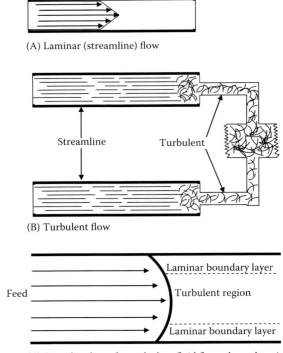

FIGURE 7.1 (A) Laminar (streamline) flow. (B) Turbulent flow. (C) Hydraulic boundary layer formed when fluid flows through a pipe.

Desalination

107

> **DID YOU KNOW?**
>
> The fouling of a reverse osmosis membrane is almost inevitable. Particulate matter will be retained and is an ideal nutrient for biomass, resulting in biofouling.

MEMBRANE FOULING

Membrane fouling is a process where a loss of membrane performance occurs due to the deposition of suspended or dissolved substances on its external surfaces, at its pore openings, or within its pores, forming a fouling layer. It can also be caused by internal changes in the membrane material. Both forms of fouling can cause membrane permeability to decline.

MEMBRANE SCALING

Membrane scaling is a form of fouling on the feed-concentrate side of the membrane that occurs when dissolved species are concentrated in excess of their solubility limit. Scaling is exacerbated by low cross-flow velocity and high membrane flux (Kucera, 2010).

SILT DENSITY INDEX

The silt density index (SDI) is a dimensionless value resulting from an empirical test used to measure the level of suspended and colloidal material in water. It is calculated from the time it takes to filter 500 mL of the test water through a 0.45-μm pore diameter filter at 30 psi pressure at the beginning and at the end of a specified test duration. The lower the SDI, the lower the potential for fouling a membrane with suspended solids. Visually, the deposited foulant on a filter membrane can be identified by its color. For example, foulant that is yellow could possibly indicate iron or organics, red foulant indicates iron, and black may indicate manganese (Kucera, 2010).

LANGELIER SATURATION INDEX

The Langelier Saturation Index (LSI) is a calculated value based on total dissolved solids (TDS), calcium concentration, total alkalinity, pH, and solution temperature. It indicates the tendency of a water solution to precipitate or dissolve calcium carbonate. The LSI is based on the pH and temperature of the water in question as well as the concentrations of total dissolved solids, calcium hardness, and alkalinity. The LSI generally ranges from <0.0 (no scale, very slight tendency to dissolve scale) to 3.0 (extremely severe scaling). For RO applications, a positive LSI indicates that the influent water has the tendency to form calcium carbonate scale (Kucera, 2010).

ANTISCALANTS

Antiscalants are chemical sequestering agents added to feedwater to inhibit scale formation.

GAS LAWS

Because gases can be pollutants as well as the conveyors of pollutants into various water bodies used as sources of drinking water and other types of water usage, it is important to have a fundamental understanding of the gas laws. Air (which is mainly nitrogen) is usually the main gas stream. Gas conditions are usually described in two ways: *standard temperature and pressure* (STP) and *standard conditions* (SC). STP represents 0°C (32°F) and 1 atm. The more commonly used SC value represents typical room conditions of 20°C (70°F) and 1 atm; SC is usually measured in cubic meters (m^3), normal cubic meters (Nm^3), or standard cubic feet (scf).

108 Hydraulic Fracturing Wastewater: Treatment, Reuse, and Disposal

To understand the physics of air it is imperative to have an understanding of the various physical laws that govern the behavior of pressurized gases. One of the more well-known physical laws is *Pascal's law*, which states that a confined gas (fluid) transmits externally applied pressure uniformly in all directions, without a change in magnitude. This parameter can be seen in a container that is flexible, as it will assume a spherical (balloon) shape. The reader has probably noticed that most compressed-gas tanks are cylindrical in shape; the spherical ends contain the pressure more effectively and allow the use of thinner sheets of steel without sacrificing safety.

BOYLE'S LAW

Though gases are compressible, note that, for a given mass flow rate, the actual volume of gas passing through the system is not constant within the system due to changes in pressure. This physical property (the basic relationship between the pressure of a gas and its volume) is described by Boyle's law, named for the Irish physicist and chemist Robert Boyle, who discovered this property in 1662. It states: "The absolute pressure of a confined quantity of gas varies inversely with its volume, if its temperature does not change." For example, if the pressure of a gas doubles, its volume will be reduced by a half, and *vice versa*; that is, as pressure goes up, volume goes down, and the reverse is true. This means, for example, that if 12 ft^3 of air at 14.7 psia (pounds per square inch absolute) is compressed to 1 ft^3, air pressure will rise to 176.4 psia, as long as the air temperature remains the same. This relationship can be calculated as follows:

$$P_1 \times V_1 = P_2 \times V_2 \tag{7.4}$$

where
P_1 = Original pressure (units for pressure must be absolute).
P_2 = New pressure (units for pressure must be absolute).
V_1 = Original gas volume at pressure P_1.
V_2 = New gas volume at pressure P_2.

This equation can be rewritten as

$$P_2/P_1 = V_1/V_2 \quad \text{or} \quad P_1/P_2 = V_2/V_1 \tag{7.5}$$

To allow for the effects of atmospheric pressure, always remember to convert from gauge pressure (psig, or pounds per square inch gauge) *before* solving the problem, then convert back to gauge pressure *after* solving it.

Pounds per square inch absolute (psia) = psig + 14.7 psi

and

Pounds per square inch gauge (psig) = psia − 14.7 psi

In a pressurized gas system where gas is caused to move through the system by the fact that gases will flow from an area of high pressure to that of low pressure, we will always have a greater actual volume of gas at the end of the system than at the beginning (assuming the temperature remains constant).

CHARLES'S LAW

Another physical law dealing with temperature is Charles's law, discovered by French physicist Jacques Charles in 1787. It states: "The volume of a given mass of gas at constant pressure is directly proportional to its absolute temperature." The absolute temperature is the temperature in Kelvin

Desalination 109

(273 + °C); absolute zero = −460°F, or 0°R on the Rankine scale. This is calculated by using the following equation:

$$P_2 = P_1 \times (T_2/T_1) \tag{7.6}$$

Charles's law also states: "If the pressure of a confined quantity of gas remains the same, the change in the volume (V) of the gas varies directly with a change in the temperature of the gas," as given below:

$$V_2 = V_1 \times (T_2/T_1) \tag{7.7}$$

IDEAL GAS LAW

The ideal gas law combines Boyle's and Charles's laws because air cannot be compressed without its temperature changing. The ideal gas law can be expressed as

$$(P_1 \times V_1)/T_1 = (P_2 \times V_2)/T_2 \tag{7.8}$$

Note that the ideal gas law is still used as a design equation even though the equation shows that the pressure, volume, and temperature of the second state of a gas are equal to the pressure, volume, and temperature of the first state. In actual practice, however, other factors such as humidity, heat of friction, and efficiency losses all affect the gas. Also, this equation uses absolute pressure (psia) and absolute temperatures (°R) in its calculations.

In air science practice, the importance of the ideal gas law cannot be overstated. It is one of the fundamental principles used in calculations involving gas flow in air-pollution-related work. This law is used to calculate actual gas flow rates based on the quantity of gas present at standard pressures and temperatures. It is also used to determine the total quantity of that contaminant in a gas that can participate in a chemical reaction. The ideal gas law has three important variables:

- Number of moles of gas
- Absolute temperature
- Absolute pressure

In practical applications, practitioners generally use the following standard ideal gas law equation:

$$V = nRT/P \quad \text{or} \quad PV = nRT \tag{7.9}$$

where
V = Volume.
n = Number of moles.
R = Universal gas constant.
T = Absolute temperature.
P = Absolute pressure.

SOLUTIONS

A solution is a condition in which one or more substances are uniformly and evenly mixed or dissolved. In other words, a solution is a homogeneous mixture of two or more substances. Solutions can be solids, liquids, or gases, such as drinking water, seawater, or air. Here, we are focusing primarily on liquid solutions. A solution has two components: a solvent and a solute (see Figure 7.2). The *solvent* is the component that does the dissolving; typically, the solvent is the species present in the greater quantity. The *solute* is the component that is dissolved. When water dissolves substances, it creates solutions with many impurities. Generally, a solution is usually transparent and

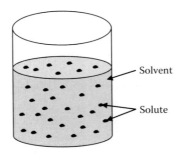

FIGURE 7.2 Solution with two components: solvent and solute.

not cloudy and visible to longer wavelength ultraviolet light. Because water is colorless, the light necessary for photosynthesis can travel to considerable depths. However, a solution may be colored when the solute remains uniformly distributed throughout the solution and does not settle with time.

SOLUTION CALCULATIONS

Remember, in chemical solutions, the substance being dissolved is called the *solute*, and the liquid present in the greatest amount in a solution (and that does the dissolving) is called the *solvent*. We should also be familiar with another term, *concentration*—the amount of solute dissolved in a given amount of solvent. Concentration is measured as

$$\% \text{ Strength} = \frac{\text{Weight of solute}}{\text{Weight of solution}} \times 100$$
$$= \frac{\text{Weight of solute}}{\text{Weight of solute} + \text{Weight of solvent}} \times 100 \qquad (7.10)$$

■ EXAMPLE 7.1

Problem: If 30 lb of chemical is added to 400 lb of water, what is the percent strength (by weight) of the solution?

Solution:

$$\% \text{ Strength} = \frac{\text{Weight of solute}}{\text{Weight of solute} + \text{Weight of solvent}} \times 100$$
$$= \frac{30 \text{ lb}}{30 \text{ lb} + 400 \text{ lb}} \times 100$$
$$= 7\%$$

Important to making accurate computations of chemical strength is a complete understanding of the dimensional units involved; for example, it is important to understand exactly what *milligrams per liter* (mg/L) signifies:

$$\text{Milligrams per liter (mg/L)} = \text{Milligrams of solute/Liters of solution} \qquad (7.11)$$

Another important dimensional unit commonly used when dealing with chemical solutions is *parts per million* (ppm):

$$\text{Parts per million (ppm)} = \text{Parts of solute/Million parts of solution} \qquad (7.12)$$

Desalination

111

Note: A part is usually a weight measurement.

For example:

$$9 \text{ ppm} = (9 \text{ lb solids})/(1,000,000 \text{ lb solution})$$

or

$$9 \text{ ppm} = (9 \text{ mg solids})/(1,000,000 \text{ mg solution})$$

This leads to two important parameters that water practitioners should commit to memory:

- $1 \text{ mg/L} = 1 \text{ ppm}$
- $1\% = 10,000 \text{ mg/L}$

When working with chemical solutions, it is also necessary to be familiar with the two chemical properties of *density* and *specific gravity*. Density is the weight of a substance per a unit of its volume—for example, pounds per cubic foot or pounds per gallon:

$$\text{Density} = \text{Mass of substance/Volume of substance} \tag{7.13}$$

Here are a few key facts about density:

- Density is often measured in units of lb/cf, lb/gal, or mg/L.
- Density of water = 62.5 lb/cf = 8.34 lb/gal.
- Density of concrete = 130 lb/cf.
- Density of liquid alum at 60°F = 11.11 lb/gal.
- Density of hydrogen peroxide (35%) = 1.13 g/mL.

Specific gravity is the ratio of the density of a substance to a standard density:

$$\text{Specific gravity} = \text{Density of substance/Density of water} \tag{7.14}$$

Here are a few facts about specific gravity:

- Specific gravity has no units.
- Specific gravity of water = 1.0.
- Specific gravity of concrete = 2.08.
- Specific gravity of liquid alum at 60°F = 1.33
- Specific gravity of hydrogen peroxide (35%) = 1.13.

When molecules dissolve in water, the atoms making up the molecules come apart (dissociate) in the water. This dissociation in water is called *ionization*. When the atoms in the molecules come apart, they do so as charged atoms (both negatively and positively charged), which, as described earlier, are called *ions*. The positively charged ions are called *cations* and the negatively charged ions are called *anions*.

The ionization that occurs when calcium carbonate ionizes is a good example:

$$CaCO_3 \quad \leftrightarrow \quad Ca^{2+} \quad + \quad CO_3^{2-}$$

| Calcium carbonate | Calcium ion (cation) | Carbonate ion (anion) |

Another good example is the ionization that occurs when table salt (sodium chloride) dissolves in water:

$$NaCl \quad\leftrightarrow\quad Na^{2+} \quad+\quad Cl^-$$

Sodium chloride   Sodium ion   Chloride ion
                          (cation)   (anion)

Some of the common ions found in water and their symbols are provided below:

Hydrogen	H^+
Sodium	Na^+
Potassium	K^+
Chloride	Cl^-
Bromide	Br^-
Iodide	I^-
Bicarbonate	HCO_3^-

Solutions serve as a vehicle to (1) allow chemical species to come into close proximity so they can react; (2) provide a uniform matrix for solid materials, such as paints, inks, and other coatings so they can be applied to surfaces; and (3) dissolve oil and grease so they can be rinsed away.

Water dissolves *polar substances* better than *nonpolar substances*. For example, polar substances such as mineral acids, bases, and salts are easily dissolved in water. Nonpolar substances such as oils and fats and many organic compounds do not dissolve as easily in water.

CONCENTRATIONS

Because the properties of a solution depend largely on the relative amounts of solvent and solute, the concentrations of each must be specified.

Note: Chemists use both relative terms, such as saturated and unsaturated, as well as more exact concentration terms, such as weight percentages, molarity, and normality.

Although polar substances dissolve better than nonpolar substances in water, polar substances dissolve in water only to a point; that is, only so much solute will dissolve at a given temperature. When that limit is reached, the resulting solution is saturated. At this point, the solution is in equilibrium—no more solute can be dissolved. A liquid/solids solution is supersaturated when the solvent actually dissolves more than an equilibrium concentration of solute (usually when heated).

Specifying the relative amounts of solvent and solute, or specifying the amount of one component relative to the whole, usually gives the exact concentrations of solution. Solution concentrations are sometimes specified as weight percentages.

MOLES

To understand the concepts of molarity, molality, and normality, we must first understand the concept of a mole. The mole is defined as the amount of a substance that contains exactly the same number of items (i.e., atoms, molecules, or ions) as 12 g of carbon-12. By experiment, Avogadro determined this number to be 6.02×10^{23} (to three significant figures). If 1 mole of carbon atoms equals 12 g, for example, what is the mass of 1 mole of hydrogen atoms? Note that carbon is 12 times heavier than hydrogen; therefore, we need only 1/12 the weight of hydrogen to equal the same number of atoms of carbon.

Note: One mole of hydrogen equals 1 gram.

Desalination 113

By the same principle:

- One mole of $CO_2 = 12 + 2(16) = 44$ g.
- One mole of $Cl^- = 35.5$ g.
- One mole of $Ra = 226$ g.

In other words, we can calculate the mass of a mole if we know the formula of the substance.

Molarity (*M*) is defined as the number of moles of solute per liter of solution. The volume of a solution is easier to measure in the lab than its mass:

$$M = (\text{No. of moles of solute})/(\text{No. of liters of solution})$$

Molality (*m*) is defined as the number of moles of solute per kilogram of solvent:

$$m = (\text{No. of moles of solute})/(\text{No. of kilograms of solutions})$$

Note: Molality is not used as frequently as molarity, except in theoretical calculations.

Especially for acids and bases, the *normality* (*N*) rather than the molarity of a solution is often reported. Normality is the number of equivalents of solute per liter of solution (1 equivalent of a substance reacts with 1 equivalent of another substance):

$$N = (\text{No. of equivalents of solute})/(\text{No. of liters of solution})$$

In acid/base terms, an *equivalent* (or gram equivalent weight) is the amount that will react with 1 mole of H^+ or OH^-; for example,

- One mole of HCl will generate 1 mole of H^+; therefore, 1 mole HCl = 1 equivalent.
- One mole of $Mg(OH)_2$ will generate 2 moles of OH^-; therefore, 1 mole of $Mg(OH)_2 = 2$ equivalents.

$$HCl \Rightarrow H^+ + Cl^-$$

$$Mg(OH)^{2+} \Rightarrow Mg^{2+} + 2OH^-$$

By the same principle:

- A 1-*M* solution of H_3PO_4 is 3 *N*.
- A 2-*N* solution of H_2SO_4 is 1 *M*.
- A 0.5-*N* solution of NaOH is 0.5 *M*.
- A 2-*M* solution of HNO_3 is 2 *N*.

Chemists titrate acid/base solutions to determine their normality. An endpoint indicator is used to identify the point at which the titrated solution is neutralized.

Note: If it takes 100 mL of 1-*N* HCl to neutralize 100 mL of NaOH, then the NaOH solution must also be 1 *N*.

PREDICTING SOLUBILITY

Predicting solubility is difficult, but there are a few general rules of thumb, such as "like dissolves like." Following are three to keep in mind when predicting solubility.

- *Liquid–liquid solubility*—Liquids with similar structure and hence similar intermolecular forces will be completely miscible. For example, we would correctly predict that methanol and water are completely soluble in any proportion.
- *Liquid–solid solubility*—Solids *always* have limited solubilities in liquids, in general because of the difference in magnitude of their intermolecular forces. Therefore, the closer the temperature is to its melting point, the better the match between a solid and a liquid.

Note: At a given temperature, lower melting solids are more soluble than higher melting solids. Structure is also important; for example, nonpolar solids are more soluble in nonpolar solvents.

- *Liquid–gas solubility*—As with solids, the more similar the intermolecular forces, the higher the solubility. Therefore, the closer the match between the temperature of the solvent and the boiling point of the gas, the higher the solubility. When water is the solvent, an additional *hydration* factor promotes solubility of charged species. Other factors that can significantly affect solubility are temperature and pressure. In general, raising the temperature typically increases the solubility of solids in liquids.

Note: Dissolving a solid in a liquid is usually an endothermic process (i.e., heat is absorbed), so raising the temperature will fuel this process. In contrast, dissolving a gas in a liquid is usually an exothermic process (i.e., it evolves heat), so lowering the temperature generally increases the solubility of gases in liquids.

Note: Thermal pollution is a problem because of the decreased solubility of O_2 in water at higher temperatures.

Pressure has only an appreciable effect on the solubility of gases in liquids. For example, carbonated beverages such as soda water are typically bottled at significantly higher atmospheres. When the beverage is opened, the decrease in the pressure above the liquid causes the gas to bubble out of solution. When shaving cream is used, dissolved gas comes out of solution, bringing the liquid with it as foam.

COLLIGATIVE PROPERTIES

Properties of a solution that depend on the concentrations of the solute species rather than their identity include the following:

- Lowering vapor pressure
- Raising boiling point
- Decreasing freezing point
- Osmotic pressure

True colligative properties are directly proportional to the concentration of the solute but entirely independent of its identity.

REVERSE OSMOSIS

Reverse osmosis membranes work on the premise that a pressure greater than the osmotic pressure of the feed solution must be applied to the system to force water through the membrane and reject the salt. The osmotic pressure is a function of the salinity of the water. It is difficult for most people to gain an understanding of reverse osmosis unless they first understand the principles of natural biological osmosis. In the simplest terms, osmosis can be defined as the naturally occurring process whereby water is transported through a membrane from a solution with a low salt content to a solution with a high salt content in order to equalize the salt concentration (see Figure 7.3).

Desalination

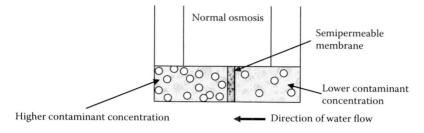

FIGURE 7.3 Osmosis.

> **DID YOU KNOW?**
>
> Reverse osmosis (RO) is not properly a filtration method. In RO, an applied pressure is used to overcome osmotic pressure, a colligative property, that is driven by chemical potential (Pure Water Products, 2014).

OSMOTIC PRESSURE

For water with very high salinity, the osmotic pressure and required system operating pressure are very high. In a well-practiced experimental demonstration of osmosis, water moves spontaneously from an area of high vapor pressure to an area of low vapor pressure (Figure 7.4). If this experiment is allowed to continue, in the end all of the water would move to the solution. A similar process occurs when pure water is separated from a concentrated solution by a *semipermeable membrane* (i.e., a membrane that only allows the passage of water molecules) (Figure 7.5). The osmotic pressure is the pressure that is just adequate to prevent osmosis (Figure 7.6). In dilute solutions, the osmotic pressure is directly proportional to the solute concentration and is independent of its identity.

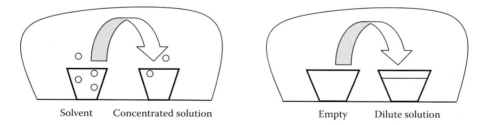

FIGURE 7.4 Osmotic pressure.

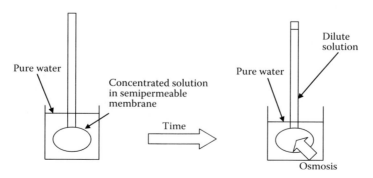

FIGURE 7.5 Colligative properties: passage of water molecules only.

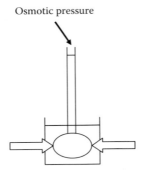

FIGURE 7.6 Osmotic pressure is the pressure just adequate to prevent osmosis.

> **DID YOU KNOW?**
>
> The energy consumption of reverse osmosis is directly related to the salt concentration, because a high salt concentration has a high osmotic pressure.

Reverse Osmosis Process[*]

Reverse osmosis (RO) is used to remove salt from produced wastewater. In order to present the principles and operations of RO in an understandable from the terms associated with the technology are presented, in plain English, in this section. Reverse osmosis is a separation or purification process (not properly a filtration process) that uses pressure to force a solvent through a semipermeable membrane that retains the solute on one side and allows the pure solvent to pass to the other side, forcing it from a region of high solute concentration through a membrane to a region of low solute concentration by applying a pressure in excess of osmotic pressure. The difference between normal osmosis and reverse osmosis is shown in Figure 7.7. Although many solvents (liquids) may be used and many applications are described in this book, the primary application of RO discussed here is water-based systems. Therefore, after an explanation of the RO process and its many different applications, the major emphasis of the discussion will focus on water as the liquid solvent; that is, drinking water purification, wastewater reuse, and desalination processes will be discussed in detail.

Process Description[†]

In the RO process, water passes through a membrane, leaving behind a solution with a smaller volume and a higher concentration of solutes. The solutes can be contaminants or useful chemicals or reagents, such as copper, nickel, and chromium compounds, which can be recycled for further use in metals plating or other metal finishing processes. The recovered water can be recycled or treated downstream, depending on the quality of the water and the needs of the plant. As shown in Figure 7.8, the water that passes through the membrane is the *permeate*, and the concentrated solution left behind is the *retentate* (or concentrate).

The RO process does not require thermal energy, only an electrically driven feed pump. RO processes have simple flow sheets and a high energy efficiency. However, RO membranes can be fouled or damage. This can result in holes in the membrane and passage of the concentrated solution to clean water—and thus a release to the environment. In addition, some membrane materials are susceptible to attack by oxidizing agents, such as free chlorine.

---

[*] Based on Spellman, F.R, *The Science of Water: Concepts and Applications*, 3rd ed., CRC Press, Boca Raton, FL, 2015.
[†] Based on USEPA, *Reverse Osmosis Process*, EPA 625/R-96/009, U.S. Environmental Protection Agency, Washington, DC, 1996; Spellman, F.R., *Physics for the Nonphysicist*, Government Institutes, Lanham, MD, 2009.

Desalination

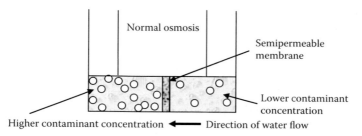

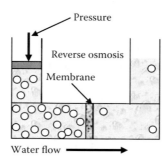

FIGURE 7.7 Normal osmosis and reverse osmosis.

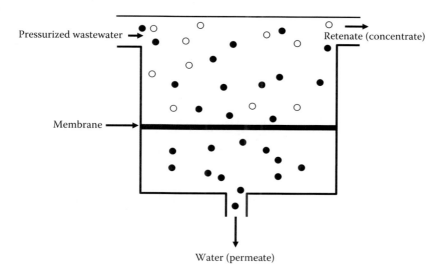

FIGURE 7.8 Reverse osmosis process.

The flux of a component A (recall that RO flux is the rate of water flow across a unit surface area) through an RO membrane is given by Equation 7.15:

$$N_A = P_A(\Delta\Phi/L) \tag{7.15}$$

where
- N_A = Flux of component A through the membrane (mass/time length^2).
- P_A = Permeability of A (mass-length/time-force).
- $\Delta\Phi$ = Driving force (DF) of A across the membrane, either pressure difference or concentration difference (force/length^2 or mass/length^2).
- L = Membrane thickness (length).

At equilibrium, the pressure difference between the two sides of the RO membrane equals the osmotic pressure difference. At low solute concentration, the osmotic pressure (π) of a solution is given by Equation 7.16:

$$\pi = C_S RT \tag{7.16}$$

where
π = Osmotic pressure (force/length^2).
C_S = Concentration of solutes in solution (moles/length^3).
R = Ideal gas constant (force-length/mass-temperature).
T = Absolute temperature (K or °R).

As a mixture is concentrated by passing water through the membrane, osmotic pressure of the solution increases, thereby reducing the driving force for further water passage. An accurate characterization of the pressure to drive the RO process must be based on an osmotic pressure computed from the average of the feed and retentate stream compositions. The water recovery of an RO process is expressed by Equation 7.17:

$$REC = (Q_p/Q_F) \times 100 \tag{7.17}$$

where
REC = Water recovery (%).
Q_p = Permeate flow rate (length^2/time).
Q_F = Feed flow rate (length^2/time).

Water recovery is determined by temperature, operating pressure, and membrane surface area. Rejection of contaminants determines permeate purity, whereas water recovery primarily determines the volume reduction of the feed or amount of permeate produced. Generally, for concentrations of waters from the metal finishing industry, greater water recoveries are desirable to obtain overall greater volume reduction.

REVERSE OSMOSIS EQUIPMENT

Membrane Materials

The membrane material refers to the substance from which the membrane itself is made. Normally, the membrane material is manufactured from a synthetic polymer, although other forms, including ceramic and metallic "membranes," may be available. Currently, almost all membranes manufactured for drinking water production are made of polymeric material, because they are significantly less expensive than membranes constructed of other materials.

The material properties of the membrane may significantly impact the design and operation of the filtration system. For example, membranes constructed of polymers that react with oxidants commonly used in drinking water treatment should not be used with chlorinated feed water. Mechanical

---

DID YOU KNOW?

It is not uncommon to confuse reverse osmosis with filtration; however, there are key differences between the two. The predominant removal mechanism in membrane filtration is straining, or size exclusion, so the process can theoretically achieve perfect exclusion of particles regardless of operational parameters such as influent pressure and concentration. On the other hand, reverse osmosis involves a diffusive mechanism so that separation efficiency is dependent on solute concentration, pressure, and water flux rate (Pure Water Products, 2014).

Desalination

strength is another consideration, as a membrane with greater strength can withstand larger transmembrane pressure (TMP) levels, allowing for greater operational flexibility and the use of higher pressures with pressure-based direct integrity testing. Similarly, a membrane with bidirectional strength may allow cleaning operations or integrity testing to be performed from either the feed or the filtrate side of the membrane. Material properties influence the exclusion characteristics of a membrane as well. A membrane with a particular surface charge may achieve enhanced removal of particulate or microbial contaminants of the opposite surface charge due to electrostatic attraction. In addition, a membrane can be characterized as being hydrophilic (i.e., water attracting or, as the author defines it, water loving) or hydrophobic (i.e., water repelling or water hating). These terms describe the ease with which membranes can be wetted, as well as the propensity of the material to resist fouling to some degree.

Reverse osmosis membranes are generally manufactured from cellulose acetate or polyamide materials (and their respective derivatives), and there are various advantages and disadvantages associated with each. Although cellulose membranes are susceptible to biodegradation and must be operated within a relatively narrow pH range of about 4 to 8, they do have some resistance to continuous low-level oxidant exposure. In general, for example, chlorine doses of 0.5 mg/L or less may control biodegradation as well as biological fouling without damaging the membrane. Polyamide (PA) membranes, by contrast, can be used under a wide range of pH conditions and are not subject to biodegradation. Although PA membranes have very limited tolerance for the presence of strong oxidants, they are compatible with weaker oxidants such as chloramines. PA membranes require significantly less pressure to operate and have become the predominant material used for RO applications.

In a symmetric membrane, the membrane is uniform in density or pore structure throughout the cross-section, whereas in an asymmetric membrane the density of the membrane material changes across the cross-sectional area. Some asymmetric membranes have a graded construction, in which the porous structure gradually decreases in density from the free to the filtrate side of the membrane. In other asymmetric membranes, there may be a distinct transition between the dense filtration layer (i.e., the skin) and the support structure. The more densely skinned layer is exposed to the feed water and acts as the primary filtration barrier, while the thicker and more porous understructure serves primarily as mechanical support. Some hollow fibers may be manufactured as single- or double-skinned membranes, with the double skin providing filtration at both the outer and inner walls of the fibers. Like the asymmetric skinned membranes, composite membranes also have a thin, dense layer that serves as the filtration barrier. However, in composite membranes the skin is a different material than the porous substructure onto which it is cast. This surface layer is designed to be thin so as to limit the resistance of the membrane to the flow of water, which passes more freely through the porous substructure. RO membrane construction is typically either asymmetric or composite.

Membrane Modules

The module is the housing that contains the membrane. Membrane modules are commercially available in four configurations:

- Plate-and-frame
- Spiral-wound
- Hollow-fiber
- Tubular

Plate-and-Frame Modules

The plate-and-frame configuration is one of the earliest membrane models developed. As shown in Figure 7.9, plate-and-frame modules use flat sheet membranes that are layered between spacers and supports. The supports also form a flow channel for the permeate water. The feed water flows across the flat sheets and from one layer to the next. Because of the very low surface area-to-volume ratio,

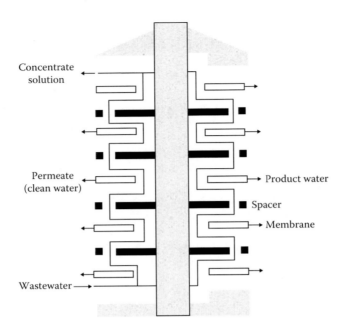

FIGURE 7.9 Plate-and-frame reverse osmosis module.

> **DID YOU KNOW?**
> Plate-and-frame modules are relatively easy to clean, which makes them ideal for use in high suspended solids applications. The best cleaning technique involves removing the plates and hand-cleaning each individual sheet of the membrane.

the plate-and-frame configuration is considered inefficient and is therefore seldom used in drinking water applications. Recent innovations have increased the packing densities for new designs of plate-and-frame modules. Maintenance on plate-and-frame modules is possible due to the nature of their assembly. They offer high recoveries with their long feed channels and are used to treat feed streams that often cause fouling problems. Advanced designs of plate-and-frame modules capable of operating at up to 25% dissolved solids and operating pressures up to 4500 psia (pounds per square in absolute) have been placed in operation in Germany (Tiwari et al., 2004). This development opens new opportunities for the use of reverse osmosis for concentration of metal-finishing wastewaters.

Spiral-Wound Modules

Spiral-wound modules were developed as an efficient configuration for the use of semipermeable membranes to remove dissolved solids, and thus are most often associated with RO processes. The basic unit of a spiral-wound module is a sandwich arrangement of flat membrane sheets, called a *leaf*, which is wound around a central perforated tube (Figure 7.10). One leaf consists of two membrane sheets placed back to back and separated by a fabric spacer called a *permeate carrier*. The layers of the leaf are glued along three edges, and the unglued edge is sealed around the perforated central tub. A single spiral-wound module 8 inches in diameter may contain up to approximately 20 leaves, each separated by a layer of plastic mesh, a *spacer*, that serves as the feed water channel. Feed water enters the spacer channels at the end of the spiral-wound element in a path parallel to the central tube. As the feed water flows across the membrane surface through the spaces, a portion permeates through either of the two surrounding membrane layers and into the permeate carrier,

Desalination

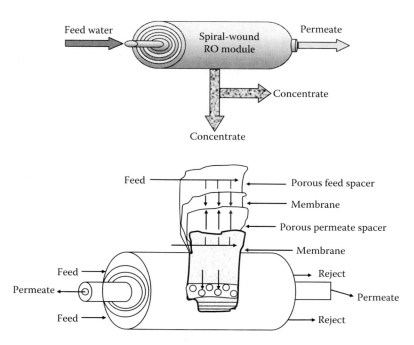

FIGURE 7.10 (Top) Cutaway view of a spiral-wound RO module consisting of internal-wound product spacers, RO membranes, feed spacers, and RO membranes. (Bottom) Internal construction of spiral-wound module.

> **DID YOU KNOW?**
>
> The spiral-bound membrane is the most commonly used module in RO systems.

leaving behind any dissolved and particulate contaminants that are rejected by the semipermeable membrane. The filtered water in the permeate carrier travels spirally inward around the element toward the central collector tube, while the water in the feed spacer that does not permeate through the membrane layer continues to flow across the membrane surface, becoming increasingly concentrated in rejected contaminants. This concentrated stream exits the element parallel to the central tube through the opposite end from which the feed water entered.

Hollow-Fiber Modules

Most hollow-fiber modules used in drinking water treatment applications are manufactured to accommodate porous membranes and designed to filter particulate matter. As the name suggests, these modules are comprised of hollow-fiber membranes (see Figure 7.11), which are long and very narrow tubes that may be constructed of any of the various membrane materials described earlier. The fibers may be bundled in one of several different arrangements. In one common configuration used by many manufacturers, the fibers are bundled together longitudinally, potted in a resin on both ends, and encased in a pressure vessel that is included as a part of the hollow-fiber module. These modules are typically mounted vertically, although horizontal mounting may also be utilized. One alternative configuration is similar to spiral-wound modules in that both are inserted into pressure vessels that are independent of the module itself. These modules (and the associated pressure vessels) are mounted horizontally. Another configuration in which the bundled hollow fibers are mounted vertically and submerged in a basin does not utilize a pressure vessel. A typical commercially available hollow-fiber module may consist of several hundred to over 10,000 fibers. Hollow-fiber modules offer the greatest packing densities of the other module configurations described in

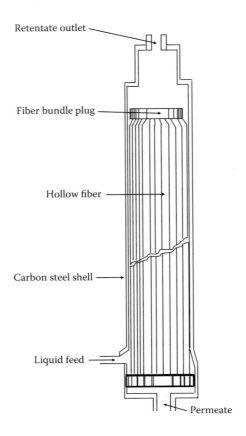

FIGURE 7.11 Hollow-fiber module.

> **DID YOU KNOW?**
> Hollow-fiber modules are relatively inexpensive because of the high surface area per unit volume achievable with this configuration.

this section. Figure 7.11 shows a hollow-fiber module. Although specific dimensions vary by manufacturer, approximate ranges for hollow-fiber construction are as follows:

- Outside diameter—0.5 to 2.0 mm
- Inside diameter—0.3 to 1.0 mm
- Fiber wall thickness—0.1 to 0.6 mm
- Fiber length—1 to 2 m

Tubular Modules

Tubular membranes are essentially a larger, more rigid version of hollow-fiber membranes. Tubular module have membranes supported within the inner part of tubes. The operator can easily service feed and permeate channels to remove fouling layers. Tubular modules are somewhat resistant to fouling when operated with a turbulent feed flow. This is accomplished with larger flow channels than those used with hollow-fiber and spiral-wound modules. The drawbacks of tubular modules are their high energy requirements for pumping large volumes of water, high capital costs, and low membrane surface area per unit volume of module (see Figure 7.12).

Desalination

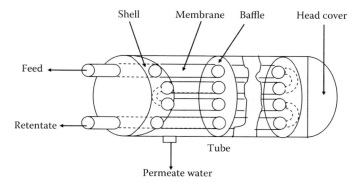

FIGURE 7.12 Tubular module.

System Configuration

Figure 7.13 illustrates a schematic of an early 1990s RO system with four modules in parallel, chemical pretreatment, and an upfront filtration step. Figure 7.14 illustrates the typical symbol used in membrane schematics. Figure 7.15 illustrates a typical RO membrane system with one influent stream (i.e., feed) and two effluent streams (i.e., permeate and concentration) that is commonly used today. As shown in the figure, a typical RO membrane systems consists of three separate subsystems: pretreatment, membrane process, and posttreatment.

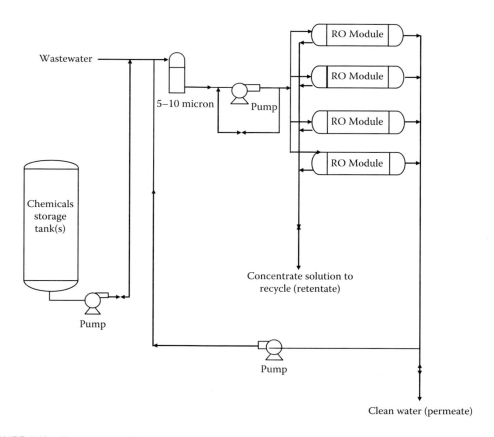

FIGURE 7.13 Reverse osmosis system.

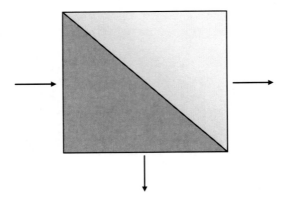

FIGURE 7.14 Typical membrane symbol.

All sources of input to an RO system must undergo some type and level of pretreatment. Pretreatment is necessary because RO thin-film composite membranes are subject to fouling by many substances:

- *Biological fouling*—Bacteria, microorganisms, viruses, and protozoans; pretreatment accomplished by chlorination
- *Particle fouling*—Suspended solids, sand, clay, and turbidity ingredients; pretreatment accomplished by filtration
- *Colloidal fouling*—Organic and inorganic complexes, colloidal particles, and microalgae; pretreatment accomplished by coagulation and filtration, with flocculation and sedimentation typically included
- *Organic fouling*—Natural organic matter (NOM), including humic and fulvic acids; pretreatment accomplished by coagulation, filtration, and activated carbon adsorption

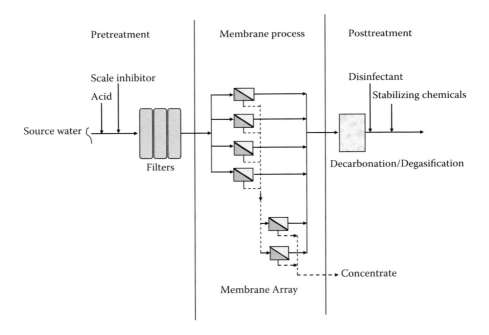

FIGURE 7.15 Typical RO membrane system.

Desalination

> **DID YOU KNOW?**
>
> For waters with high salinity, reverse osmosis is not a practical solution. Reverse osmosis generally is considered a cost-effective treatment technology to use with seawater or a salinity up to 40,000 ppm TDS.

- *Mineral fouling*—Calcium, magnesium, barium, or strontium sulfates and carbonates; pretreatment accomplished by acidification and antiscalant dosing
- *Oxidant fouling*—Chlorine, ozone, potassium permanganate ($KMNO_4$); pretreatment accomplished by oxidant scavenger dosing with sodium (metabisulfite and granulated activated carbon)

Pretreatment processes usually involve adding acid, scale inhibitor, or both to prevent precipitation of sparingly soluble salts as the reject ions become more concentrated, followed by cartridge filtration (5-micron) as the last step to protect the RO membranes from damage and particulate fouling (i.e., debris, sand, and piping materials). In some cases, additional pretreatment is required upstream of the filter cartridges for those input waters with higher fouling potential. Posttreatment usually includes unit processes common to conventional drinking water treatment, such as aeration, degasification, pH adjustment, addition of corrosion-control chemicals, fluoridation, and disinfection, and other unit processes that are discussed later.

NANOFILTRATION

Like reverse osmosis, nanofiltration (NF) is a cross-flow membrane filtration process. Nanofiltration membranes have a higher rate of salt transport and preferentially retain higher charged ions such as calcium and sulfate than reverse osmosis membranes. Depending on the specific membrane properties, the rejection of monovalent ions, such as nitrate, sodium, and chloride, will be as low as 45%. Both reverse osmosis and nanofiltration act on charged molecules. Uncharged particles will pass through the membrane, depending on the size. Some molecules may be too large to pass through the membrane and are retained in the concentrate. Experience has shown that the use of RO and NF membranes for produced wastewater treatment is hampered by organic fouling problems, which reduce process efficiency and increase treatment costs. Because both RO and NF are harmed by particulates, biological materials, and high concentrations of slightly soluble salts, pretreatment is recommended to provide acceptable feed water.

ELECTRODIALYSIS AND ELECTRODIALYSIS REVERSAL

Electrodialysis (ED) is a mature, robust, electrically driven process for brackish water desalination. For produced water, electrodialysis has been tested for produced wastewater at laboratory scale. The process consists of a stack of alternating cation-transfer membranes and anion-transfer membranes between an anode and cathode (see Figure 7.16). An electrical current is passed through the water. The dissolved salts in the water exist as ions and migrate toward the oppositely charged electrode. The anion-transfer membrane only allows passage of negatively charged ions, and the cation-transfer membrane only allows passage of positively charged ions. The alternating anion- and cation-transfer membranes are arranged in a stack. The membranes are impermeable to water. These systems are operated at very low pressure, usually below 25 psi. Electrodialysis reversal can also be implemented, such that the charge on the electrodes is frequently reversed. This prevents the buildup of scale, biofilm, and other foulants on the membrane surface. Thus, these systems have relatively lower fouling propensity and higher recovery as compared to RO systems. With regard to the energy required for ED treatment, it is related to the total dissolved solids (TDS) of

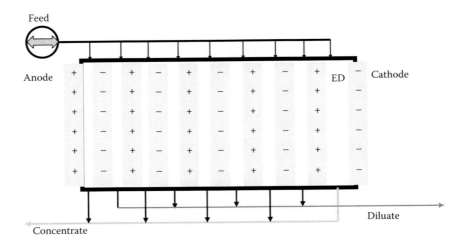

FIGURE 7.16 Electrodialysis (ED).

the water—the higher the TDS, the more energy required for treatment. Current research suggests the ED is not cost competitive for treating water with a TDS greater than 1500 mg/L. Sirivedhin et al. (2004) tested ED on five simulated produced water types of high and low TDS using Neospeta® membranes. They found that, at 6.5 volts per stack, ED was not capable of producing water with a sodium absorption ratio (SAR) that would be suitable for irrigation because ED removes divalent ions to a greater extent than monovalent ions. If ED, using divalent selective membranes, is to be used to treat produced water for beneficial use as irrigation water, calcium and/or magnesium will have to be added back to the water to lower the SAR.

FORWARD OSMOSIS

Forward osmosis is a membrane-driven process (see Figure 7.17) that offers high rejection of all contaminants. Forward osmosis uses the feed water to be treated as the dilute process stream, and water is moved across the membrane from the dilute feed water stream to a concentrated brined stream with a high osmotic pressure. The concentrated brine stream is called a *draw solution*, which requires reconcentration. To allow forward osmosis to be cost effective, the components that contribute to the high osmotic pressure in the brine stream must be removed (periodic membrane cleaning is necessary) to leave behind the fresh water product.

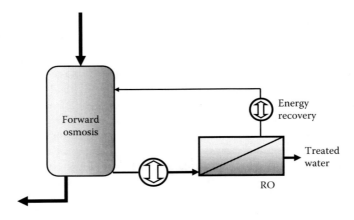

FIGURE 7.17 Forward osmosis for produced wastewater.

Desalination

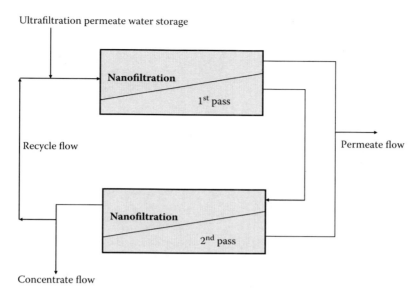

FIGURE 7.18 Two-pass nanofiltration membrane process for produced wastewater treatment.

HYBRID MEMBRANE PROCESSES

TWO-PASS NANOFILTRATION

The two-pass nanofiltration (NF^2) technique (Figure 7.18), developed and patented by California's Long Beach Water Department, involves treating produced wastewater with nanofiltration and then further treating the permeate water with nanofiltration again. This process is used to obtain a permeate stream with even lower TDS than a single-pass NF process and is less energy intensive than reverse osmosis. The process can operate at lower pressures than reverse osmosis, which would allow a lower operating cost. The two-pass nanofiltration process can provide an additional physical barrier for contaminant removal. Another important feature of the NF^2 process is that the second-pass concentrate recycle dilutes the feed water, allowing lower feed pressures to be used. Western Environmental pilot tested this process for produced wastewater (USDOI, 2013).

DUAL RO WITH CHEMICAL PRECIPITATION

Dual RO with chemical precipitation consists of both physical (purification through an impermeable membrane) and chemical precipitation methods to enhance water recovery beyond that of a single-stage RO process. The concentrate from the first RO process is further treated with lime softening ($Ca(OH)_2$) or caustic soda (NaOH) and is then fed to a second-stage RO process. The permeate streams from both RO processes are collected and provide the product water from this process. This method is used when high recovery is desired when desalting produced wastewater. Reported recoveries using this process are 95% and higher for brackish water applications. Utilizing this process enhances the recovery of the RO process but requires additional chemicals, additional equipment, and an increased footprint (USDOI, 2013). Figure 7.19 is an illustration of the process.

DUAL RO WITH SOFTENING PRETREATMENT AND OPERATION AT HIGH pH

Called HERO™, this patented system consists of chemical softening as a pretreatment step, primary RO, and ion exchange, degasification, and pH increase on the concentrate from the first RO stage. The treated concentrate stream then is treated with a secondary RO. The product water from

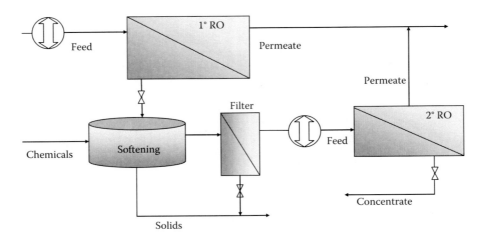

FIGURE 7.19 Dual RO with chemical precipitation.

the primary and secondary RO units is combined to make up the product water for this process. As with the dual RO with chemical precipitation, this process is designed to increase the product wastewater recovery of the process. Reported recovery rates range from 90 to 95% (USDOI, 2011).

Dual RO with Seeded Slurry Precipitation and Recycling RO

Water recovery in reverse osmosis (RO) processes may be enhanced by creating a seeded crystalline slurry (e.g., adding gypsum) to precipitate sparingly soluble salts from the water. The seeded crystals provide a preferential growth site for silicates, sulfate, and calcium. The crystals are then separated from the concentrate process stream using a cyclone separator, and the remaining water then is recycled back to the RO feed. The combined water recovery of the process is greater than 94%. Figure 7.20 is an illustration of the process.

High-Efficiency Electrodialysis

High-efficiency electrodialysis (HEED®) is an electromembrane process in which the ions are transported through a membrane from one section to another under the influence of an electrical potential. High-efficiency electrodialysis consists of dual or multiple side-by-side ion exchange membranes and features an improved gasket design that results in greater efficiency than traditional ED processes (EET Corp., 2011). High-efficiency electrodialysis is more resistant to fouling than RO or NF.

Electrodeionization

Electrodeionization (EDI) involves ion exchange resins, ion exchange membranes, and a direct electrical current (DC) (Figure 7.21). The distinguishing characteristic of an EDI system, as compared to a standard ED system, is that its desalting compartments are filled with an ion exchange resin. Ions are transported to the ion exchange resin by diffusion; they are then transported through the resin by the current. The current flows through the ion exchange resin because this is the path of least resistance. This process is capable of removing weakly ionized species and desalting water to very low concentrations. The advantage of EDI over desalting technologies is that its use decreases chemical usage by as much as 90% and the volume of the chemical wastestream by approximately 50%. EDI also has a smaller footprint and reduced operating and capital costs.

Desalination

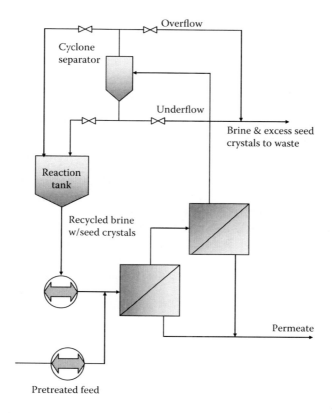

FIGURE 7.20 Dual RO with slurry precipitation and recycling RO.

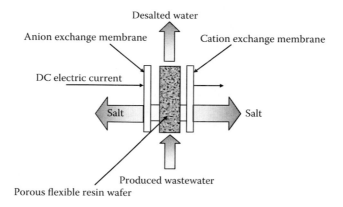

FIGURE 7.21 Electrodeionization (EDI).

THERMAL DESALINATION TECHNOLOGIES

MEMBRANE DISTILLATION

Membrane distillation is a thermally driven membrane process that uses the vapor pressure gradient between the feed solution and the product solution as the driving force. The membrane is hydrophobic and microporous. The flux and salt rejection of this process are independent of feed water salinity. There are many different configurations for the application of membrane distillation (USDOI, 2013).

MULTISTAGE FLASH DISTILLATION

Water, in a vacuum, is converted to steam at low temperatures in a multistage flash distillation system. At vacuum pressures, the boiling point of water is lower than at atmospheric pressure, requiring less energy. The water is preheated and then subjected to a vacuum pressure that causes vapor to flash off the warm liquid. The vapor is then condensed to form fresh water, and the remaining concentrated brine that does not flash is sent to the next chamber where a similar process takes place. The multiple stages are designed to improve the recovery of the process. Many of the older seawater desalination plants use the multistage flash distillation process.

MULTIPLE-EFFECT DISTILLATION

Often used for seawater desalination, multiple-effect distillation (MED) consists of multiple "effects," or stages in series. In each effect, the feed water is heated by steam in tubes. Vapor from the first effect is condensed in the second effect, and the heat of condensation is used to evaporate the water in the second effect. Each effect is connected in series and essentially reuses the energy from the previous effect. Thus, this system is a low energy consumer compared to other thermal processes. MED has the advantages of simple operation and low maintenance costs.

VAPOR COMPRESSION

Vapor compression desalination (considered a clean process) refers to a distillation process where the evaporation of saline water is obtained by the application of heat delivered by compressed vapor. In the process, the feed water is preheated in a heat exchanger by the product and reject streams from the process. The process uses a still that contains tubes. The water is then fed to the inside of the tubes, and the vapors are fed to the outside of the tubes to condense. The gases that do not condense are removed from the steam-condensation space by a vent pump or ejector. The mechanic pump or ejector is a requirement of this process and is necessary to increase the pressure of the vapor to cause condensation. The vapor compression process is not new; it has been used for produced water treatment, and commercially available products currently are marketed for this application. In terms of energy consumption and water recovery ratio, the vapor compression desalination process is more efficient than any other system in the market.

ALTERNATIVE DESALINATION PROCESSES

CAPACITIVE DEIONIZATION

Capacitive deionization is a novel and emerging desalination technology where water is passed through pairs of high-surface-area carbon electrodes (see Figure 7.22) that are held at a potential difference of 1.2 volts. Ions and other charged particles are attracted to the oppositely charged electrode.

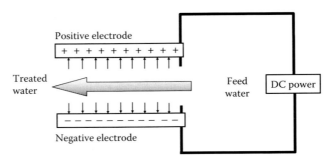

FIGURE 7.22 Capacitive deionization.

Desalination

The carbon electrodes have a relatively high surface area (500 m²/g) and provide high electrical conductivity, and they have high ion permeability. When the electrodes have become saturated with ions, they must be regenerated by removing the applied potential and rinsing the ions out of the system.

SOFTENING

Softening can be used to remove hardness and silica from the water. Hardness in water is caused by the presence of certain positively charged metallic ions in solution in the water. The most common of these hardness-causing ions are calcium and magnesium; others include iron, strontium, and barium. The two primary constituents of water that determine the hardness of water are calcium and magnesium. If the concentration of these elements in the water is known, the total hardness of the water can be calculated. To make this calculation, the equivalent weights of calcium, magnesium, and calcium carbonate must be known; the equivalent weights are given below:

Calcium (Ca)	20.04
Magnesium (Mg)	12.15
Calcium carbonate ($CaCO_3$)	50.045

Calculating Calcium Hardness as $CaCO_3$

The hardness (in mg/L as $CaCO_3$) for any given metallic ion is calculated using Equation 7.18:

$$\frac{\text{Calcium hardness (mg/L) as } CaCo_3}{\text{Equivalent weight of } CaCo_3} = \frac{\text{Calcium (mg/L)}}{\text{Equivalent weight of calcium}} \tag{7.18}$$

■ EXAMPLE 7.2

Problem: A water sample has a calcium content of 51 mg/L. What is this calcium hardness expressed as $CaCO_3$?

Solution:

$$\frac{\text{Calcium hardness (mg/L) as } CaCo_3}{\text{Equivalent weight of } CaCo_3} = \frac{\text{Calcium (mg/L)}}{\text{Equivalent weight of calcium}}$$

$$\frac{x \text{ mg/L}}{50.045} = \frac{51 \text{ mg/L}}{20.04}$$

$$x = \frac{51 \times 50.045}{20.04} = 127.4 \text{ mg/L Ca as } CaCo_3$$

■ EXAMPLE 7.3

Problem: The calcium content of a water sample is 26 mg/L. What is this calcium hardness expressed as $CaCO_3$?

Solution:

$$\frac{\text{Calcium hardness (mg/L) as } CaCo_3}{\text{Equivalent weight of } CaCo_3} = \frac{\text{Calcium (mg/L)}}{\text{Equivalent weight of calcium}}$$

$$\frac{x \text{ mg/L}}{50.045} = \frac{26 \text{ mg/L}}{20.04}$$

$$x = \frac{26 \times 50.045}{20.04} = 64.9 \text{ mg/L Ca as } CaCo_3$$

Calculating Magnesium Hardness as $CaCO_3$

To calculate magnesium hardness, we use Equation 7.19:

$$\frac{\text{Magnesium hardness (mg/L) as } CaCo_3}{\text{Equivalent weight of } CaCo_3} = \frac{\text{Magnesium (mg/L)}}{\text{Equivalent weight of magnesium}} \quad (7.19)$$

■ EXAMPLE 7.4

Problem: A sample of water contains 24 mg/L magnesium. Express this magnesium hardness as $CaCO_3$.

Solution:

$$\frac{\text{Magnesium hardness (mg/L) as } CaCo_3}{\text{Equivalent weight of } CaCo_3} = \frac{\text{Magnesium (mg/L)}}{\text{Equivalent weight of magnesium}}$$

$$\frac{x \text{ mg/L}}{50.045} = \frac{24 \text{ mg/L}}{12.15}$$

$$x = \frac{24 \times 50.045}{12.15} = 98.9 \text{ mg/L Mg as } CaCo_3$$

■ EXAMPLE 7.5

Problem: The magnesium content of a water sample is 16 mg/L. Express this magnesium hardness as $CaCO_3$.

Solution:

$$\frac{\text{Magnesium hardness (mg/L) as } CaCo_3}{\text{Equivalent weight of } CaCo_3} = \frac{\text{Magnesium (mg/L)}}{\text{Equivalent weight of magnesium}}$$

$$\frac{x \text{ mg/L}}{50.045} = \frac{16 \text{ mg/L}}{12.15}$$

$$x = \frac{16 \times 50.045}{12.15} = 65.9 \text{ mg/L Mg as } CaCo_3$$

Calculating Total Hardness

Calcium and magnesium ions are the two constituents that are the primary cause of hardness in water. To find total hardness, we simply add the concentrations of calcium and magnesium ions, expressed in terms of calcium carbonate ($CaCO_3$), using Equation 7.20:

$$\text{Total hardness (mg/L) as } CaCO_3 = \text{Ca hardness (mg/L) as } CaCO_3 \\ + \text{Mg hardness (mg/L) as } CaCO_3 \quad (7.20)$$

■ EXAMPLE 7.6

Problem: A sample of water has a calcium content of 70 mg/L as $CaCO_3$ and a magnesium content of 90 mg/L as $CaCO_3$.

Solution:

$$\text{Total hardness (mg/L) as } CaCO_3 = \text{Ca hardness (mg/L) as } CaCO_3 \\ + \text{Mg hardness (mg/L) as } CaCO_3 \\ = 70 \text{ mg/L} + 90 \text{ mg/L} = 160 \text{ mg/L as } CaCo_3$$

Desalination 133

■ EXAMPLE 7.7

Problem: Determine the total hardness as $CaCO_3$ of a sample of water that has a calcium content of 28 mg/L and a magnesium content of 9 mg/L.

Solution: Express calcium and magnesium in terms of $CaCO_3$:

$$\frac{\text{Calcium hardness (mg/L) as } CaCo_3}{\text{Equivalent weight of } CaCo_3} = \frac{\text{Calcium (mg/L)}}{\text{Equivalent weight of calcium}}$$

$$\frac{x \text{ mg/L}}{50.045} = \frac{28 \text{ mg/L}}{20.04}$$

$$x = \frac{28 \times 50.045}{20.04} = 69.9 \text{ mg/L Ca as } CaCo_3$$

$$\frac{\text{Magnesium hardness (mg/L) as } CaCo_3}{\text{Equivalent weight of } CaCo_3} = \frac{\text{Magnesium (mg/L)}}{\text{Equivalent weight of magnesium}}$$

$$\frac{x \text{ mg/L}}{50.045} = \frac{9 \text{ mg/L}}{12.15}$$

$$x = \frac{9 \times 50.045}{12.15} = 37.1 \text{ mg/L Mg as } CaCo_3$$

Now, total hardness can be calculated:

$$\text{Total hardness (mg/L) as } CaCO_3 = \text{Ca hardness (mg/L) as } CaCO_3$$
$$+ \text{Mg hardness (mg/L) as } CaCO_3$$
$$= 69.9 \text{ mg/L} + 37.1 \text{ mg/L}$$
$$= 107 \text{ mg/L as } CaCo_3$$

Calculating Carbonate and Noncarbonate Hardness

As mentioned, total hardness is comprised of calcium and magnesium hardness. When total hardness has been calculated, it is sometimes used to determine another expression of hardness—carbonate and noncarbonate. When hardness is numerically greater than the sum of bicarbonate and carbonate alkalinity, the amount of hardness equivalent to the total alkalinity (both in units of mg $CaCO_3$/L) is called the *carbonate hardness*; the amount of hardness in excess of this is the *noncarbonate hardness*. When the hardness is numerically equal to or less than the sum of carbonate and noncarbonate alkalinity, all hardness is carbonate hardness, and noncarbonate hardness is absent. Again, the total hardness is comprised of carbonate hardness and noncarbonate hardness:

$$\text{Total hardness} = \text{Carbonate hardness} + \text{Noncarbonate hardness} \qquad (7.21)$$

When the alkalinity (as $CaCO_3$) is greater than the total hardness, all of the hardness is carbonate hardness:

$$\text{Total hardness (mg/L) as } CaCO_3 = \text{Carbonate hardness (mg/L) as } CaCO_3 \qquad (7.22)$$

When the alkalinity (as $CaCO_3$) is less than the total hardness, then the alkalinity represents carbonate hardness, and the balance of the hardness is noncarbonate hardness:

$$\text{Total hardness (mg/L) as CaCo}_3 = \text{Carbonate hardness (mg/L) as CaCo}_3 \\ + \text{Noncarbonate hardness (mg/L) as CaCo}_3 \tag{7.23}$$

When carbonate hardness is represented by the alkalinity, we use Equation 7.24:

$$\text{Total hardness (mg/L) as CaCo}_3 = \text{Alkalinity (mg/L) as CaCo}_3 \\ + \text{Noncarbonate hardness (mg/L) as CaCo}_3 \tag{7.24}$$

■ EXAMPLE 7.8

Problem: A water sample contains 110 mg/L alkalinity as $CaCO_3$ and 105 mg/L total hardness as $CaCO_3$. What is the carbonate and noncarbonate hardness of the sample?

Solution: Because the alkalinity is greater than the total hardness, all of the hardness is carbonate hardness:

$$\text{Total hardness (mg/L) as CaCo}_3 = \text{Carbonate hardness (mg/L) as CaCo}_3 \\ + \text{Noncarbonate hardness (mg/L) as CaCo}_3$$

$$105 \text{ mg/L as CaCO}_3 = \text{Carbonate hardness}$$

No noncarbonate hardness is present in this water.

■ EXAMPLE 7.9

Problem: The alkalinity of a water sample is 80 mg/L as $CaCO_3$. If the total hardness of the water sample is 112 mg/L as $CaCO_3$, what is the carbonate and noncarbonate hardness in mg/L as $CaCO_3$?

Solution: Alkalinity is less than total hardness; therefore, both carbonate and noncarbonate hardness will be present in the hardness of the sample:

$$\text{Total hardness (mg/L) as CaCo}_3 = \text{Carbonate hardness (mg/L) as CaCo}_3 \\ + \text{Noncarbonate hardness (mg/L) as CaCo}_3$$

$$112 \text{ mg/L} = 80 \text{ mg/L} + x \text{ mg/L}$$

$$x = 112 \text{ mg/L} - 80 \text{ mg/L} = 32 \text{ mg/L noncarbonate hardness}$$

ION EXCHANGE

Ion exchange, or the Higgins Loop™, is a robust sodium ion exchange technology for water with a high concentration of sodium. The ion exchange process is a reversible chemical reaction that removes and replaces dissolved ions in solution with less troublesome ions of the same charge. This process is beneficial if there are no other ions of concern besides sodium and if sodium absorption ratio (SAR) adjustment is necessary. In operation, the cation exchange resin in the Higgins Loop™ process exchanges sodium ions for hydrogen ions; up to 90% exchange levels are achieved. As the resin becomes loaded with sodium, the flows to the adsorption portion of the process temporarily are interrupted. The resin is then advanced by a pulsing action through the loop in the opposite direction of the liquid flow. The loaded resin is then regenerated with hydrochloric acid and rinsed before being advanced back into the adsorption portion of the loop. Treated water is slightly acidic because H^+ ions are added to the water and the pH is raised. Calcium is added by passing the treated water through a limestone bed in the pH-controlling process step. For many produced wastewaters, removing the sodium ions will have a large effect on the total dissolved solids concentration and render the water suitable for beneficial use. Other ion exchange resins also may be employed to target specific ions for removal. The resin may or may not be able to be regenerated. For the resins that can be regenerated, they must be recharged periodically when the target ions begin to pass through the system. A concentrated (20% by weight) sodium chloride solution is used to recharge anion or cation resin in the salt phase. In the acid–base phase, hydrochloric acid can be used to recharge the cation resin and sodium hydroxide to recharge the anion resin.

COMMERCIAL DESALINATION PROCESSES

Numerous companies currently market produced wastewater management technologies. Most of these companies tailor their package plants to meet the specific treatment needs for each individual application. Although numerous proven commercial processes are available for use in produced wastewater operations, the focus here is on the CDM (Helena, MT) technology, which is widely used and proven. CDM is marketed for treating flow-back water from subsurface hydraulic fracturing (USDOI, 2013).

CDM Produced Wastewater Technology

The CDM treatment process is comprised of a train of different technologies in series to meet site-specific treatment goals (Figure 7.23). The specific processes included in the treatment train are dictated by the feed water quality and the desired product water quality. Some of the technologies that may be used include advanced filtration, weak acid cation IX softener, UV disinfection, low-pressure RO, antiscalant addition, seawater/high-pressure RO, evaporation, and crystallization. The feed stream is kept anoxic to minimize oxidation of iron and other metals and to reduce the fouling potential of the water. Depending on the feed water quality, the process can achieve more than 97% recovery. A computer program was developed that assists in selecting the required technologies and predicts the performance and scale formation within the system based on feed water quality (USDOI, 2013).

Pretreatment for the process consists of media filters and polymeric hollow-fiber ultrafiltration membranes to remove particulates, silt, oil, grease, coal fines, clay, and bacteria. The filtration system is backwashed using RO permeate. A weak acid cation (WAC) IX softener is used to reduce hardness and other metals. The resin is regenerated using hydrochloric acid. The water is then

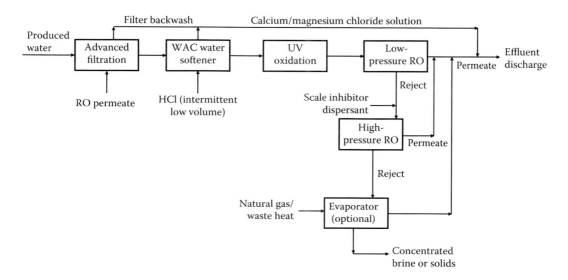

FIGURE 7.23 Schematic diagram of CDM produced water treatment process.

disinfected using UV. The calcium- and magnesium-rich WAC regeneration solution is combined with the filter backwash and is either treated separately or combined with the product streams from the membrane processes and discharged, depending on the scenario and the feed water quality.

Low-pressure RO (capable of achieving 85% recovery) is employed after pretreatment. The train size and type of membrane employed are tailored based on the feed water quality. An antiscalant (approximately 10 mg/L) is added to the concentrate stream to stabilize the silica and to prevent scale formation in the next high-pressure RO stage. The second RO stage consists of high-pressure or seawater RO membranes that can achieve 80% water recovery. The RO permeate is combined with the low-pressure RO permeate for discharge or beneficial use. The concentrate, approximately 2 to 3% of the initial feed volume, either is disposed of as a waste or can be treated for zero liquid discharge (ZLD). Because many produced wastewaters contain high levels of sodium and low levels of divalent ions, the SAR may be too high, even after treatment, for beneficial use of the water. In these cases, a limestone bed is used to add calcium to the water and to lower the SAR.

THOUGHT-PROVOKING QUESTIONS

1. Desalination has adverse impacts on the environment. Explain.
2. When brackish groundwater is desalinated this causes an overdraft of groundwater. What are the possible ramifications?
3. Which of the hybrid desalination techniques is best? Why?
4. Which of the hybrid desalination techniques is least effective? Explain.

REFERENCES AND RECOMMENDED READING

AWWA. (2007). *Reverse Osmosis and Nanofiltration*, 2nd ed. American Water Works Association, Denver, CO.

Bailer, H.H., Lykins, Jr., B.W., Fronk, C.A., and Kramer, S.H. (1987). Using reverse osmosis to remove agricultural chemicals from groundwater. *J. AWWA*, 79(8):55–60.

Bird, R.B, Stewart, W.E., and Lightfoot, E.N. (1960). *Transport Phenomena*. John Wiley & Sons, New York.

EET Corp. (2011). *HEED® Summary*. EET Corporation, Harriman, TN (http://www.eetcorp.com/lts/HEEDsumm.htm).

Jost, N.J. (1992). Surface and ground water pollution control technology, in Knowles, P.C., Ed., *Fundamentals of Environmental Science and Technology*. Government Institutes, Rockville, MD.

Kucera, J. (2010). *Reverse Osmosis: Industrial Applications and Processes*. John Wiley & Sons, New York.

Ludzack, F.J. and Noran, D.K. (1965). *Tolerance of High Salinities by Conventional Wastewater Treatment Processes*. Water Environment Federation, Arlington, VA.

NETL. (2016a). *Produced Water Management Technology Descriptions*. National Energy Technology Laboratory, Washington, DC (www.netl.doe.gov/research/coal/crosscutting/pwmis/tech-desc).

NETL. (2016b). *Fact Sheet—Industrial Use*. National Energy Technology Laboratory, Washington, DC (www.netl.doe.gov/research/coal/crosscutting/pwmis/tech-desc/induse).

Pure Water Products. (2014). *Reverse Osmosis Rejection Percentages*. Pure Water Products, LLC, Denton, TX (http://www.purewaterproducts.com/articles/ro-rejection-rates).

Puretec. (2014). *Filtration Spectrum*, puretecwater.com/resources/filtration-spectrum.pdf.

Sirivedhin, T. et al. (2004). Reclaiming produced water for beneficial use: salt removal by electrodialysis. *Journal of Membrane Science*, 243(1–2):335–343.

Spellman, F.R. (2015). *The Science of Water: Concepts and Applications*, 3rd ed. CRC Press, Boca Raton, FL.

Sperelakis, N. (2012). *Cell Physiology Source Book: Essentials of Membrane Biophysics*, 4th ed. Academic Press, New York.

Tiwari, S.A., Bhattacharyya, K.P., Goswami, D., Srivastava, V.K., and Hanra, M.S. (2004). Hydrodynamic considerations of reverse osmosis membrane modules and their merits and demerits with respect to their applications. *BARC Newsletter*, 249:220–225 (http://www.barc.gov.in/publications/nl/2004/200410-35.pdf).

USDOI. (2011). *Oil and Gas Produced Water Management and Beneficial Use in the Western United States*. U.S. Department of the Interior, Washington, DC.

USDOI. (2013). *Two-Pass Nanofiltration Seawater Desalination Prototype Testing and Evaluation*, Desalination and Water Purification Research and Development Report No. 158. U.S. Department of the Interior, Washington, DC.

USDOI. (2014). *Produced Water Treatment Primer: Case Studies of Treatment Applications.* U.S. Department of the Interior, Washington, DC.

USEPA. (1980). *Onsite Wastewater Treatment and Disposal Systems Design Manual*, EPA 625/1-80-012. U.S. Environmental Protection Agency, Washington, DC.

Veil, J.A., Puder, M.G., Elcock, D., and Redweik, Jr., R.J. (2004). *A White Paper Describing Produced Water from Production of Crude Oil, Natural Gas, and Coal Bed Methane.* U.S. Department of Energy, Washington, DC.

Wachinski, A.M. (2013). *Membrane Processes for Water Reuse.* McGraw-Hill, New York.

Wagner, K. et al. (1986). *Remedial Action Technology for Waste Disposal Sites.* William Andrew, Inc., Norwich, NY.

Weekley, E. (1967). *An Etymological Dictionary of Modern English.* Dover, New York.

Section IV

Produced Wastewater Disposal

8 Surface Water Disposal Options

Based on experience, observation, and years of research, the author has come to many conclusions about life in general and the machinations of humans relative to our interface with the environment. It is quite apparent that in the United States produced wastewater volume generation and management are not well characterized or documented and are rarely even thought about. Obviously, to gain an understanding of these issues a compilation of data on produced wastewater associated with oil and gas production is necessary and is long overdue to better understand the production volumes and management of this water. This is especially the case when it comes to the production and management of produced wastewater to be disposed of, one way or another, into the environment.

INTRODUCTION*

Produced wastewater is the largest volume byproduct or wastestream associated with oil and gas exploration and production. The cost of managing such a large volume of water is a key consideration for oil and gas producers. National produced wastewater volume estimates are in the range of 15 to 20 billion barrels (1 barrel = 42 U.S gallons) generated each year in the United States, which is a huge amount of produced wastewater. This high volume of produced wastewater raises numerous questions for fracking operators, fracking managers, and local regulators and residents. For example, what is to be done with it? Where do we put it? Where does one put billions of barrels of produced wastewater? One option is to treat the produced wastewater and reuse it in one way or another, but if the produced wastewater is not being treated then more questions are raised: What is to be done with produced wastewater that is too contaminated to be treated and reused for any purpose? What is to be done with produced wastewater that is too expensive to treat? What is to be done with produced wastewater in areas where treatment is impractical because of the remoteness of location or other footprint issues? How about formation water? Produced wastewater contains human-added contaminants from the fracking process itself. But, formation water contamination, though naturally added from its formation source, can also contain contaminants from the subsurface geological formations. Whether or not formation contaminants are harmful to the environment depends on what those contaminants are; thus, testing of the formation water must be conducted to determine what to do with it.

These are the types of questions, along with many others, that fracking personnel ask themselves and then make decisions about each and every day. Some readers might be wondering if the produced wastewater could be dumped back into the water source from which it was taken in the first place. If the source water originated from an above-ground source such as a lake, pond, stream, river, or creek, why not dump the produced wastewater back there? The old adage that *dilution is the solution to pollution* could apply, but the receiving body can eventually become overwhelmed. This practice has actually been applied by many in the past and to some extent still today. U.S. Environmental Protection Agency (USEPA) and local regulations usually prohibit discharge into local surface waters unless the discharged water meets certain permit standards. The problem with the discharge of untreated produced wastewater is that many of the surface water bodies where the fracking operations take place are in remote locations. This remoteness may also

---

* Much of the information in this chapter is from ANL, *Produced Water Volumes and Management Practices in the United States*, ANL/EVS/R-09/1, Argonne National Laboratory, Argonne, IL, 2009; Spellman, F.R., *Environmental Impacts of Hydraulic Fracturing*, CRC Press, Boca Raton, FL, 2013.

affect the size of surface water bodies in the area. Dilution is never the solution to pollution in small water bodies. Dumping millions of gallons of produced wastewater into a small stream or lake is not advised and should not be practiced.

So, if it is not feasible to treat onsite produced wastewater or it cannot be discharged into a local water body, what is to be done with it? There are other options available. An alternative surface disposal option is evaporation in a waste pit or pond. Two other disposal practices include commercial disposal and subsurface injection. Injection may or may not require treatment and is the number one disposal method (98% of produced wastewater) used today. This chapter primarily considers disposal via discharge to surface waters.

DISCHARGE TO SURFACE WATERS

In many parts of the world, discharge to surface water bodies represents one of the principal options for disposing of produced water from oil and gas exploration and production operations. Subject to certain exceptions, the following trends can be observed (NETL, 2016):

- Most U.S. onshore oil and gas operators inject their produced water for enhanced oil and gas recovery or final disposal. This reflects the prevailing regulatory situation of prohibiting discharges from most onshore wells.
- Many U.S. coalbed methane (CBM) well operators prefer to discharge produced water to surface water bodies if authorized by the regulator. Tightening treatment requirements for CBM produced water under increasingly restricted discharge standards may change the mix of management options in the future.
- Most U.S. offshore operators discharge produced water to the ocean subject to all applicable regulatory requirements. Offshore produced water is also typically discharged in other parts of the world. Different countries employ different discharge standards.

In the United States, discharge activities are subject to all applicable regulatory controls required by the USEPA and state agencies. USEPA national discharge standards—the effluent limitation guidelines (ELGs) for the oil and gas extraction point source category—include various subcategories:

- The *onshore* subcategory generally prohibits produced waste discharge from onshore wells, subject to limited exceptions made available by two other subcategories for onshore wells.
- Under the *stripper* subcategory, states decide whether to authorize produced water discharge from very small oil wells. Because low oil production volumes do not contribute much income to stripper well operators, they are not able to undertake complicated or expensive treatment. Adewumi et al. (1992) described a simple, low-cost system used for produced water treatment in Pennsylvania that involved separation, pH adjustment, aeration, solids separation, and filtration.
- The *agricultural and wildlife* subcategory allows discharges of produced water that are clean enough (and with sufficiently low salinity). The discharge must meet a limit for oil and grease of 35 mg/L and must actually be put to a beneficial agricultural or wildlife reuse. Little information is available on the treatment methods used before discharging produced water from oil and conventional gas wells under this subcategory.
- The *offshore* subcategory governs produced water discharges from as many as 4000 U.S. platforms. Most offshore produced water is discharged to the ocean.
- Wells in the *coastal* subcategory are generally prohibited from discharging produced water, subject to an exception for wells located in Alaska's Cook Inlet. Discharges from these wells must meet all applicable offshore standards.

Surface Water Disposal Options

DID YOU KNOW?

Oil and grease are key constituents of produced water and are therefore subject to regulation in nearly all permits authorizing produced water discharges. Oil and grease do not occur as a single chemical compound; rather, they serve as "indicator pollutants" and provide a measure of many different types of organic materials that respond to a particular analytical procedure.

WATER QUALITY REGULATIONS

To this point in the presentation, it has been made clear that anyone working at any level of fracking operations must deal in one way or another with produced wastewater. Thus, it follows that those dealing with produced wastewater must be aware of and abide by various federal, state, and local water quality regulations (potential impacts to water quality are primarily regulated under several federal statues and the accompanying state programs). The primary federal statutes governing water quality issues related to shale gas development are the Clean Water Act (CWA), the Safe Drinking Water Act (SDWA), and the Oil Pollution Act. These statutes and their relationships to shale gas development are discussed below, but first we present a short history of clean water reform.

GENESIS OF CLEAN WATER REFORM

To help the reader better understand the history of (and thus the impetus behind) the reform movement generated to clean up our water supplies, a chronology of some of the significant events precipitated by environmental organizations and citizens' groups that have occurred since the mid-1960s is provided here.* This chronology of events presents only a handful of the significant actions taken by Congress (with helpful prodding and guidance provided by the Sierra Club and the National Resources Defense Council, as well as others) with regard to enacting legislation and regulations to protect our nation's waters. No law has been more important to furthering this effort than the Clean Water Act, which is discussed in the next section.

1969

Americans came face to face with the grim condition of the nation's waterways when the industrial-waste-laden Cuyahoga River caught on fire. That same year, waste from food processing plants killed almost 30 million fish in Lake Thonotosassa, Florida.

1972

Congress enacted the Clean Water Act (after having overridden President Nixon's veto). The passage of the Clean Water Act has been called "literally a life-or-death proposition for the Nation." The Act set the goals of achieving water quality levels that were "fishable and swimmable" by 1983, receiving zero discharges of pollutants by 1985, and prohibiting the discharge of toxic pollutants in toxic amounts.

1974

The Safe Drinking Water Act (SDWA) passed, requiring the USEPA to establish national standards for contaminants in drinking water systems, underground wells, and sole-source aquifers, as well as several other requirements.

---

* Chronology adapted from Sierra Club, Clean water timeline, *The Planet Newsletter*, 4(8), 1997 (http://vault.sierraclub.org/planet/199710/time1.asp).

> **DID YOU KNOW?**
>
> There are approximately 155,000 public water systems in the United States. The USEPA classifies these water systems according to the number of people they serve, the source of their water, and whether they serve the same customers year-round or on an occasional basis.

1984

An alliance of the Natural Resources Defense Council, the Sierra Club, and others successfully sued Philips, a New York industrial polluter that had dumped waste into the Seneca River. According to the Sierra Club's water committee chair, Samuel Sage, the case "tested the muscles of citizens against polluters under the Clean Water Act." During this same time frame, the Clean Water Act reauthorization bill drew the wrath of environmental groups, who dubbed it the "Dirty Water Act" after lawmakers added last-minute pork and weakened wetland protection and industrial pretreatment provisions. Because of grassroots actions, most of these pork provisions were dropped. That same year, the highest environmental penalty to date—$70,000—was imposed against Alcoa Aluminum in Messina, NY (for polluting the St. Lawrence River) as a result of a suit filed by the Sierra Club.

1986

Tip O'Neill, Speaker of the House of Representatives, stated that he would not let a Clean Water Act reauthorization bill on the floor without the blessing of environmental groups. Later, after the bill was crafted and passed by Congress, President Reagan vetoed the bill. Also, amendments to the Safe Drinking Water Act directed the USEPA to publish a list of drinking water contaminants requiring legislation.

1987

The Clean Water Act was reintroduced and became law after Congress overrode President Reagan's veto. A new provision established the National Estuary Program.

1995–1996

The House passed H.R. 961 (again dubbed the "Dirty Water Act"), which in some cases eliminated standards for water quality, wetlands protection, sewage treatment, and agricultural and urban runoff. The Sierra Club collected over 1 million signatures supporting the Environmental Bill of Rights and released "Danger on Tap," a report that showed polluter contributions to friends in Congress who wanted to gut the Clean Water Act. Due in part to these efforts, the bill was stopped in the Senate.

CLEAN WATER ACT[*]

Concern with the disease-causing pathogens residing in many of our natural waterways and otherwise filthy water was not the initial lightning rod that got Joe or Nancy Citizen's attention regarding the condition and health of our country's waterways. Instead, it was their aesthetic qualities. Americans in general have a strong emotional response to the beauty of nature and have acted to prevent the pollution and degradation of our nation's waterways simply because many of us expect rivers, waterfalls, and mountain lakes to be natural and naturally beautiful—in the state they were intended to be, pure and clean.

---

[*] This section is adapted from USEPA, *Summary of the Clean Water Act*, U.S. Environmental Protection Agency, Washington, DC, 2012 (www.epa.gov/lawsregs/laws/cwa.html); Spellman, F.R., *The Science of Water*, 2nd ed., CRC Press, Boca Raton, FL, 2007.

Surface Water Disposal Options

Much of this emotional attachment to the environment was generated from the sentimentality of the popular literature and art of early 19th-century American writers and painters. From Longfellow's *Song of Hiawatha* to *Huckleberry Finn* to the vistas of the Hudson River School of Winslow Homer, American culture abounds with expressions of this singularly strong attachment. As the saying goes: "Once attached, detachment is never easy."

Federal water pollution legislation dates back to the turn of the century, to the Rivers and Harbors Act of 1899, though the Clean Water Act (CWA) stems from the Federal Water Pollution Control Act, which was originally enacted in 1948 to protect surface waters such as lakes, rivers, and coastal areas. The Clean Water Act is the primary federal law in the United States governing pollution of surface water. Established to protect water quality, the Act includes regulation of pollutant limits on the discharge of oil- and gas-related produced water. Regulation is achieved through the National Pollutant Discharge Elimination System (NPDES) permitting process.

The Clean Water Act was significantly expanded and strengthened in 1972 in response to growing public concern for serious and widespread water pollution problems. This 1972 legislation provided the foundation for subsequent dramatic progress in reducing water pollution. Amendments to the 1972 Clean Water Act were made in 1977, 1981, and 1987.

The Clean Water Act focuses on improving water quality by maintaining and restoring the physical, chemical, and biological integrity of the nation's waters. It provides a comprehensive framework of standards, technical tools, and financial assistance to address the many stressors that can cause pollution and adversely affect water quality, including municipal and industrial wastewater discharges, polluted runoff from urban and rural areas, and habitat destruction.

The Clean Water Act requires national performance standards for major industries (such as iron and steel manufacturing and petroleum refining) that provide a minimum level of pollution control based on the best technologies available. These national standards result in the removal of over a billion pounds of toxic pollution from our waters every year. The Clean Water Act also establishes a framework whereby states and Indian tribes survey their waters, determine an appropriate use (such as recreation or water supply), then set specific water quality criteria for various pollutants to protect those uses. These criteria, together with the national industry standards, are the basis for permits that limit the amount of pollution that can be discharged to a water body. Under the National Pollutant Discharge Elimination System, sewage treatment plants and industries that discharge wastewater are required to obtain permits and to meet the specified limits in those permits.

Note: The Clean Water Act requires the USEPA to set effluent limitations. All dischargers of wastewaters to surface waters are required to obtain NPDES permits, which require regular monitoring and reporting.

DID YOU KNOW?

The Clean Water Act made it unlawful to discharge any pollutant from a point source into the navigable waters of the United States, unless done in accordance with a specific approved permit. The NPDES permit program controls discharges from point sources that are discrete conveyances, such as pipes or manmade ditches. Industrial, municipal, and other facilities such as shale gas production sites or commercial facilities that handle the disposal or treatment of shale gas produced water must obtain permits if they intend to discharge directly into surface water (USEPA, 2008a,b). Large facilities usually have individual NPDES permits. Discharge from some smaller facilities may be eligible for inclusion under general permits that authorize a category of discharge under the CWA within a geographic area. A general permit is not specifically tailored to an individual discharger. Most oil and gas production facilities with related discharges are authorized under general permits because there are typically numerous sites with common discharges in a geographic area.

The Clean Water Act also provides federal funding to help states and communities meet their clean water infrastructure needs. Since 1972, federal funding has provided more than $66 billion in grants and loans, primarily for building or upgrading sewage treatment plants. Funding is also provided to address another major water quality problem—polluted runoff from urban and rural areas.

Protecting valuable aquatic habitat—wetlands, for example—is another important component of this law. American waterways have suffered loss and degradation of biological habitat, a widespread cause of the decline in the health of aquatic resources. When Europeans colonized this continent, North America held approximately 221 million acres of wetlands. Today, most wetlands are lost. Roughly 22 states have lost 50% or more of their original acreage of wetlands, and 10 states have lost about 70% of their wetlands. The Clean Water Act sections dealing with wetlands have become extremely controversial. Although wetlands are among our nation's most fragile ecosystems and provide a valuable role in maintaining regional ecology and preventing flooding, while serving as home to numerous species of insects, birds, and animals, wetlands also possess potential expandable monetary value in the eyes of private landowners and developers. Herein lies the major problem. Many property owners feel they are being unfairly penalized by a Draconian regulation that restricts their right to develop their own property. Alternative methods that do not involve destroying the wetlands do exist. These methods include wetlands mitigation and mitigation banking. Since 1972, when the Clean Water Act was passed, permits from the Army Corps of Engineers have been required to work in wetland areas. To obtain these permits, builders must agree to restore, enhance, or create an equal number of wetland acres (generally in the same watershed) as those damaged or destroyed in the construction project. Landowners are given the opportunity to balance the adverse affects by replacing environmental values that are lost. This concept is known as *wetlands mitigation.*

Mitigation banking allows developers or public bodies that seek to build on wetlands to make payments to a "bank" for use in the enhancement of other wetlands at a designated location. The development entity purchases credits from the bank and transfers full mitigation responsibility to an agency or environmental organization that runs the bank. Environmental professionals design, construct, and maintain a specific natural area using these funds.

Note that a state that meets the federal primacy requirements is allowed to set more stringent state-specific standards for this program. Because individual states can acquire primacy over their respective programs, it is not uncommon to have varying requirements from state to state. This variation is important to oil and gas industry managers because it can affect how they manage produced water within a drainage basin located within two or more states, such as the Marcellus Shale in the Appalachian Basin. Effluent limitations serve as the primary mechanism under NPDES permits for controlling discharges of pollutants to receiving waters. When developing effluent limitations for NPDES permits, the permit writers must consider limitations based on both the technology available to control the pollutants (i.e., technology-based effluent standards) and the regulations that protect the water quality standards of the receiving water (i.e., water quality-based effluent standards).

The intent of technology-based effluent limits in NPDES permits is to require treatment of effluent concentrations to less than a maximum allowable standard for point source discharges to the specific surface water today. This is based on available treatment technologies, while allowing the discharger to use any available control technique to meet the limits. For industrial (and other nonmunicipal) facilities, technology-based effluent limits are derived by (1) using national effluent limitations guidelines and standards established by the USEPA, or (2) using best professional judgment (BPJ) on a case-by-case basis in the absence of national guidelines and standards.

Prior to the granting of a permit, the authorizing agency must consider the potential impact of every proposed surface water discharge on the quality of the receiving water, not just individual dischargers. If the authorizing agency determines that technology-based effluent limits are not sufficient to ensure that water quality standards will be attained in the receiving water, the CWA (Section 303(b)(1)(c)) and NPDES (40 CFR 122.44(d)) regulations require that more stringent limits be imposed as part of the permit (USEPA, 2008b). USEPA establishes effluent limitation guidelines

Surface Water Disposal Options

(ELGs) and standards for different non-municipal (i.e., industrial) categories. These guidelines are developed based on the degree of pollutant reduction attainable by an industrial category through the application of pollution control technologies. The CWA requires the USEPA to develop specific effluent guidelines that represent the following:

1. *Best conventional technology* (BCT) for control of conventional pollutants and applicable to existing dischargers
2. *Best practicable technology* (BPT) currently available for control of conventional, toxic, and nonconventional pollutants and applicable to existing dischargers
3. *Best available technology* (BAT) economically achievable for control of toxic and nonconventional pollutants and applicable to existing dischargers
4. *New source performance standards* (NSPS) for conventional pollutants and applicable to new sources

To date, the USEPA has established national guidelines and standards for wastewater discharges to surface waters and publicly owned treatment works. At present, more than 50 different industrial categories are listed (USEPA, 2008c) (Table 8.1). The EGLs for oil and gas extraction, which were published in 1979, can be found at 40 CFR Part 435. The onshore subcategory, Subpart C, is applicable to discharges associated with shale gas development and production.

The Clean Water Act also includes a program to control stormwater discharges. The 1987 Water Quality Act (WQA) added Section 402(p) to the CWA that required the USEPA to develop and implement a stormwater permitting program. The USEPA developed this program in two phases (in 1990 and 1999). The regulations establish NPDES permit requirements for municipal, industrial, and construction site stormwater runoff. The WQA also added Section 402(1)(2) to the CWA which specified that the USEPA and states shall not require NPDES permits for unconventional stormwater discharges from oil and gas exploration, production, processing, or treatment operations or from transmission facilities. This exemption applies where the runoff is not contaminated by contact with raw materials or wastes.

The USEPA had previously interpreted the 402(1)(2) exemption as not applying to construction activities of oil and gas development, such as building roads and pads (i.e., an NPDES permit was required) (USEPA, 2008d); however, the Energy Policy Act of 2005 modified Section 402(1)(2) by defining the excluded oil and gas sector operations as including all oil and gas field activities and operations, including those necessary to prepare a site for drilling and for the movement and placement of drilling equipment. The USEPA promulgated a rule to implement this exemption. On May 23, 2008, the U.S. Court of Appeals for the Ninth Circuit released a decision vacating the permitting exemption for discharges of sediment from oil and gas construction activities that contribute to violations of the CWA (*NRDC v. USEPA*, 9th Cir. P. 5947, 2008).

The court based its decision on the fact that the new rule exempted runoff contaminated with sediment, while the CWA does not exempt such runoff. As a result of the court's decision, stormwater discharges contaminated with sediment resulting in a water quality violation require permit coverage under the NPDES stormwater permitting program. Although the USEPA stormwater permitting rule now contains a broad exclusion of oil and gas sector construction activities, it is important to note that individual states and Indian tribes may still regulate stormwater associated with these activities. The USEPA has clarified its position that states and tribes may not regulate such stormwater discharges under their CWA authority but are free to regulate under their own independent authorities: "This final rule is not intended to interfere with the ability of states, tribes, or local governments to regulate any discharges through a non-NPDES permit program" (71 FR 33635). In addition to state and tribal regulation, the industry has a voluntary program of Reasonable and Prudent Practices for Stabilization (RAPPS) of oil and gas construction sites (IPAA, 2004). Producers use RAPPS to control erosion and sedimentation associated with stormwater runoff from areas disturbed by clearing, grading, and excavating activities related to site preparation.

TABLE 8.1

Existing Effluent Guidelines

Industry Category	40 CFR Part	Industry Category	40 CFR Part
Aluminum Forming	467	Meat and Poultry Products	432
Asbestos Manufacture	427	Metal Finishing	433
Battery Manufacturing	461	Metal Molding and Casting (Foundries)	464
Canned and Preserved Fruits and Vegetable Processing	407	Metal Products and Machinery	438
Canned and Preserved Seafood (Seafood Processing)	408	Mineral Mining and Processing	436
Carbon Black Manufacturing	458	Nonferrous Metals Forming and Metal Powders	471
Cement Manufacturing	411	Nonferrous Metals Manufacturing	421
Centralized Waste Treatment	437	Oil and Gas Extraction	435
Coal Mining	434	Ore Mining and Dressing (Hard Rock Mining)	440
Coal Coating	465	Organic Chemicals, Plastics, and Synthetic Fibers (OCPSF)	414
Concentrated Animal Feeding Operations (CAFO)	412	Paint Formulating	446
Concentrated Aquatic Animal Production	451	Paving and Roofing Materials (Tars and Asphalt)	443
Copper Forming	468	Pesticide Chemicals Manufacturing, Formulating, and Packaging	455
Dairy Products Processing	405	Petroleum Refining	419
Electrical and Electronic Components	469	Pharmaceutical Manufacturing	439
Electroplating	413	Phosphate Manufacturing	422
Explosives Manufacturing	457	Photography	459
Ferroalloy Manufacturing	424	Porcelain Enameling	466
Fertilizer Manufacturing	414	Pulp, Paper, and Paperboard	430
Glass Manufacturing	426	Rubber Manufacturing	428
Grain Mills Manufacturing	406	Soap and Detergent Manufacturing	417
Gum and Wood Chemicals	454	Steam Electric Power Generating	423
Hospitals	460	Sugar Processing	409
Ink Formulating	447	Textile Mills	410
Inorganic Chemicals	415	Timber Products Processing	429
Iron and Steel Manufacturing	420	Transportation Equipment Cleaning	442
Landfills	445	Waste Combustors	444
Leather Tanning and Finishing	425		

DID YOU KNOW?

Point-source water pollution is any discernible, defined, and discrete conveyance, including but not limited to any pipe, ditch, channel, tunnel, conduit, well, discrete fissure, container, rolling stock, vessel, or other floating craft from which pollutants are or may be discharged into a river or other surface water body. *Non-point-source pollution* consists of runoff from irrigated agricultural land.

The history of the CWA is much like that of the environmental movement itself. Once widely supported and buoyed by its initial success, the CWA has since encountered increasingly difficult problems—polluted stormwater runoff, for example, and non-point-source pollution, as well as unforeseen legal challenges, such as the debate on wetlands and property rights. Unfortunately, the CWA is only part of the way toward achieving its goal. At least a third of U.S. rivers, half of U.S. estuaries, and more than half of U.S. lakes are still not safe for such uses as swimming or fishing. Thirty-one states reported toxins in fish exceeding the action levels set by the Food and Drug Administration (FDA). Every pollutant in a USEPA study on chemicals in fish showed up in at least one location. Water quality is seen as deteriorated and viewed as the cause of the decreasing number of shellfish in the waters.

SAFE DRINKING WATER ACT[*]

When we get the opportunity to travel the world, one of the first things we learn to ask is whether or not the water is safe to drink. Unfortunately, in most of the places in the world, the answer is "no." As much as 80% of all sickness in the world is attributable to inadequate water or sanitation (Masters, 2007). In a speech in Racine, Wisconsin, in 1998, environmentalist William C. Clark probably summed it up best: "If you could tomorrow morning make water clean in the world, you would have done, in one fell swoop, the best thing you could have done for improving human health by improving environmental quality." An estimated three fourths of the population in Asia, Africa, and Latin America lack a safe supply of water for drinking, washing, and sanitation (Morrison, 1983). Money, technology, education, and attention to the problem are essential for improving these statistics, to solving the problem this West African proverb succinctly states: "Filthy water cannot be washed." Left alone, Nature provides for us. Left alone, Nature feeds us. Left alone, Nature refreshes and sustains us with untainted air. Left alone, Nature provides and cleans the water we need to drink to survive. As Norse (1985) put it, "In every glass of water we drink, some of the water has already passed through fishes, trees, bacteria, worms in the soil, and many other organisms, including people. ...Living systems cleanse water and make it fit, among other things, for human consumption." Left alone, Nature performs at a level of efficiency and perfection we cannot imagine. The problem, of course, is that our human populations have grown too large, demanding, and intrusive to allow Nature to be left alone.

Our egos allow us to think that humans are the real reason why Nature exists at all. In our eyes, our infinite need for water is why Nature works its hydrologic cycle—to provide the constant supply of drinking water we need to sustain life. But the hydrologic cycle itself is unstoppable, human activity or not. Bangs and Kallen (1985) summed it up best: "Of all our planet's activities—geological movements, the reproduction and decay of biota and even the disruptive propensities of certain species (elephants and humans come to mind)—no force is greater than the hydrologic cycle."

---

[*] This section is adapted from USEPA, *Understanding the Safe Drinking Water Act*, EPA 816-F-04-030, U.S. Environmental Protection Agency, Washington, DC, 2004.

Nature, through the hydrologic cycle, provides us with an endless (we hope) resupply of water; however, we find that developing and maintaining an adequate supply of safe drinking water requires the coordinated efforts of scientists, technologists, engineers, planners, water treatment plant operators, and regulatory officials. In this section, we concentrate on the regulations that have been put into place in the United States to ensure that the water supplies developed are protected and are kept safe, fresh, and palatable.

Legislation to protect drinking water quality in the United States (the nation's first water quality standards) began with the Public Health Service Act of 1912. With time, the Act evolved, but not until the passage of the Safe Drinking Water Act (SDWA) in 1974 was federal responsibility extended beyond intestate carriers to include all community water systems serving 15 or more outlets, or 25 or more customers. Prompted by public concern over findings of harmful chemicals in drinking water supplies, the law established the basic federal–state partnership for drinking water used today. It focuses on ensuring safe water from public water supplies and on protecting the nation's aquifers from contamination. The law was amended in 1986 and 1996 and requires many actions to protect drinking water and its sources, including rivers, lakes, reservoirs, springs, and groundwater wells. Before we examine the basic tenets of the SDWA, we must define several of the terms used in the Act.

SDWA Definitions*

Action level (AL)—The amount required to trigger treatment or other action.

Best management practices (BMPs)—Schedules of activities, prohibitions of practices, maintenance procedures, and other management practices to prevent or reduce the pollution of waters of the United States.

Contaminant—Any physical, chemical, biological, or radiological substance or matter in water.

Consumer Confidence Report (CCR)—Annual water quality report that a community water system is required to provide to its customers. The CCR helps people make informed choices about the water they drink. They let people know what contaminants, if any, are in their drinking water and how these contaminants may affect their health. CCRs also give the system a chance to tell customers what it takes to deliver safe drinking water.

Discharge of a pollutant—Any addition of any pollutant to navigable waters from any point source.

Exemption—A document for water systems having technical and financial difficulty meeting the National Primary Drinking Water Regulations; it is effective for one year and is granted by the USEPA due to compelling factors.

Likely source—Where a contaminant could come from.

Maximum contaminant level (MCL)—The maximum permissible level of a contaminant in water that is delivered to any user of a public water system.

Maximum contaminant level goal (MCLG)—The level at which no known or anticipated adverse effects on the health of persons occur and which allows an adequate margin of safety.

Maximum residual disinfectant level (MRDL)—The highest level of a disinfectant allowed in drinking water.

Maximum residual disinfectant level goal (MRDLG)—The level of a drinking water disinfectant below which there is no known or expected risk to health.

Microbiological contaminants—Microbes used as indicators that other, potentially harmful bacteria may be present.

National Pollutant Discharge Elimination System (NPDES)—The national program for issuing, modifying, revoking and reissuing, terminating, monitoring, and enforcing permits, in addition to imposing and enforcing pretreatment requirements, under Sections 307, 402, 318, and 405 of the Clean Water Act.

---

* This section is adapted from 40 CFR Part 122.2; Safe Drinking Water Act Section 1401; Clean Water Act Section 502.

Surface Water Disposal Options

Navigable waters—Waters of the United States, including territorial seas.

pCi/L—Picocuries per liter (a measure of radioactivity).

Person—An individual, corporation, partnership, association, state, municipality, commission, or political subdivision of a state, or any interstate body.

Point source—Any discernible, confined, and discrete conveyance, including but not limited to any pipe, ditch, channel, tunnel, conduit, well, discrete fissure, container, rolling stock, concentrated animal feeding operation, or vessel, or other floating craft, from which pollutants are or may be discharged. This term does not include agricultural stormwater discharges and return flows from irrigated agriculture.

Pollutant—Dredged soil, solid waste, incinerator residue, filter backwash, sewage, garbage, sewage sludge, munitions, chemical wastes, biological materials, radioactive materials (except those regulated under the Atomic Energy Act of 1954), heat, wrecked or discarded equipment, rock, sand, cellar dirt, and industrial, municipal, and agricultural waste discharged into water. It does not mean (a) sewage from vessels or (b) water, gas, or other material injected into a well to facilitate production of oil or gas, or water derived in association with oil and gas production and disposal of in a well, if the well used either to facilitate production or for disposal purposes is approved by authority of the state in which the well is located, and if the state determines that the injection or disposal will not result in the degradation of ground or surface water sources.

Public water system—A system for the provision to the public of piped water for human consumption, if such system has at least 15 service connections or regularly serves at least 25 individuals.

Publicly owned treatment works (POTW)—Any device or system used in the treatment of municipal sewage or industrial wastes of a liquid nature which is owned by a state or municipality. This definition includes sewer, pipes, or other conveyances only if they convey wastewater to a POTW providing treatment.

Recharge zone—The area through which water enters a sole or principal source aquifer.

Regulated substances—Substances that are regulated by the USEPA; they cannot be present at levels above the MCL.

Significant hazard to public health—Any level of contaminant that causes or may cause the aquifer to exceed any maximum contaminant level set forth in any promulgated National Primary Drinking Water Regulations at any point where the water may be used for drinking purposes or which may otherwise adversely affect the health of persons, or which may require a public water system to install additional treatment to prevent such adverse effect.

Sole or principal source aquifer—An aquifer that supplies 50% or more of the drinking water for an area.

Streamflow source zone—Upstream headwaters area that drains into an aquifer recharge zone.

Toxic pollutants—Pollutants that, after discharge and upon exposure, ingestion, inhalation, or assimilation into any organism, will cause death, disease, behavioral abnormalities, cancer, genetic mutations, physiological malfunctions, or physical deformations in such organisms or their offspring.

Treatment technique (TT)—A required process intended to reduce the level of a substance in drinking water.

Turbidity—A measure of the cloudiness of water; turbidity is not necessarily harmful but can interfere with the disinfection of drinking water.

Unregulated monitored substances—Substances that are not regulated by the USEPA but must be monitored so information about their presence in drinking water can be used to develop limits.

Variance—A document for water systems having technical and financial difficulty meeting National Primary Drinking Water Regulations that postpones compliance when such postponement will not result in an unreasonable risk to health.

DID YOU KNOW?

State agencies are the principal organizations for enforcing water quality regulations. They have inspectors, usually located at regional offices throughout the state, who visit oil and gas well sites to ensure compliance with regulations.

Waters of the United States—(1) All waters that are currently used, were used in the past, or may be susceptible to use in interstate or foreign commerce, including all waters that are subject to the ebb and flow of the tide. (2) All interstate waters, including interstate wetlands. (3) All other waters such as interstate lakes, rivers, streams, mudflats, sandflats, wetlands, sloughs, prairie potholes, wet meadows, playa lakes, or natural ponds, the use, degradation, or destruction of which would affect interstate or foreign commerce.

Wetlands—Areas that are inundated or saturated by surface water or groundwater at a frequency and duration sufficient to support a prevalence of vegetation typically adapted for life in saturated soil conditions. Wetlands generally include swamps, marshes, bogs, and similar areas.

SDWA Specific Provisions

To ensure the safety of public water supplies, the Safe Drinking Water Act requires the USEPA to set safety standards for drinking water. Standards are now in place for over 80 different contaminants. The USEPA sets a maximum level for each contaminant; however, in cases where making this distinction is not economically or technologically feasible, the USEPA specifies an appropriate treatment technology instead. Water suppliers must test their drinking water supplies and maintain records to ensure quality and safety. Most states carry the responsibility for ensuring that their public water supplies are in compliance with the national safety standards. Provisions also authorize the USEPA to conduct basic research on drinking water contamination, to provide technical assistance to states and municipalities, and to provide grants to states to help them manage their drinking water programs. To protect groundwater supplies, the law provides a framework for managing underground injection compliance. As part of that responsibility, the USEPA may disallow new underground injection wells based on concerns over possible contamination of a current or potential drinking water aquifer.

Each state is expected to administer and enforce the SDWA regulations for all public water systems. Public water systems must provide water treatment, ensure proper drinking water quality through monitoring, and provide public notification of contamination problems. As mentioned, the 1986 amendments to the SDWA significantly expanded and strengthened its protection of drinking water. Under the 1986 provisions, the SDWA required the following five basic activities:

- *Establishment and enforcement of maximum contaminant levels (MCLs)*—These are the maximum levels of certain contaminants that are allowed in drinking water from public systems. Under the 1986 amendments, the USEPA has set numerical standards or treatment techniques for an expanded number of contaminants.
- *Monitoring*—The USEPA requires monitoring of all regulated and certain unregulated contaminants, depending on the number of people served by the system, the source of the water supply, and the contaminants likely to be found.
- *Filtration*—The USEPA has criteria for determining which systems are obligated to filter water from surface water sources.
- *Use of lead materials*—The use of solder or flux containing more than 0.2% lead or pipes and pipe fittings containing more than 8% lead is prohibited in public water supply systems. Public notification is required where lead is used in construction materials of the public water supply system, or where water is sufficiently corrosive to cause leaching of lead from the distribution system or lines.

DID YOU KNOW?

If monitoring the contaminant level in drinking water is not economically or technically feasible, the USEPA must specify a treatment technique that will effectively remove the contaminant from the water supply or reduce its concentration. The MCLs currently cover a number of volatile organic chemicals, organic chemicals, inorganic chemicals, and radionuclides, as well as microbes and turbidity (cloudiness or muddiness). The MCLs are based on an assumed human consumption of 2 liters (roughly 2 quarts) of water per day.

- *Well head protection*—The 1986 amendments require all states to develop well head protection programs. These programs are designed to protect public water supplies from sources of contamination.

The *National Drinking Water Standards* were developed by the USEPA to meet the requirements of the SDWA. Found in CFR 40, these regulations are subdivided into National Primary Drinking Water Regulations (40 CFR 141), which specify maximum contaminant levels (MCLs) based on health-related criteria, and National Secondary Drinking Water Regulations (40 CFR 143), which are unenforceable guidelines based on both aesthetic qualities such as taste, odor, and color of drinking water, as well as non-aesthetic qualities such as corrosivity and hardness. In setting MCLs, the USEPA is required to balance the public health benefits of the standard against what is technologically and economically feasible. In this way, MCLs are different from other set standards, such as the National Ambient Air Quality Standards (NAAQS), which must be set at levels that protect public health regardless of cost or feasibility (Masters, 2007).

The USEPA also creates unenforceable maximum contaminant level goals (MCLGs) set at levels that present no known or anticipated health effects and include a margin of safety, regardless of technological feasibility or cost. The USEPA is also required (under SDWA) to periodically review the actual MCLs to determine whether they can be brought closer to the desired MCLGs.

National Primary Drinking Water Regulations

Categories of primary contaminants include *organic chemicals*, *inorganic chemicals*, *microorganisms*, *turbidity*, and *radionuclides*. Except for some microorganisms and nitrate, water that exceeds the listed MCLs will pose no immediate threat to public health; however, all of these substances must be controlled, because drinking water that exceeds the standards over long periods of time may be harmful.

Organic Chemicals

Organic contaminants for which MCLs are being promulgated are classified using the following three groupings: *synthetic organic chemicals* (SOCs), *volatile organic chemicals* (VOCs), and *trihalomethanes* (THMs). Table 8.2 shows a partial list of maximum allowable levels for several selected

DID YOU KNOW?

For non-carcinogens, MCLGs are determined by a three-step process. The first step is calculating the reference dose (RfD) for each specific contaminant. The RfD is an estimate of the amount of a chemical that a person can be exposed to on a daily basis that is not anticipated to cause adverse systemic health effects over the person's lifetime. A different assessment system is used for chemicals that are potential carcinogens. If toxicological evidence leads to the classification of the contaminant as a human or probable human carcinogen, the MCLG is set at zero (Boyce, 1997).

TABLE 8.2
Selected Primary Standard MCLs and MCLGs for Organic Chemicals

Contaminant	Health Effects	MCL–MCLG (mg/L)	Sources
Aldicarb	Nervous system effects	0.003–0.001	Insecticide
Benzene	Possible cancer risk	0.005–0	Industrial chemicals, paints, plastics, pesticides
Carbon tetrachloride	Possible cancer risk	0.005–0	Cleaning agents, industrial wastes
Chlordane	Possible cancer	0.002–0	Insecticide
Endrin	Nervous system, liver, kidney effects	0.002–0.002	Insecticide
Heptachlor	Possible cancer	0.0004–0	Insecticide
Lindane	Nervous system, liver, kidney effects	0.0002–0.0002	Insecticide
Pentachlorophenol	Possible cancer risk, liver, kidney effects	0.001–0	Wood preservative
Styrene	Liver, nervous system effects	0.1–0.1	Plastics, rubber, drug industry
Toluene	Kidney, nervous system, liver, circulatory effects	1–1	Industrial solvent, gasoline additive chemical manufacturing
Total trihalomethanes (TTHM)	Possible cancer risk	0.1–0	Chloroform, drinking water chlorination byproduct
Trichloroethylene (TCE)	Possible cancer risk	0.005–0	Waste from disposal of dry cleaning material and manufacture of pesticides, paints, waxes; metal degreaser
Vinyl chloride	Possible cancer	0.002–0	May leach from PVC pipe
Xylene	Liver, kidney, nervous system effects	10–10	Gasoline refining byproduct, paint ink, detergent

Source: USEPA, *Is Your Drinking Water Safe?*, EPA 810-F-94-002, U.S. Environmental Protection Agency, Washington, DC, 1994; USEPA, *National Primary Drinking Water Regulations*, EPA 816-F-09-0004, U.S. Environmental Protection Agency, Washington, DC, 2009.

Surface Water Disposal Options

DID YOU KNOW?

In water, VOCs are particularly dangerous. VOCs are absorbed through the skin through contact with water—for example, every shower or bath. Hot water allows these chemicals to evaporate rapidly; they are harmful if inhaled. VOCs can be present in any tapwater, regardless of location or water source. If tapwater contains significant levels of these chemicals, they pose a health threat from skin contact, even if the water is not ingested (Ingram, 1991).

organic contaminants. As we learn more from research about the health effects of various contaminants, the number of regulated organics is likely to grow. Public drinking water supplies must be sampled and analyzed for organic chemicals at least once every three years.

Synthetic organic chemicals (SOCs) are manmade and are often toxic to living organisms. SOCs are compounds used in the manufacture of a wide variety of agricultural and industrial products, including pesticides and herbicides. This group includes PCBs, carbon tetrachloride, 2.4-D, aldicarb, chlordane, dioxin, xylene, phenols, and thousands of other synthetic chemicals. A 1995 study of 29 Midwestern cities and towns by the Washington, DC-based nonprofit Environmental Working Group found pesticide residues in the drinking water in nearly all of them. In Danville, Illinois, the level of the weed killer cyanazine (made by DuPont) was 34 times the federal standard. In Fort Wayne, Indiana, one glass of tap water contained nine kinds of pesticides. The fact is, each year approximately 2.6 billion pounds of pesticides are used in the United States (Lewis, 1996). These pesticides find their way into water supplies and thus present increased risk to public health.

Volatile organic chemicals (VOCs) are synthetic chemicals that readily vaporize at room temperature. These include degreasing agents, paint thinners, glues, dyes, and some pesticides—more specifically, benzene, carbon tetrachloride, 1,1,1-trichloroethane (TCA), trichloroethylene (TCE), and vinyl chloride.

Trihalomethanes (THMs) are created in the water itself as byproducts of water chlorination. Chlorine (present in essentially all U.S. tapwater) combines with organic chemicals to form THMs. They include chloroform, bromodichloromethane, dibormochloromethane, and bromoform. THMs are known carcinogens—substances that increase the risk of getting cancer—and they are present at varying levels in all public tapwater.

Inorganic Chemicals

Several inorganic substances (particularly lead, arsenic, mercury, and cadmium) are of public health importance. These inorganic contaminants and others contaminate drinking water supplies as a result of natural processes, environmental factors, or, more commonly, human activity. Some of these chemicals are listed in Table 8.3. For most organics, MCLs are the same as MCLGs, but the MCLG for lead is zero. Note that in Table 8.3 the nitrate level is set at 10 mg/L, because nitrate levels above 10 mg/L pose an immediate threat to children under 1 year old. Excessive levels of nitrate can react with hemoglobin in blood to produce an anemic condition known as "blue babies." Treated water is sampled and tested for inorganics at least once per year (Nathanson, 1997).

DID YOU KNOW?

The abbreviation mg/L stands for milligrams per liter. In metric units, this is the weight of the chemical dissolved in 1 liter of water. One liter is about equal to 1 quart, and 1 ounce is equal to about 28,500 milligrams, so 1 milligram is a very small amount. About 25 grains of sugar weigh 1 milligram.

TABLE 8.3
Selected Primary Standard MCLs for Inorganic Chemicals

Contaminant	Health Effects	MCL (mg/L)	Sources
Arsenic	Nervous system effects	0.010	Geological, pesticide residues, industrial waste, smelter operations
Asbestos	Possible cancer	7 MFL[a]	Natural mineral deposits, A/C pipe
Barium	Circulatory system effects	2	Natural mineral deposits, paint
Cadmium	Kidney effects	0.005	Natural mineral deposits, metal finishing
Chromium	Liver, kidney, digestive system effects	0.1	Natural mineral deposits, metal finishing, textile and leather industries
Copper	Digestive system effects	TT[b]	Corrosion of household plumbing, natural deposits, wood preservatives
Cyanide	Nervous system effects	0.2	Electroplating, steel, plastics, fertilizer
Fluoride	Dental fluorosis, skeletal effects	4	Geological deposits, drinking water additive, aluminum industries
Lead	Nervous system and kidney effects, toxic to infants	TT	Corrosion of lead service lines and fixtures
Mercury	Kidney, nervous system effects	0.002	Industrial manufacturing, fungicide, natural mineral deposits
Nickel	Heart, liver effects	0.1	Electroplating, batteries, metal alloys
Nitrate	Blue-baby effect	10	Fertilizers, sewage, soil and mineral deposits
Selenium	Liver effects	0.05	Natural deposits, mining, smelting

Source: USEPA, *Is Your Drinking Water Safe?*, EPA 810-F-94-002, U.S. Environmental Protection Agency, Washington, DC, 1994; USEPA, *National Primary Drinking Water Regulations*, EPA 816-F-09-0004, U.S. Environmental Protection Agency, Washington, DC, 2009.

[a] Million fibers per liter.

[b] Treatment techniques have been set for lead and copper because the occurrence of these chemicals in drinking water usually results from corrosion of plumbing materials. All systems that do not meet the action level at the tap are required to improve corrosion control treatment to reduce the levels. The action level for lead is 0.015 mg/L, and for copper it is 1.3 mg/L.

Surface Water Disposal Options

Microorganisms (Microbiological Contaminants)

This group of contaminants includes bacteria, viruses, and protozoa, which can cause typhoid, cholera, and hepatitis, as well as other waterborne diseases. Bacteria are closely monitored in water supplies because they can be dangerous and because their presence can be easily detected. Because tests designed to detect individual microorganisms in water are difficult to perform, in actual practice a given water supply is not tested by individually testing for specific pathogenic microorganisms. Instead, a simpler technique is used that is based on testing water for evidence of any fecal contamination. Coliform bacteria are used as indicator organisms, whose presence suggests that the water is contaminated. The number of monthly samples required to test for total coliforms is based on the population served and on the size of the distribution system.

Because the number of coliform bacteria excreted in feces is on the order of 50 million per gram, and because the concentration of coliforms in untreated domestic wastewater is usually several million per 100 mL, that water contaminated with human wastes would have no coliforms is highly unlikely. That conclusion is the basis for the drinking water standard for microbiological contaminants, which specifies in essence that, on the average, water should contain no more than 1 coliform per 100 mL. The SDWA standards now require that coliforms not be found in more than 5% of the samples examined during a 1-month period. Known as the *presence/absence concept*, it replaces previous MCLs based on the number of coliforms detected in the sample. Viruses are very common in water. If we removed a teaspoonful of water from an unpolluted lake, over a billion viruses would be present in the water. The two most common and troublesome protozoans found in water are called *Giardia* and *Cryptosporidium* (or *Crypto*). In water, these protozoans occur in the form of hard-shelled cysts. Their hard covering makes them resistant to chlorination and chlorine residual that kills other organisms.

Turbidity

Turbidity is the measure of fine suspended matter in water, which is mostly caused by clay, silt, organic particulates, plankton, and other microscopic organisms, ranging in size from collcidal to coarse dispersion. Turbidity in the water is measured in *nephelometric turbidity units* (NTUs), which measure the amount of light scattered or reflected from the water. Officially reported in standard units or equivalent to milligrams per liter of silica of diatomaceous earth that could cause the same optical effect, turbidity testing is not required for groundwater sources.

Radionuclides

Radioactive contamination of drinking water is a serious matter. Radionuclides (the radioactive metals and minerals that cause this contamination) come from both natural and manmade sources. Naturally occurring radioactive minerals move from underground rock strata and geologic formations into the underground streams flowing through them and primarily affect groundwater. In water, radium-226, radium-228, radon-222, and uranium are the natural radionuclides of most concern. Uranium is typically found in groundwater and, to a lesser degree, in some surface waters. Radium in water is found primarily in groundwater. Radon is a colorless, odorless gas and a known cancer-causing agent. It is created by the natural decay of minerals. Radon is an unusual contaminant in water, because the danger arises not from drinking radon-contaminated water but from breathing the gas after it has been released into the air. Radon dissipates rapidly when exposed to air. When radon is present in household water, it evaporates easily into the air, where household members may inhale it.

Some experts believe that the effects of radon inhalation are more dangerous than those of any other environmental hazard. Manmade radionuclides (more than 200 are known) are believed to be potential drinking water contaminants. Manmade sources of radioactive minerals in water are nuclear power plants, nuclear weapons facilities, radioactive materials disposal sites, and docks for nuclear-powered ships.

TABLE 8.4

National Secondary Drinking Water Regulations

Contaminants	Suggested Levels	Contaminant Effects
Aluminum	0.05–0.2 mg/L	Discoloration of water
Chloride	250 mg/L	Salty taste; corrosion of pipes
Color	15 color units	Visible tint
Copper	1.0 mg/L	Metallic taste; blue-green staining of porcelain
Corrosivity	Noncorrosive	Metallic taste; fixture staining corroded pipes (corrosive water can leach pipe materials, such as lead, into drinking water)
Fluoride	2.0 mg/L	Dental fluorosis (brownish discoloration of the teeth)
Foaming agents	0.5 mg/L	Aesthetic: frothy, cloudy, bitter taste, odor
Iron	0.3 mg/L	Bitter metallic taste; staining of laundry, rusty color, sediment
Manganese	0.05 mg/L	Taste; staining of laundry, black to brown color, black staining
Odor	3 TON[a]	Rotten egg, musty, or chemical smell
pH	6.5–8.5	Low pH: bitter metallic taste, corrosion; High pH: slippery feel, soda taste, deposits
Silver	0.1 mg/L	Argyria (discoloration of skin), graying of eyes
Sulfate	250 mg/L	Salty taste; laxative effects
Total dissolved solids	500 mg/L	Taste and possible relation between low hardness and cardiovascular disease; also an indicator of corrosivity (related to lead levels in water); can damage plumbing and limit effectiveness of soaps and detergents
Zinc	5 mg/L	Metallic taste

Source: USEPA, *Secondary Drinking Water Regulations: Guidance for Nuisance Chemicals*, EPA 816-F-10-079, U.S. Environmental Protection Agency, Washington, DC, 2012.

[a] Threshold odor number.

Nation Secondary Drinking Water Regulations

The National Secondary Drinking Water Regulations are non-enforceable guidelines regulating contaminants that may cause cosmetic effects (such as skin or tooth discoloration) or aesthetic effects (such as taste, odor, or color) in drinking water. A range of concentrations is established for substances that affect water only aesthetically and have no direct effect on public health. Secondary regulations are provided in Table 8.4.

1996 Amendments to SDWA

After more than 3 years of effort, the Safe Drinking Water Act Reauthorization (one of the most significant pieces of environmental legislation passed to date) was adopted by Congress and signed into law by President Clinton on August 6, 1996. The new streamlined version of the original SDWA gives states greater flexibility in identifying and considering the likelihood for contamination in potable water supplies and in establishing monitoring criteria. It establishes increased reliance on "sound science" instead of "feel-good science," paired with more consumer information presented in readily understandable form, and calls for increased attention to assessment and protection of source waters. The significance of the 1996 SDWA amendments lies in the fact that they are a radical rewrite of the law that the USEPA, states, and water systems had been trying to implement for the past 10 years. In contrast to the 1986 amendments (which were crafted with little substantive input from the regulated community and embraced a command-and-control approach with compliance costs rooted in water rates), the 1996 amendments were developed with significant

Surface Water Disposal Options

159

contributions from water suppliers and state and local officials and embody a partnership approach that includes major new infusions of federal funds to help water utilities—especially the thousands of smaller systems—comply with the law. Table 8.5 provides a summary of many of the major provisions of the new amendments, which are as complex as they are comprehensive.

Implementing SDWA

On December 3, 1998, at the oceanfront of Fort Adams State Park, Newport, Rhode Island, in remarks by President Clinton to the community of Newport, a significant part of the 1996 SDWA and amendments were announced—the expectation being that the new requirements would protect most of the nation from dangerous contaminants while adding only about $2 to many monthly water bills. The rules require approximately 13,000 municipal water suppliers to use better filtering systems to screen out *Cryptosporidium* and other microbes, ensuring that U.S. community water supplies are safe from microbial contamination. In his speech, President Clinton said:

> This past summer I announced a new rule requiring utilities across the country to provide their customers regular reports on the quality of their drinking water. When it comes to the water our children drink, Americans cannot be too vigilant.
>
> Today I want to announce three other actions I am taking. First, we're escalating our attack on the invisible microbes that sometimes creep into the water supply. ...Today, the new standards we put in place will significantly reduce the risk from *Cryptosporidium* and other microbes, to ensure that no community ever has to endure an outbreak like the one Milwaukee suffered.
>
> Second, we are taking steps to ensure that when we treat our water, we do it as safely as possible. One of the great health advances to the 20th century is the control of typhoid, cholera, and other diseases with disinfectants. Most of the children in this audience have never heard of typhoid and cholera, but their grandparents cowered in fear of it, and their great-grandparents took it as a fact of life that it would take away significant numbers of the young people of their generation. But as with so many advances, there are trade-offs. We now see that some of the disinfectants we use to protect our water can actually combine with natural substances to create harmful compounds. So today I'm announcing standards to significantly reduce our exposure to these harmful byproducts, to give our families greater peace of mind with their water.
>
> The third thing we are doing today is to help communities meet higher standards, releasing almost $800 million to help communities in all 50 states to upgrade their drinking water systems...to give 140 million Americans safer drinking water.

OIL POLLUTION ACT OF 1990

The Oil Pollution Act (OPA) was signed into law in 1990, largely in response to rising public concern following the *Exxon Valdez* incident. The CWA and the OPA include both regulatory and liability provisions that are designed to reduce damage to natural resources from oil spills. Congress added Section 311 to the CWA, which in part authorized the President to issue regulations establishing procedures, methods, equipment, and other requirements to prevent discharges of oil from vessels and facilities (Section 311(j)(1)(C)). The OPA amended Section 311 of the CWA and contains provisions applicable to onshore facilities and operations. Section 311, as amended by the OPA, provides for spill prevention requirements, spill reporting obligations, and spill response planning. It regulates the prevention of and response to accidental release of oil and hazardous substances into navigable waters, on adjoining shorelines, or affecting natural resources belonging to or managed by the United States. This authority is primarily carried out through the creation and implementation of facility and response plans. These plans are intended to establish measures that will prevent the discharge of oil into navigable waters of the United States or adjoining shorelines as opposed to response and cleanup after a spill occurs.

A cornerstone of the strategy to prevent oil spills from reaching the nation's waters is the Spill Prevention, Control, and Countermeasure (SPCC) plan. The USEPA promulgated regulations to implement this part of the OPA:

TABLE 8.5

Summary of Major Amendment Provisions of the 1996 SDWA Regulations

Definition	Constructed conveyances such as cement ditches used primarily to supply substandard drinking water to farm workers are now SDWA protected.
Contaminant Regulation	Deletes old contamination selection requirement (USEPA regulate 25 new contaminants every 3 years).
	Requires the USEPA to evaluate at least five contaminants for regulation every 5 years, addressing the most risky first, and considering vulnerable populations. The USEPA must issue a *Cryptosporidium* rule (Enhanced Surface Water Treatment Rule) and disinfection byproduct rules within agreed deadlines.
	Deletes the Senate provision giving industry veto power over the USEPA's expediting the rules. The USEPA is authorized to address "urgent threats to health" using an expedited, streamlined process.
	No earlier than 3 years after enactment, no later than the date the USEPA adopts the State II DBP rule, the USEPA must adopt a rule requiring disinfection of certain groundwater systems and provide guidance on determining which systems must disinfect. The USEPA may use cost/benefit provisions to establish this regulation.
Risk Assessment, Management, and Communication	Requires cost/benefit analysis, risk assessment, vulnerable population impact assessment, and development of public information materials for USEPA rules.
	Standard setting provision allows but does not require the USEPA to use risk assessment and cost/benefit analysis in setting standards.
Standard Setting	Reduces the Senate's process to issue standards from three to two steps, deleting the requirement for advanced notice of proposed rule making.
	Requires risks to vulnerable populations to be considered.
	Makes considering costs/benefits and risks/risks a discretionary USEPA authority. "Sound science" provision is limited to standard setting and scientific decisions.
	Requires standard to be reevaluated every 6 years instead of every 3 years.
Treatment Technologies for Small Systems	Establishes new guidelines for the USEPA to identify the best treatment technology for meeting specific regulations.
	For each new regulation, the USEPA must identify affordable treatment technologies that achieve compliance for three categories of small systems: (1) those serving 3301–10,000, (2) those serving 501–3000, and (3) those serving 500 or fewer.
	For all contaminants other than microbials and their indicators, the technologies can include package systems as well as point-of-use and point-of-entry units owned and maintained by water systems.
	The USEPA has 2 years to list such technologies for current regulations, and 1 year to list such technologies for the surface water treatment rule.
	The USEPA must identify the best treatment technologies for the same system categories for use under variances. Such technologies do not have to achieve compliance but must achieve maximum reduction, be affordable, and protect public health.
	The USEPA has 2 years to identify variance technologies for current regulations.

(continued)

Limited Alternative to Filtration	Allows systems with fully controlled pristine watersheds to avoid filtration if the USEPA and state agree that health is protected through other effective inactivation of microbial contaminants.
	The USEPA has 4 years to regulate recycling of filter backwash.
Effective Date of Rules	Extends compliance time from 18 months (current law) to 3 years, with available extensions of up to 5 years total.
Arsenic, Sulfate, and Radon	*Arsenic*—Requires the USEPA to set new standard by 2001 using new standard setting language, after more research and consultation with the National Academy of Sciences (NAS). The law authorizes $2.5 million/year for 4 years for research.
	Sulfate—The USEPA has 30 months to complete a joint study with the Centers for Disease Control and Prevention (CDC) to establish a reliable dose–response relationship. Must consider sulfate for regulation within 5 years. If the USEPA decides to regulate sulfate, it must include public notice requirements and allow alternative supplies to be provided to at-risk populations.
	Radon—Requires the USEPA to withdraw its proposed radon standard and to set a new standard in 4 years, after NAS conducts a risk assessment and a study of risk-reduction benefits associated with various mitigation measures. Authorizes cost/benefit analysis for radon, taking into account the costs and benefits of indoor air radon control measures. States or water systems obtaining USEPA approval of a multimedia radon program in accordance with USEPA guidelines would only have to comply with a weaker "alternative maximum contaminant level" for radon that would be based on the contribution of outdoor radon to indoor air.
State Primacy	Primacy states have 2 years to adopt new or revised regulations no less stringent than federal ones and allows 2 years or more if the USEPA finds it necessary and justified.
	Provides states with interim enforcement authority between the time they submit their regulations to the USEPA and USEPA approval
Enforcement and Judicial Review	Streamlines USEPA administrative enforcement, increases civil penalties, clarifies enforceability of the lead ban and other previously ambiguous requirements, allows enforcement to be suspended in some cases to encourage system consolidation or restructuring, requires states to have administrative penalty authority, and clarifies provisions for judicial review of final USEPA actions.
Public Right to Know	"Consumer confidence reports" provision requires consumers be provided at least annually: (1) the levels of regulated contaminants detected in tap water; (2) what the enforceable maximum contaminant levels and the health goals are for the contaminants (and what those levels mean); (3) the levels found of unregulated contaminants required to be monitored; (4) information on the system's compliance with health standards and other requirements; (5) information on the health effects of regulated contaminants found at levels above enforceable standards; (6) information on health effects of up to three regulated contaminants found at levels below USEPA enforceable health standards where health concerns may still exist; and (7) USEPA's toll-free hotline for further information.
	Governors can waive the requirement to mail these reports for systems serving under 10,000 people, but systems must still publish the report in the paper.
	Systems serving 500 or fewer people need only prepare the report and tell their customers it is available.
	States can later modify the content and form of the reporting requirements.
	The public information provision modestly improves public notice requirements for violations (such as requiring "prominent" newspaper publication instead of buried classified ads). States and the USEPA must prepare annual reports summarizing violations.

(continued)

TABLE 8.5 (continued)

Summary of Major Amendment Provisions of the 1996 SDWA Regulations

Variances and Exemptions	Provisions for small system variances make minor changes to current provisions regarding exemption criteria and schedules.
	States are authorized to grant variances to systems serving 3300 or fewer people but need USEPA approval to grant variances to systems serving between 3301 and 10,000 people. Such variances are available only if the USEPA identifies an applicable variance technology and systems install it.
	Variances are only granted to systems that cannot afford to comply (as defined by state criteria that meet USEPA guidelines) through treatment, alternative sources, or restructuring, and when states determine that the terms of the variance ensure adequate health protection. Systems granted such variances have 3 years to comply with its terms and may be granted an extra 2 years if necessary; states must review the eligibility of such variances every 5 years thereafter.
	Variances are not allowed for regulations adopted prior to 1986 for microbial contaminants or their indicators.
	The USEPA has 2 years to adopt regulations specifying procedures for granting or denying such variances and for informing consumers of proposed variances and pertinent public hearings. They also must describe proper operation of variance technologies and eligibility criteria. The USEPA and the federal Rural Utilities Service have 18 months to provide guidance to help states define affordability criteria.
	The USEPA must periodically review state small-system variance programs, may object to proposed variances, and may overturn issued variances if objections are not addressed. Also, customers of a system for which a variance is proposed can petition the USEPA to object.
	New York may extend deadlines for certain small, unfiltered systems in nine counties to comply with federal filtration requirements.
Capacity Development	States must acquire authority to ensure that community and nontransient/noncommunity systems beginning operation after October 1, 1999, have the technical, managerial, and financial capacity to comply with SDWA regulations. States that fail to acquire authority lose 20% of their annual state revolving loan fund grants.
	States have 1 year to send the USEPA a list of systems with a history of significant noncompliance and 5 years to report on the success of enforcement mechanisms and initial capacity development efforts. State primacy agencies must also provide progress reports to governors and the public.
	States have 4 years to implement strategy to help systems acquire and maintain capacity before losing portions of their SRLF grants.
	The USEPA must review existing capacity programs and publish information within 18 months to help states and water systems implement such programs. The USEPA has 2 years to provide guidance for ensuring capacity of new systems and must describe likely effects of each new regulation on capacity.
	The law authorizes $26 million over 7 years for grants to establish small water systems technology assistance centers to provide training and technical assistance. The law also authorizes $1.5 million/year through 2003 for the USEPA to establish programs to provide technical assistance aimed at helping small systems achieve and maintain compliance.

(continued)

Surface Water Disposal Options

Operator Certification	Requires all operators of community and nontransient/noncommunity systems be certified. The USEPA has 30 months to provide guidance specifying minimum standards for certifying water system operators, and states must implement a certification program within 2 years or lose 20% of Sustainability Revolving Loan Fund (SRLF) grants.
	States with such programs can continue to use them as long as the USEPA determines that they are substantially equivalent to its program guidelines.
	The USEPA must reimburse states for the cost of certification training for operators of systems serving 3300 or fewer people, and the law authorizes \$30 million/year through 2003 for such assistance grants
State Supervision Program	Authorizes \$100 million/year through 2003 for public water system supervision grants to states
	Allows the USEPA to reserve a state's grant should the USEPA assume primacy and, if needed, use SRLF resources to cover any shortfalls in Public Water System Supervision (PWSS) appropriations.
Drinking Water Research	Gives the USEPA authorization to conduct drinking water and groundwater research and is required to develop a strategic research plan and to review the quality of all such research.
Water Return Flows	Repeals the provision in current law that allows businesses to withdraw water from a public water system (such as for industrial cooling purposes), then to return the used water—perhaps with contamination—to the water system's pipe.
Enforcement	Expands and clarifies the USEPA's enforcement authority in primacy and nonprimacy states and provides for public hearings regarding civil penalties ranging from \$5000 to \$25,000.
	Provides enforcement relief to systems that submit a plan to address problems by consolidating facilities or management or transferring ownership.
	Requires states to obtain authority to issue administrative penalties, which cannot be less than \$1000/day for systems serving over 10,000 people.
	The USEPA can assess civil penalties as high as \$15,000/day under its emergency powers authority.

Source: Adapted from Drinan, J.E., *Water and Wastewater Treatment: A Guide for the Nonengineering Professional*, CRC Press, Boca Raton, FL, 2000, pp. 289–295.

DID YOU KNOW?

In addition to implementing federal statutes for the NPDES, UIC, and stormwater programs, states and tribes may impose their own requirements to protect their water resources, both surface and underground. For example, they can establish water quality standards for some or all of their surface waters. These standards are approved by the USEPA and become the baseline for CWA permits (USEPA, 2008i).

1. SPCC plans must be prepared, certified (by a professional engineer), and implemented by facilities that store, process, transfer, distribute, use, drill for, produce, or refine oil.
2. Facilities must establish procedures and methods and install proper equipment to prevent an oil release.
3. Facilities must train personnel to properly respond to an oil spill by conducting drills and training sessions.
4. Facilities must have a plan that outlines steps to contain, clean up, and mitigate any effects of an oil spill on waterways (USEPA, 2008h).

Before a facility is subject to the SPCC rule, it must meet three criteria:

1. It must be non-transportation-related.
2. It must have an aggregate aboveground storage capacity greater than 1320 gal (31.4 bbl) and completely buried storage capacity greater than 42,000 gal (1000 bbl).
3. There must be a reasonable expectation of a discharge into or upon navigable waters of the United States or adjoining shorelines.

An SPCC plan is site specific and describes the measures the facility owner has taken to prevent oil spills and what measures are in place to contain and clean up spills. It includes information about the facility, the oil storage containment, inspections, and a site diagram showing locations of tanks (above and below ground) and drainage, and other pertinent details. Prevention measures include secondary containment around tanks and oil-containing equipment. The SPCC program is not as applicable to shale gas operations as it is to oil production sites. Shale gas operators may have to prepare plans if they store large amounts of fuel (exceeding the volumes stated above) onsite or if oil-filled equipment is present and there is a risk of that oil impacting U.S. waters.

THOUGHT-PROVOKING QUESTIONS

1. Is the deposit of produced wastewater into surface waters a good idea?
2. Is the deposit of treated produced wastewater into surface waters a good idea?
3. Is the deposit of produced wastewater into the oceans a good idea?

REFERENCES AND RECOMMENDED READING

Adewumi, M.A., Erb, J.E., and Watson, R.W. (1992). Initial design considerations for a cost effective treatment of stripper oil well produced water, in Ray, J.P. and Engelhart, F.R., Eds., *Produced Water*. Plenum Press, New York, pp. 511–522.

ALL Consulting and Montana Board of Oil and Gas Conservation (MBOGC). (2002). *Handbook on Best Management Practices and Mitigation Strategies for Coal Bed Methane in the Montana Portion of the Powder River Basin*, prepared for National Energy Technology Laboratory, National Petroleum Technology Office, U.S. Department of Energy, Tulsa, OK, 44 pp.

ALL Consulting and Montana Board of Oil and Gas Conservation (MBOGC). (2004). *Coal Bed Natural Gas Handbook: Resources for the Preparation and Review of Project Planning Elements and Environmental Documents*, prepared for National Petroleum Technology Office, U.S. Department of Energy, Washington, DC, 182 pp.

Arms, K. (1994). *Environmental Science*, 2nd ed. Saunders College Publishing, Fort Worth, TX.

AWWA. (1995). *Water Quality*, 2nd ed. American Water Works Association, Denver, CO.

Baden, J. and Stroup, R.C., Eds. (1981). *Bureaucracy vs. Environment*. University of Michigan Press, Ann Arbor.

Bangs, R. and Kallen, C. (1985). *Rivergods: Exploring the World's Great Wild Rivers*. Sierra Club Books, San Francisco, CA.

Bird, R.B, Stewart, W.E., and Lightfoot, E.N. (1960). *Transport Phenomena*. John Wiley & Sons, New York.

Botkin, D.B. (1995). *Environmental Science: Earth as a Living Planet*. John Wiley & Sons, New York.

Boyce, A. (1997). *Introduction to Environmental Technology*. Van Nostrand Reinhold, New York.

Caldeira, K. and Wickett, M.E. (2003). Anthropogenic carbon and ocean pH. *Nature*, 425(6956):365.

Carson, J.E. and Moses, H. (1969). The validity of several plume rise formulas. *Journal of the Air Pollution and Control Association*, 19(11):862–866.

CDPHE. (2007). *Pit Monitoring Data for Air Quality*. Colorado Department of Public Health and Environment, Denver.

CH2M HILL. (2007). *Review of Oil and Gas Operation Emissions and Control Options, Final Report*, prepared for Air Pollution Control Division, Colorado Department of Public Health and Environment, Denver, CO.

Cobb, R.W. and Elder, C.D. (1983). *Participation in American Politics*, 2nd ed. The Johns Hopkins University Press, Baltimore, MD.

Davis, M.L. and Cornwell, D.A. (1991). *Introduction to Environmental Engineering*. McGraw-Hill, New York.

DRBC. (2010). *Administrative Manual—Part III: Water Quality Regulations: 18 CFR Part 410*. Delaware River Basin Commission, West Trenton, NJ.

Easterbrook, G. (1995). *A Moment on the Earth: The Coming Age of Environmental Optimism*. Viking Penguin, Bergenfield, NJ.

Freedman, B. (1989). *Environmental Ecology*. Academic Press, New York.

Franck, I. and Brownstone, D. (1992). *The Green Encyclopedia*. Prentice Hall, New York.

Henry, J.G. and Heinke, G.W. (1995). *Environmental Science and Engineering*, 2nd ed. Prentice Hall, New York.

Ingram, C. (1991). *The Drinking Water Book*. Ten Speed Press, Berkeley, CA.

IOGCC. (1996). *Review of Existing Reporting Requirements for Oil and Gas Exploration and Production Operators in Five Key States*. Interstate Oil and Gas Compact Commission, Washington, DC.

IOGCC. (2008). *Issues: States' Rights*. Interstate Oil and Gas Compact Commission, Washington, DC.

IPAA. (2000). *IPAA Opposes EPA's Possible Expansion of TRI*, Environment and Safety Fact Sheets. Independent Petroleum Association of America, Washington, DC (www.ipaa.org/issues/factsheets/environment_safety/tri.php).

IPAA. (2004). *Reasonable and Prudent Practices for Stabilization (RAPPS) of Oil and Gas Construction Sites*, Guidance Document. Independent Petroleum Association of America, Washington, DC.

Jackson, A.R. and Jackson, J.M. (1996). *Environmental Science: The Natural Environment and Human Impact*. Longman, New York.

Jost, N.J. (1992). Surface and ground water pollution control technology, in Knowles, P.C., Ed., *Fundamentals of Environmental Science and Technology*. Government Institutes, Rockville, MD.

Karliner, J. (1998). Earth predators. *Dollars and Sense*, 218:7.

Keller, E.A. (1988). *Environmental Geology*. Merrill, Columbus, OH.

Kimberlin, J. (2009). That reeking paper mill keeps Franklin running. *The Virginian-Pilot*, February 1.

Leopold, A. (1970). *A Sand County Almanac*. Ballentine Books, New York.

Lewis, S.A. (1996). *Safe Drinking Water*. Sierra Club Books, San Francisco, CA.

Masters, G.M. (1991). *Introduction to Environmental Engineering and Science*. Prentice-Hall, Englewood Cliffs, NJ.

Masters, G.M. (2007). *Introduction to Environmental Engineering and Science*, 3rd ed. Prentice Hall, Englewood Cliffs, NJ.

McHibben, B. (1995). *Hope, Human and Wild: True Stories of Living Lightly on the Earth*. Little, Brown & Company, Boston, MA.

Miller, G.T. (2004). *Environmental Science*, 10 ed. Brooks/Cole, Belmont, CA.

Morrison, A. (1983). In Third World villages, a simple handpump saves lives. *Civil Engineering*, 52:68–72.

Nathanson, J.A. (1997). *Basic Environmental Technology: Water Supply, Waste Management, and Pollution Control*. Prentice Hall, Upper Saddle River, NJ.

NETL. (2016). *Fact Sheet—Discharge*. National Energy Technology Laboratory, Washington, DC (https://www.netl.doe.gov/research/coal/crosscutting/pwmis/tech-desc/discharge).

Norse, E.A. (1985). The value of animal and plant species for agriculture, medicine, and industry, in Hoage, R.J., Ed., *Animal Extinctions*. Smithsonian Institution Press, Washington, DC, pp. 59–70.

Peavy, H.S., Rowe, D.R., and Tchobanoglous, G. (1985). *Environmental Engineering*. McGraw-Hill, New York.

Pepper, I.L., Gerba, C.P., and Brusseau, M. L. (1996). *Pollution Science*. Academic Press, San Diego, CA.

Santoro, R.L., Howarth, R.W., and Ingraffea, A.R. (2011). *Indirect Emissions of Carbon Dioxide from Marcellus Shale Gas Development*, Technical Report. Agriculture, Energy, and Environment Program, Cornell University, New York.

Spellman, F.R. (1996). *Stream Ecology and Self-Purification: An Introduction for Wastewater and Water Specialists*. Technomic, Lancaster, PA.

Spellman, F.R. (2015). *The Science of Water: Concepts and Applications*, 3rd ed. CRC Press, Boca Raton, FL.

Spellman, F.R. and Whiting, N. (2006). *Environmental Science and Technology: Concepts and Applications*. CRC Press, Boca Raton, FL.

STRONGER. (2008a). *History of STRONGER—Helping to Make an Experiment Work*. State Review of Oil and Natural Gas Regulations, Oklahoma City, OK (www.strongerinc.org/about/history.asp).

STRONGER. (2008b). *List of State Reviews*. State Review of Oil and Natural Gas Regulations, Oklahoma City, OK (www.strongerinc.org/reviews/reviews.asp).

Tower, E. (1995). *Environmental and Natural Resource Economics*. Eno River Press, New York.

USEPA. (1999). *USEPA's Program to Regulate the Placement of Waste Water and Other Fluids Underground*, EPA 810-F-99-019. U.S. Environmental Protection Agency, Washington, DC.

USEPA. (2002a). *Exemption of Oil and Gas Exploration and Production Wastes from Federal Hazardous Waste Regulations*. U.S. Environmental Protection Agency, Washington, DC (epa.gov/osw/nonhaz/industrial/special/oil/oil-gas.pdf).

USEPA. (2002b). *Introduction to Underground Injection Control Program*. U.S. Environmental Protection Agency, Washington, DC.

USEPA. (2004). *Understanding the Safe Drinking Water Act*, EPA 816-F-04-030. U.S. Environmental Protection Agency, Washington, DC (http://water.epa.gov/lawsregs/guidance/sdwa/basicinformation.cfm).

USEPA. (1988). *Regulatory Determination for Oil and Gas and Geothermal Exploration, Development and Production Wastes*. U.S. Environmental Protection Agency, Washington, DC.

USEPA. (2008a). *Summary of the Clean Water Act*. U.S. Environmental Protection Agency, Washington, DC (www.epa.gov/lawsregs/laws/cwa.html).

USEPA. (2008b). *Water Quality and Technology-Based Permitting*. U.S. Environmental Protection Agency, Washington, DC (http://cfpub.epa.gov/npdes/generalissues/watertechnology.cfm).

USEPA. (2008c). *Effluent Limitations Guidelines and Standards*. U.S. Environmental Protection Agency, Washington, DC (http://cfpub.epa.gov/npdes/techbasedpermitting/effguide.cfm).

USEPA. (2008d). *Final Rule: Amendments to the Storm Water Regulations for Discharges Associated with Oil and Gas Construction Activities*. U.S. Environmental Protection Agency, Washington, DC (www.epa.gov/npdes/regulations/final_oil_gas_factsheet.pdf).

USEPA. (2008e). *The Green Book Nonattainment Areas for Criteria Pollutants*. U.S. Environmental Protection Agency, Washington, DC (www.epa.gov/oar/oaqps/greenbk/).

USEPA. (2008f). *Air Trends: Basic Information*. U.S. Environmental Protection Agency, Washington, DC (www.epa.gov/airtrends/sixpoll.html).

USEPA. (2008g). *Region 2. Solid Waste; RCRA Subtitle D*. U.S. Environmental Protection Agency, Washington, DC (www.epa.Gov/region2/waste/dsummary.htm).

USEPA. (2008h). *Region 9. Superfund*. U.S. Environmental Protection Agency, Washington, DC (http://www.epa.gov/region9/superfund/).

USEPA. (2008i). *Water. State, Tribal and Territorial Standards*. U.S. Environmental Protection Agency, Washington, DC (www.epa.gov/waterscience/standards/wqslibrary/).

USEPA. (2008j). *Summary of the Oil Pollution Act*. U.S. Environmental Protection Agency, Washington, DC (www.epa.gov/lawsregs/laws/opa.html).

9 Disposal by Evaporation

If Henry David Thoreau were around today and living next to a fracking produced wastewater pond or pit, and if he were tasked to write a tome about such a holding pond or pit, I am confident that he would not describe such ponds or pits as being "lovelier than diamonds" as he did when he described the ponds in his classic *Walden; or, Life in the Woods*. So, the question becomes how would Thoreau apply his descriptive genius to describing a produced wastewater pond or pit? Well, using some literary license, he might describe it as a large cocktail made with a stinky, oily, greasy, and chemical-laced mixer.

PRODUCED WASTEWATER EVAPORATION PONDS

During my travels in northern Appalachia, I often found fracking operations in progress or completed and abandoned. In some, I readily identified excavations in the ground where produced wastewater and flowback had been deposited for storage or for evaporation. Originally, many of the early evaporation ponds were nothing more than bulldozed holes without liners. This practice was modified as time passed and as the appropriate regulators insisted. Today, many produced wastewater and flowback ponds or pits are used only to evaporate liquid contents. This is a very attractive, natural disposal method for two main reasons: low maintenance and low cost. Evaporation itself is driven by local conditions; because these conditions vary with location, evaporation rates and efficiency of operation differ. Thus, drier climates generally favor evaporation as a waste management tool. Ponds and pits with very large surface areas are the most efficient. This is especially the case where precipitation rates do not exceed inflow rates.

Evaporation ponds and pits do have a few drawbacks; for example, waterfowl are often attracted to them. Our feathered friends are not compatible with oil, grease, and assorted hydrocarbon chemicals. One pond I observed was covered with netting to prevent fouling the fowl, and one of the operators said that the netting was effective in preventing waterfowl landings.

At another active fracking site, the produced wastewater pond had spray nozzles installed to produce a fine mist of water droplets that hovered like a low cloud over the pond area. An onsite operator reported that the nozzles were quite effective in speeding up evaporation: "We just spray those droplets of water into the air and the contaminants evaporate more quickly." He also said that the site operators had learned the technique from a southwestern operation where they employed mist sprayers to accomplish the same increased evaporation rate. One issue that became immediately apparent to me was the tendency for the wind to pick up the airborne water mist droplets and move them out of the pond and onto the surrounding grounds and foliage. The operator agreed: "Well, that can be a nasty thing for sure. That stuff contained in the water—you know, the salt, oily stuff, chemicals, and all that all over the place—that's bad." He went on to say, "My boss, the engineer, gets quite upset when that happens, and he insists that we turn off the sprayers when there is a steady, strong wind and any time the site is unmanned."

Later, after digesting the operator's comments about shutting down the spray nozzles whenever the wind was dispersing the suspended droplets out of the pond and onto the surrounding landscape, I recalled from my research that whenever water molecules leave a liquid surface they produce a vapor pressure in the air over the surface. Water temperature is a measure of the energy of the water molecules—the higher the temperature, the greater the rate at which water molecules will escape. Consequently, vapor pressure at the surface is directly related to the temperature of the surface. It is important to point out that some molecules of water vapor already in the air will move in the other

167

direction and condense on the liquid surface. The net gain or loss of water at the surface depends on the differences in vapor pressure between the atmosphere and the surface. The rate of evaporation or condensation is directly proportional to the magnitude of those vapor pressure differences. In semi-arid regions, evaporation is much more common than condensation, and the very term "semi-arid" suggests that evaporation usually proceeds at a fairly rapid rate.

Now, the part of the operator's statement that I had some issue with was that the spray nozzles were turned off whenever wind conditions were brisk. I understood the need to prevent saline and chemical spray from leaving the pond area and contaminating the surrounding landscape, vegetation, and wildlife, but I also understood that if the air above a water surface is still and if evaporation from the surface continues long enough then the air above the surface becomes saturated. When this occurs, there are no longer vapor pressure differences, and evaporation stops. If evaporation is to continue, the surface air layer must be constantly removed and replaced with unsaturated air. Over the surface of produced wastewater ponds or pits, wind is usually responsible for this removal and replacement of air. The faster the wind is blowing, the faster evaporation will take place. Thus, wind blowing across a pond or pit is not always a bad thing, depending on your point of view or on the possibility of degradation of the surrounding area by windblown pond contaminants. Over ponds and pits, wind is usually responsible for this removal and replacement of air. The faster the wind is blowing, the faster evaporation will take place. After moving on to another site, I noticed equipment that looked like a snow-making machine near the evaporation pond; it was in operation and was producing fine water droplets to aid evaporation.

At a couple of fracking sites in North Dakota, where the climate can get quite cold, I observed equipment that combined natural evaporation with a freeze crystallization process. This freeze–thaw/evaporation (FTE®) process was described in detail by Boysen et al. (1999). Freeze crystallization processes are increasingly being recognized as low-cost, energy-efficient means of treating water containing a wide variety of undesirable chemical constituents, including salts. Water purification using freeze crystallization processes has been shown to simultaneously and significantly remove salts, organics, and heavy metals from impure aqueous solutions. In addition, freeze crystallization processes have demonstrated the ability to produce significant quantities of water suitable for industrial, agricultural, and municipal uses. Although freeze crystallization is not a new technology, recent technical advances have made it an increasingly attractive option for the treatment of a wide variety of waters to produce water for beneficial uses.

In the natural freeze–thaw process, freezing is a crystallization process that can be used to purify water. When salts or other constituents are dissolved in water, the freezing point of the solution is lowered below 32°F, the freezing point of pure water. Partial freezing occurs when the solution is cooled to below 32°F but not below the freezing point of the solution. Relatively pure ice crystals form, and an unfrozen solution, or brine, containing elevated concentrations of the chemical constituents also forms. Because of the presence of these chemical constituents in the brine, it has a higher density than that of the purified ice and therefore readily flows from the ice; thus, the purified ice and the brine are naturally separated.

The advantages of natural freezing for water purification are that the required refrigeration is provided at no cost and the ice pack is repeatedly subjected to freeze–thaw cycling. This repeated freeze–thaw cycling promotes the formation of large ice crystals, which, in turn, increase the permeability of the ice pack. This increased permeability allows the brine to flow more readily through the purified ice pack.

Research and observation have shown that if an ice pack is tightly frozen by ambient temperatures well below 0°F, pure ice is formed first. The remaining solution, which is initially unfrozen, again contains elevated concentrations of chemical constituents. As more of this solution freezes, the concentrations of chemical constituents in the unfrozen solution continue to increase until the entire solution freezes. A tightly frozen ice pack, created by freezing under these types of atmospheric conditions, contains zones of ice with elevated concentrations of chemical constituents and zones of relatively pure ice. When this type of ice pack begins to melt during thawing periods, such

Disposal by Evaporation

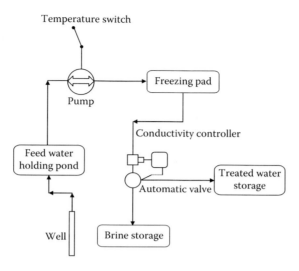

FIGURE 9.1 The natural freeze–thaw water purification process.

as in the spring or early summer, the concentrated zones in the ice melt first, and the initial runoff from the ice contains elevated concentrations of chemical constituents that were incorporated in the ice. Again, the concentrated brine and purified ice are naturally separated (Stinson, 1976).

Figure 9.1 is a simplified block flow diagram of the freeze–thaw process, in which impure water (feed water) is pumped from a holding pond or groundwater well. When the ambient air temperature is below 32°F, the feed water is sprayed or dripped onto a freezing pad to create an ice pile. During subfreezing conditions, runoff from the ice pile will have high concentrations of chemical constituents. This runoff is automatically diverted to a brine storage pond or back to the feed water holding pond or well for recycle based on the conductivity of the runoff. When temperatures promote melting or thawing, the runoff from the freezing pad will be highly purified water that is automatically diverted, based on its conductivity, to a treated water storage pond for later beneficial uses or surface discharge.

EVAPORATION PONDS AND PITS

It was pointed out earlier that, primarily in the past, fracking ponds or pits were commonly bulldozed into existence and that was that, job complete. Today, such ponds or pits are constructed more scientifically with proper engineering techniques because of various environmental regulations. The science of ponds and pits is a bit more complicated than just digging a hole in the ground, which will become apparent in this section.

Environmental professionals and fracking managers involved with constructing and overseeing the operation of produced wastewater ponds, pits, impoundments, or lagoons are at first generally concerned with determining and measuring pond or pit morphometric data, which is commonly recorded on pre-impoundment topographic maps. Determining and maintaining impoundment systems is also a major area of concern for environmental engineers. Mapping the impoundment pond or pit should be the centerpiece of any comprehensive study on the topic. Calculations made from the map allow the engineer to accumulate and relate a lot of data concerning the impoundment unit. Impoundment modeling, a direct measurement method, is undertaken to help the environmental engineer organize an extended project.

Various calculations used in the design and operation of produced wastewater ponds and pits are discussed below; however, before they are presented it is important for the reader to become familiar with the various parameters used in obtaining data for pond and pit measurements:

DID YOU KNOW?

Leaks of flowback and produced wastewater from onsite ponds, pits, impoundments, or lagoons have caused releases as large as 57,000 gal (220,000 L) that impact surface water and groundwater (PA DEP, 2010). Damage from illegal disposal of the range of wastes associated with drilling and production is by far the most common problem. Results of illegal discharge include fish kills, vegetation kills, and death of livestock from drinking polluted water. Discharged fluids include oil, brines of up to 180,000-ppm chlorides, drilling fluids containing detergent and bentonite mud, and fracking fluids that can have a pH as low as 3.0 (highly acidic). Illegal discharges take many forms, including drainage of saltwater holding tanks into creeks or streams, breaching of reserve ponds into streams, siphoning of ponds or pits into streams, or dumping of vacuum truck contents into fields or streams.

- *Physical*—The only physical measurements (size and location) required by the various procedures are latitude, elevation, and surface area.
- *Meteorological*—The various methods may require measurements or estimates of temperature, vapor pressure, wind, radiation, or temperature-dependent constants.
- *Water temperature*—This measurement is needed primarily to estimate saturation vapor pressure at the pond surface. Small, shallow ponds usually mix well, resulting in little thermal stratification. The suggested procedure for measuring average daily pond temperature is to use a mercury-in-glass thermometer held a foot or two below the water surface as far as possible from the water's edge. Temperature should be measured in the morning and in the late afternoon and averaged to provide the estimate of mean daily water temperature.
- *Air temperature (maximum, minimum, mean daily)*—Air temperatures may be measure with recording thermographs or with mercury in-glass max-min thermometers housed in standard weather shelters, preferably situated on the prevailing downwind side of the pond. A recording thermograph (or preferably a well-calibrated hygrothermograph) will allow determination of a weighted mean daily temperature and maintain a permanent record. The simple average of daily maximum and minimum temperatures will provide an estimate of mean daily temperature.
- *Vapor pressure (relative humidity, saturation, ambient)*—Estimates of relative humidity at the mean daily temperature may be made directly with a well-calibrated hygrothermograph. Ambient vapor pressure is an air mass property and therefore changes slowly over the course of a day unless there is an obvious frontal passage. Saturation vapor pressure is determined by temperature. A sling psychrometer used in the morning and afternoon may be used alternatively. In either case, familiarity with psychrometric tables or charts allows the estimate of any vapor pressure measure required by the various procedures.
- *Dew point temperature*—The dew point is a conservative measure of atmospheric moisture content.
- *Wind*—The mass-transfer method requires an estimate of total wind run during the day.
- *Radiation*—Accurate radiation (water reflection related to albedo) measurements require expensive equipment.
- *Other*—Some of the measurement parameters are constants, such as the psychrometric constant (g) or variables that primarily show temperature dependency, such as the latent heat of vaporization. These temperature-dependent variables have values that can be found in standard textbooks that deal with evaporation processes and principles.

Disposal by Evaporation

WASTEWATER STABILIZATION PONDS

POND AND PIT MORPHOMETRY CALCULATIONS

Produced wastewater impoundment pond and pit volume (V), shoreline development index (SDI), and mean depth (D) can be calculated using the formulas provided by Wetzel (1975) and Cole (1994).

Volume

The volume (V) of a fracking wastewater or flowback impoundment can be calculated when the area circumscribed by each isobath (i.e., each subsurface contour line) is known. The formula for water body volume is as follows (Wetzel, 1975):

$$V = \sum_{i=0}^{n} \frac{h}{3} \left(A_i + A_{i+1} + \sqrt{A_i \times A_{i+1}} \right) \tag{9.1}$$

where
  V = Volume (ft^3, acre-ft, m^3).
  h = Depth of the stratum (ft, m).
  i = Number of depth stratum.
  A_i = Area at depth i (ft^2, acre, m^2).

The formula for the volume of water between the shoreline contour (z_0) and the first subsurface contour (z_1) is as follows (Cole, 1994):

$$V_{z_1-z_0} = \frac{1}{3} \left(A_{z_0} + A_{z_1} + \sqrt{\left(A_{z_0} + A_{z_1} \right)\left(z_1 - z_0 \right)} \right) \tag{9.2}$$

where
  z_0 = Shoreline contour.
  z_1 = First subsurface contour.
  A_{z_0} = Total area of the water body.
  A_{z_1} = Area limited by the z_1 line.

Shoreline Development Index

The shoreline development index (D_L) is a comparative figure relating the shoreline length to the circumference of a circle that has the same area as the impoundment body. The smallest possible index would be 1.0. For the following formula, both L and A must be in consistent units for this comparison—meters and square meters:

$$D_L = \frac{L}{2\sqrt{\pi A}} \tag{9.3}$$

where
  D_L = Shoreline development index.
  L = Length of shoreline (miles or m).
  A = Surface area of impoundment body (acre, ft^2, m^2).

Mean Depth

The impoundment body volume divided by its surface area will yield the mean depth. Remember to keep the units the same. If volume is in cubic meters, then area must be in square meters. The equation is as follows:

$$\bar{D} = \frac{V}{A} \tag{9.4}$$

where
  $\bar{D}$ = Mean depth (ft, m).
  V = Volume of lake (ft³, acre-ft, m³).
  A = Surface area (ft², acre, m²).

■ EXAMPLE 9.1

Problem: A large impoundment pond has a shoreline length of 8.6 miles. Its surface area is 510 acres. Its maximum depth is 8.0 feet. The areas for each foot depth are 460, 420, 332, 274, 201, 140, 110, 75, 30, and 1. Calculate the volume of the lake, shoreline development index, and mean depth of the pond.

Solution: Compute the volume of the pond:

$$V = \sum_{i=0}^{n} \frac{h}{3}\left(A_i + A_{i+1} + \sqrt{A_i \times A_{i+1}}\right)$$

$$= \frac{1}{3}\begin{bmatrix} \left(510+460+\sqrt{510\times460}\right)+\left(460+420+\sqrt{460\times420}\right)+\left(420+332+\sqrt{420\times332}\right) \\ +\left(332+274+\sqrt{332\times274}\right)+\left(274+201+\sqrt{274\times201}\right)+\left(201+140+\sqrt{201\times140}\right) \\ +\left(140+110+\sqrt{140\times110}\right)+\left(110+75+\sqrt{110\times75}\right)+\left(75+30+\sqrt{75\times30}\right) \\ +\left(30+1+\sqrt{30\times0}\right) \end{bmatrix}$$

$$= 1/3 \times 6823 = 2274 \text{ acre-ft}$$

Compute the shoreline development index:

$$A = 510 \text{ acres} = 510 \text{ acres} \times \frac{1 \text{ m}^2}{640 \text{ acres}} = 0.7969 \text{ m}^2$$

$$D_L = \frac{L}{2\sqrt{\pi A}} = \frac{8.60}{2\sqrt{3.14 \times 0.7969}} = \frac{8.60}{3.16} = 2.72$$

Compute the mean depth:

$$\bar{D} = \frac{V}{A} = \frac{2274 \text{ acre-ft}}{510 \text{ acres}} = 4.46 \text{ ft}$$

Bottom Slope

$$S = \frac{\bar{D}}{D_m} \tag{9.5}$$

where
  S = Bottom slope.
  $\bar{D}$ = Mean depth (ft, m).
  D_m = Maximum depth (ft, m).

Disposal by Evaporation

Volume Development

Another morphometric parameter is volume development (D_v) (Cole, 1994). This metric compares the shape of the impoundment basin to an inverted cone with a height equal to D_m and a base equal to the surface area of the impoundment body:

$$D_v = 3 \times \frac{\bar{D}}{D_m} \tag{9.6}$$

Water Retention Time

$$RT = \frac{\text{Storage capacity } \left(\text{acre-ft, m}^3\right)}{\text{Annual runoff } \left(\text{acre-ft/yr, m}^3/\text{yr}\right)} \tag{9.7}$$

where RT is retention time (years).

Ratio of Drainage Area to Water Body Capacity

$$R = \frac{\text{Drainage area } \left(\text{acre, m}^2\right)}{\text{Storage capacity } \left(\text{acre-ft, m}^3\right)} \tag{9.8}$$

IMPOUNDMENT SURFACE EVAPORATION

In pond, pit, and lagoon management, knowledge of evaporative processes is important to the environmental professional in understanding how water losses through evaporation are determined. Several models and empirical methods are used for calculating pond and pit evaporative processes. The following text discusses the water budget and energy budget models, as well as the Priestly–Taylor, Penman, DeBruin–Keijman, and Papadakis equations.

Water Budget Model

The water budget model for pond and pit evaporation is used to make estimations of pond or pit evaporation in some areas. It depends on an accurate measurement of the inflow and outflow of the impoundment body and is expressed as

$$\Delta S = P + R + G_I - G_O - E - T - O \tag{9.9}$$

where
  ΔS = Change in lake storage (mm).
  P = Precipitation (mm).
  R = Surface runoff or inflow (mm).
  G_I = Groundwater inflow (mm).
  G_O = Groundwater outflow (mm).
  E = Evaporation (mm).
  T = Transpiration (mm).
  O = Surface water release (mm).

Because fracking evaporation ponds and pits have little vegetation and zero groundwater inflow and outflow, impoundment body evaporation can be estimated by

$$E = P + R - O \pm \Delta S \tag{9.10}$$

Energy Budget Model

The energy budget (Lee and Swancar, 1996) is recognized as the most accurate method for determining lake evaporation. It is also the most costly and time-consuming method (Mosner and Aulenbach, 2003). The evaporation rate is given by

$$E_{EB} = \frac{Q_s - Q_r + Q_a + Q_{ar} - Q_{bs} + Q_v - Q_x}{L(1 + BR) + T_0} \tag{9.11}$$

where

E_{EB} = Evaporation (cm/day).
Q_s = Incident shortwave radiation (cal/cm^2/day).
Q_r = Reflected shortwave radiation (cal/cm^2/day).
Q_a = Incident longwave radiation from atmosphere (cal/cm^2/day).
Q_{ar} = Reflected longwave radiation (cal/cm^2/day).
Q_{bs} = Longwave radiation emitted by lake (cal/cm^2/day).
Q_v = Net energy advected by streamflow, ground water, and precipitation (cal/cm^2/day).
Q_x = Change in heat stored in water body (cal/cm^2/day).
L = Latent heat of vaporization (cal/g).
BR = Bowen ratio (dimensionless).
T_0 = Water surface temperature (°C).

Priestly–Taylor Equation

The Priestly–Taylor equation (Winter et al., 1995) is used to calculate potential evapotranspiration, which is a measure of the maximum possible water loss from an area under a specified set of weather conditions or evaporation as a function of latent heat of vaporization and heat flux in a water body. It is defined as

$$PET = \alpha \times \left(\frac{s}{s + \gamma} \right) \left[\frac{(Q_n - Q_x)}{L} \right] \tag{9.12}$$

where

PET = Potential evapotranspiration (cm/day).
α = 1.26, a Priestly–Taylor empirically derived constant (dimensionless).
s = Slope of the saturated vapor pressure gradient (dimensionless).
γ = Psychrometric constant (dimensionless).
Q_n = Net radiation (cal/cm^2/day).
Q_x = Change in heat stored in water body (cal/cm^2/day).
L = Latent heat of vaporization (cal/g).

Penman Equation

The Penman equation (Winter et al., 1995) estimates potential evapotranspiration:

$$E_0 = \frac{(\Delta/\gamma) H_e + E_a}{(\Delta/\gamma) + 1} \tag{9.13}$$

where

E_0 = Evapotranspiration.
Δ = Slope of the saturation absolute humidity curve at the air temperature.
γ = Psychrometric constant.
H_e = Evaporation equivalent of the net radiation.
E_a = Aerodynamic expression for evaporation.

Disposal by Evaporation

DeBruin–Keijman Equation

The DeBruin–Keijman equation (Winter et al., 1995) determines evaporation rates as a function of the moisture content of the air above the water body, the heat stored in the impoundment body, and the psychrometric constant, which is a function of atmospheric pressure and latent heat of vaporization:

$$PET = \left(\frac{SVP}{0.95SVP + 0.63\gamma} \right) \times (Q_n - Q_x) \tag{9.14}$$

where SVP is the saturated vapor pressure at mean air temperature (millibars/K). All other terms have been defined previously.

Papadakis Equation

The Papadakis equation (Winter et al., 1995) does not account for the heat flux that occurs in the impoundment water body to determine evaporation. Instead, the equation depends on the difference in the saturated vapor pressure above the impoundment water body at maximum and minimum air temperatures, and evaporation is defined by the following equation:

$$PET = 0.5625 \left[e_0 \max - (e_0 \min - 2) \right] \tag{9.15}$$

where all terms have been defined previously.

THOUGHT-PROVOKING QUESTIONS

1. Are evaporation ponds the answer to what to do with produced wastewater?
2. With regard to the environmental impact of produced water evaporation ponds, is the impact worth the result? Explain.
3. Do evaporation ponds contaminate underlying groundwater supplies?

REFERENCES AND RECOMMENDED READING

Boysen, J.E., Harju, J.A., Rousseau, C., Sole, J., and Stepan, D.J. (1999). *Evaluation of the Natural Freeze–Thaw Process for the Desalinization of Groundwater from the North Dakota Aquifer to Provide Water for Grand Forks, North Dakota*. U.S. Department of Interior, Washington, DC.

Camp, T.R. (1946). Grit chamber design. *Sewage Works Journal*, 14:368–389.

Cole, G.A. (1994). *Textbook of Limnology*, 4th ed. Waveland Press, Prospect Heights, IL.

Crites, R.W., Middlebrooks, E.J., and Reed, S.C. (2006). *Natural Wastewater Treatment Systems*. CRC Press, Boca Raton, FL.

Lawrence, A.W. and McCarty, P.L. (1970). Unified basis for biological treatment design and operation. *Journal of the Sanitary Engineering Division*, 96(3):757–778.

Lee, T.M. and Swancar, A. (1996). *Influence of Evaporation, Ground Water, and Uncertainty in the Hydrologic Budget of Lake Lucerne, a Seepage Lake in Polk County, Florida*. U.S. Geological Survey, Atlanta, GA.

Mancini, J.L. and Barnhart, E.L. (1968). Industrial waste treatment in aerated lagoons, in Gloyna, E.R. and Eckenfelder, Jr., W.W., Eds., *Advances in Water Quality Improvement*. University of Texas Press, Austin.

Metcalf & Eddy. (1991). *Wastewater Engineering Treatment, Disposal, and Reuse*. McGraw-Hill, New York.

Mosner, M.S. and Aulenbach, B.T. (2003). *Comparison of Methods Used to Estimate Lake Evaporation for a Water Budget of Lake Seminole, Southwestern Georgia and Northwestern Florida*. U.S. Geological Survey, Atlanta, GA.

Oswald, W.J. (1996). *A Syllabus on Advanced Integrated Pond Systems®*. University of California, Berkeley.

PA DEP. (2010). *PA DEP Fines Atlas Resources for Drilling Wastewater Spill in Washington County*. Pennsylvania Department of Environmental Protection Southwest Regional Office, Pittsburgh (http://www.prnewswire.com/news-releases/pa-dep-fines-atlas-resources-for-drilling-wastewater-spill-in-washington-county-100888514.html).

Polprasert, C. and Bhattarai, K.K. (1985). Dispersion model for waste stabilization ponds. *Journal of Environmental Engineering*, 11(EEI):45–49.

Ricther, B.C. and Kreitler, C.W. (1993). *Geochemical Techniques for Identifying Sources of Ground-Water Salinization*. CRC Press, Boca Raton, FL.

Rosenberry, D.O., Sturrock, A.M., and Winter, T.C. (1993). Evaluation of the energy budget method of determining evaporation at Williams Lake, Minnesota, using alternative instrumentation and study approaches. *Water Resources Research*, 29(8):2473–2483.

Spellman, F.R. (2007). *The Science of Water*, 2nd ed. CRC Press, Boca Raton, FL.

Spellman, F.R. (2010). *Spellman's Standard Handbook for Wastewater Operators*, Vol. 1. CRC Press, Boca Raton, FL.

Spellman, F.R. and Drinan, J. (2001). *Stream Ecology and Self-Purification*, 2nd ed. Technomic, Lancaster, PA.

Spengel, D.B. and Dzombak, D.A. (1992). Biokinetic modeling and scale-up considerations for biological contractors. *Water Environment Research*, 64(3):223–234.

Stinson, D.L. (1976). Atmospheric freezing for water desalination. *AIChE Symposium Series*, 166(73):112–118.

Thirumurthi, D. (1974). Design criteria for waste stabilization ponds. *Journal of the Water Pollution Control Federation*, 46:2094–2106.

Vaidyanthan, G. (2013). Hydraulic fracturing: when 2 wells meet, spills can often follow. *EnergyWire*, August 5, http://www.eenews.net/energywire/stories/1059985587.

Wetzel, R.G. (1975). *Limnology*. W.B. Saunders, Philadelphia, PA.

WHO. (1987). *Wastewater Stabilization Ponds: Principles of Planning and Practice*, Technical Publications Series 10. World Health Organization Regional Office for the Eastern Mediterranean, Alexandria, Egypt.

Winter, T.C., Rosenberry, D.O., and Sturrock, A.M. (1995). Evaluation of eleven equations for determining evaporation for a small lake in the north central United States. *Water Resources Research*, 31(4):983–993.

WPCF. (1985). *Sludge Stabilization*, Manual of Practice FD-9. Water Pollution Control Federation, Alexandria, VA.

10 Injection of Produced Wastewater

Subsurface injection of liquid wastes has been practiced in the United States for more than 50 years. Early use was for the return to the subsurface of oilfield brines brought to the surface during petroleum production. Application to other liquid wastes began during the 1930s but did not achieve a significant status until the introduced of comprehensive Federal water pollution control laws and regulations in the 1960s and early 1970s.

Warner (1984)

DEFINE YOUR TERMS, PLEASE!

It was Voltaire who said: "If you wish to converse with me, define your terms." This chapter defines the terms, concepts, and ideas that fracking practitioners use when applying their skills to make their technological endeavors bear fruit. Also included are important symbols and dimensions used in the mechanics of aquifer systems (and, secondarily, land subsidence). These concepts are presented early in the chapter so the reader can become familiar with the terms before the text discusses topics that involve these terms. The practicing fracking engineer, manager, operator, or student of fracking should know these concepts; otherwise, it will be difficult (if not impossible) to practice or understand the injection of produced wastewaters.

Injection of hydraulic fracturing and shale gas produced wastewater along with the mechanics of aquifer systems have an extensive and unique terminology, most with well-defined meanings; however, a few terms often are not only poorly defined but also defined from different and conflicting points of view. For the purposes of this chapter, the definitions of key terms are provided, and many of the terms that are poorly defined elsewhere are explained. Other terms not defined below but used in the text are defined when used and are listed and defined in the comprehensive glossary at the end of the book.

SYMBOLS AND DIMENSIONS

Symbol	Dimensions	Description
A	L^2	Area
b	L	Saturated thickness of aquifer
g	LT^{-2}	Acceleration due to gravity
n	Dimensionless	Porosity
n_e	Dimensionless	Effective porosity
p	FL^{-2}	Geostatic stress
p'	FL^{-2}	Effective stress
p_a	FL^{-2}	Applied stress
p_c	FL^{-2}	Preconsolidation stress
P	L^2	Coefficient of permeability
P	$ML^{-1}T^{-2}$	Pressure
P_m	L^2	Meinzer's coefficient of permeability
P_f	LT^{-1}	Field coefficient of permeability

S	Dimensionless	Storage coefficient
S_s	L^{-1}	Specific storage
u_w	FL^{-2}	Neutral stress or pressure
Y_s	Dimensionless	Specific yield
β_t	L^2F^{-1}	Compressibility of the structural skeleton of the medium (for stress changes in the elastic range of response)
β_w	L^2F^{-1}	Compressibility of water
Y_w	FL^{-3}	Unit weight of water
c_v	L^2T^{-1}	Coefficient of consolidation
H	L	Total head, $h_e + h_p + h_v$
h	L	Static head
h_e	L	Elevation head
h_p	L	Pressure head
h_v	L	Velocity head
J	FL^{-2}	Seepage stress
K	LT^{-1}	Hydraulic conductivity
K_e	LT^{-1}	Effective hydraulic conductivity
m	L	Thickness of deposit
m_v	L^2F^{-1}	Coefficient of volume compressibility of fine-grained sediments (for effective-stress change in range exceeding preconsolidation stress)
Φ	L^2T^{-2}	Fluid potential

DEFINITIONS[*]

Aquiclude

An aquiclude is an aerially extensive body of saturated but relatively impermeable material that does not yield appreciable quantities of water to wells. Aquicludes are characterized by very low values of *leakance* (ratio of vertical hydraulic conductivity to thickness), so they transmit only minor inter-aquifer flow and also have very low rates of yield from compressible storage. Therefore, they constitute boundaries of aquifer flow systems.

Aquifer System

Groundwater is found in saturated layers called *aquifers* under the Earth's surface. An aquifer is a heterogeneous body of intercalated permeable and poorly permeable material that functions regionally as a water-yielding hydraulic unit; it is comprised of two or more permeable beds separated at least locally by aquitards that impede groundwater movement but do not greatly affect the regional hydraulic continuity of the system. Three types of aquifers exist: *unconfined*, *confined*, and *springs*. Aquifers are made up of a combination of solid material such as rock and gravel and open spaces called *pores*. Regardless of the type of aquifer, the groundwater in the aquifer is in a constant state of motion. This motion is caused by gravity or by pumping.

The actual amount of water in an aquifer depends on the amount of space available between the various grains of material that make up the aquifer. The amount of space available is called *porosity*. The ease of movement through an aquifer is dependent on how well the pores are connected; for example, clay can hold a lot of water and has high porosity, but the pores are not connected, so water moves through the clay with difficulty. The ability of an aquifer to allow water to infiltrate is referred to as *permeability*.

---

[*] From USDOI, *Glossary of Selected Terms Useful in Studies of the Mechanics of Aquifer Systems and Land Subsidence Due to Fluid Withdrawal*, U.S. Department of the Interior, Washington, DC, 1974; Spellman, F.R., *Handbook of Water and Wastewater Operations*, 3rd ed. CRC Press, Boca Raton, FL, 2014.

Injection of Produced Wastewater

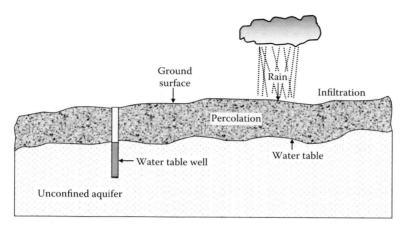

FIGURE 10.1 Unconfined aquifer.

The unconfined aquifer that lies just under the Earth's surface is called the *zone of saturation* (Figure 10.1). The top of the zone of saturation is the *water table*. An unconfined aquifer is only contained on the bottom and is dependent on local precipitation for recharge. This type of aquifer is often referred to as a *water table aquifer*. Unconfined aquifers are a primary source of shallow well water (see Figure 10.1). Because these wells are shallow they are not desirable as public drinking water sources. They are subject to local contamination from hazardous and toxic materials, such as fuel and oil, as well as septic tank and agricultural runoff that provides increased levels of nitrates and microorganisms. These wells may be classified as groundwater under the direct influence of surface water and therefore require treatment for control of microorganisms.

A confined aquifer is sandwiched between two impermeable layers that block the flow of water. The water in a confined aquifer is under hydrostatic pressure. It does not have a free water table (see Figure 10.2). Confined aquifers are referred to as *artesian aquifers*. Wells drilled into artesian

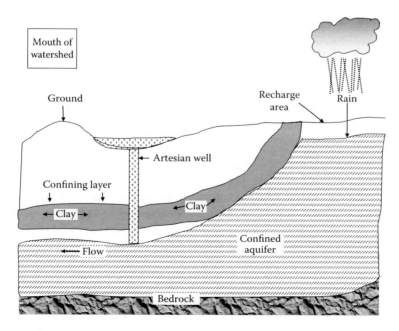

FIGURE 10.2 Confined aquifer.

aquifers are *artesian wells* and commonly yield large quantities of high-quality water. An artesian well is any well where the water in the well casing would rise above the saturated strata. Wells in confined aquifers are normally referred to as *deep wells* and are not generally affected by local hydrological events. A confined aquifer is recharged by rain or snow in the mountains where the aquifer lies close to the surface. Because the recharge area is some distance from areas of possible contamination, the possibility of contamination is usually very low; however, once contaminated, confined aquifers may take centuries to recover.

Aquitard

An aquitard is a saturated but poorly permeable bed that impedes groundwater movement and does not yield water freely to wells, but it may transmit appreciable water to or from adjacent aquifers and, where sufficiently thick, may constitute an important groundwater storage unit. Aquitards are characterized by values of leakance that may range from relatively low to relatively high. A really extensive aquitard of relatively low leakance may function regionally as boundaries of aquifer flow systems.

Coefficient of Volume Compressibility

The coefficient of volume compressibility (L^2F^{-1}; L^{-1}) is the compression of a lithologic unit, per unit of original thickness per unit increase of effective stress, in the load range exceeding preconsolidation stress (i.e., preload stress). The symbol for the coefficient of volume compressibility is m_v (Terzaghi and Peck, 1948).

Compaction

Compaction has been defined as the "decrease in volume of sediments, as a result of comprehensive stress, usually resulting from continued deposition above them" (AGI, 1957). In this book, compaction is defined as the decrease in thickness of sediments that results from an increase in vertical compressive stress, and the term is synonymous with *one-dimensional consolidation* as used by engineers. The term "compaction" is applied to both the process and the measured change in thickness. Compaction of sediments in response to an increase in applied stress is *elastic* if the applied stress increase is in the stress range less than preconsolidation stress, and it is *virgin* if the applied stress increase is in the stress range greater than preconsolidation stress.

Elastic compaction (or expansion) is approximately proportional to the change in effective stress over a moderate range of stress and is fully recoverable if the stress reverts to the initial condition. Elastic changes occur almost instantaneously in permeable sediments and, for stresses less than preconsolidation stress, with a relatively small time delay in strata of low permeability. Virgin compaction has two components: an inelastic component that is not recoverable upon a decrease in stress and a recoverable elastic component. Virgin compaction of aquitards is usually roughly proportional to the logarithm of effective stress increase. In aquitards (fine-grained beds), virgin compaction in response to a manmade increase in applied stress beyond the preconsolidation stress is a delayed process involving the slow expulsion of pore water and the gradual conversion of the increased applied stress to an increased effective stress. Until sufficient time has passed for excess pore pressure to decrease to zero, measured values of compaction are less than ultimate values. In virgin compaction of aquitards, the inelastic component commonly is many times larger than the elastic component. In coarse-grained beds, on the other hand, the inelastic component may be small compared to the elastic component.

Compaction, Residual

Residual compaction is compaction that would ultimately occur if a given increase in applied stress were maintained until steady-state pore pressures were achieved but has not occurred as of a specified time because excess pore pressure still exists in beds of low diffusivity in the compacting system. It can also be defined as the difference between (1) the amount of compaction that will occur ultimately for a given increase in applied stress, and (2) that which has occurred at a specified time.

Injection of Produced Wastewater

Compaction, Specific

Specific compaction (L^3F^{-1}) is the decrease in thickness of deposits per unit of increase in applied stress during a specified time period.

Compaction, Specific Unit

Specific unit compaction (L^2F^{-1}; L^{-1}) is the compaction of deposits, per unit of thickness per unit of increase in applied stress, during a specified time period. Ultimate specific unit compaction is attained when pore pressures in the aquitards have reached hydraulic equilibrium with pore pressures in contiguous aquifers; at that time, specific unit compaction equals gross compressibility of the system.

Compaction, Unit

Unit compaction is the compaction per unit thickness of the compacting deposits. It is usually computed as the measured compaction in a given depth interval during a specified period of time, divided by the thickness of the interval.

Consolidation

In soil mechanics, consolidation is the adjustment of a saturated soil in response to increased load, involving the squeezing of water from the pores and a decrease in void ratio (ASCE, 1962). In our reports, the geologic term "compaction" is used in preference to consolidation, except to report and discuss results of laboratory consolidation tests made in accordance with soil-mechanics techniques.

Excess Pore Pressure

Excess pore pressure (FL^{-2}; L) is transient pore pressure at any point in an aquitard or aquiclude in excess of the pressure that would exist at that point if steady-flow conditions had been attained through the bed.

Expansion, Specific

Specific expansion (L^3F^{-1}) is the increase in thickness of deposits per unit of decrease in applied stress. Specific expansion is a net specific expansion if compaction is continuing in parts of the interval being measured.

Expansion, Specific Unit

Specific unit expansion (L^2F^{-1}; L^{-1}) is the expansion of deposits per unit of thickness per unit decrease in applied stress. Specific unit expansion is a net value if compaction is occurring in parts of the interval being measured during the period of decrease in applied stress.

Hydraulic Diffusivity

Hydraulic diffusivity (L^2T^{-1}) is the ratio of the hydraulic conductivity (K) of a porous medium to its unit water-storage capacity (specific storage, S_s), namely K/S_s. The specific storage (S_s) may be defined as the volume of water released from a unit volume of a saturated medium as the result of a unit decline in head. Within the regions of the aquifer system that remain saturated, S_s is comprised of two principal components: (1) expansion of the pore water as head is reduced, and (2) reduction in pore volume as the skeletal structure of the medium compresses under increasing effective stress. In a confined system, the increase in effective stress is equivalent to the decline in head if the position of the overlying water table remains unchanged. Under these conditions, it may be shown that

$$S_s = \gamma_w \beta_w n + \gamma_w \beta_t$$

and

$$K/S_s = K/[\gamma_w(\beta_w n + \beta_t)]$$

where γ_w is the unit weight of water, β_w is the compressibility of water (reciprocal of the bulk modulus of elasticity), n is the porosity, and β_t is the compressibility of the skeletal structure of the medium for stress changes in the elastic rang of response.

In highly compressible fine-grained sediments subjected to stresses exceeding the preconsolidation stress (preload stress), the component due to compressibility of water becomes relatively insignificant; therefore, in the terminology of soil mechanics, the diffusivity is

$$K/\gamma_w m_v = c_v$$

where c_v is the coefficient of consolidation, and m_v is the coefficient of volume compressibility of the fine-grained sediment. At any given point within a saturated porous medium, the rate at which the head changes in response to a change in head imposed at some other fixed point in the medium is a function of the hydraulic diffusivity. Thus, the hydraulic diffusivity determines the rate at which a head change of specified magnitude migrates through a porous medium.

Hydrocompaction

Hydrocompaction is the process of volume decrease and density increase that occurs when moisture-deficient deposits compact as they are wetted for the first time since burial (Lofgren, 1969; Prokopovich, 1963). The vertical downward movement of the land surface that results from the process has been called *shallow subsidence* (Inter-Agency Committee on Land Subsidence in the San Joaquin Valley, 1958) and *near-surface subsidence* (Bull, 1964; Lofgren, 1960).

Piezometric Surface

The piezometric surface is an imaginary surface that coincides with the level of the water to which water in a system would rise in a *piezometer* (an instrument used to measure pressure). The surface of water that is in contact with the atmosphere is known as *free water surface*. Many important hydraulic measurements are based on the difference in height between the free water surface and some point in the water system. The piezometric surface is used to locate this free water surface in a vessel where it cannot be observed directly.

To understand how a piezometer actually measures pressure, consider the following example. If a clear, see-through pipe is connected to the side of a clear glass or plastic vessel, the water will rise in the pipe to indicate the level of the water in the vessel. Such a see-through pipe—a piezometer—allows us to see the level of the top of the water in the pipe; this is the piezometric surface. In practice, a piezometer is connected to the side of a tank or pipeline. If the water-containing vessel is not under pressure (as is the case in Figure 10.3), the piezometric surface will be the same as the free water surface in the vessel, just as when a drinking straw (the piezometer) is left standing in a glass of water. When a tank and pipeline system is pressurized, as is often the case, the pressure will

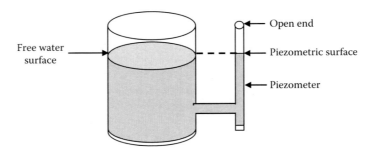

FIGURE 10.3 A container not under pressure where the piezometric surface is the same as the free water surface in the vessel.

Injection of Produced Wastewater

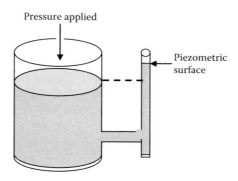

FIGURE 10.4 A container under pressure where the piezometric surface is above the level of the water in the tank.

cause the piezometric surface to rise above the level of the water in the tank. The greater the pressure, the higher the piezometric surface (see Figure 10.4). An increased pressure in a water pipeline system is usually obtained by elevating the water tank.

Note: In practice, piezometers are not installed on pipelines or on water towers because water towers are hundreds of feet high. Instead, pressure gauges are used that record pressure in feet of water or in psi.

Stress, Applied

Applied stress (FL^{-2}; L) is the downward stress imposed at an aquifer boundary. At any given boundary, the applied stress is the weight (per unit area) of sediments and moisture above the water table, plus the submerged weight (per unit area) of the saturated sediments overlying the boundary, plus or minus the net seepage stress (hydrodynamic drag) generated by downward or upward components, respectively, of flow within the specified saturated sediments.

Applied stress differs from effective stress in that it defines only the external stress tending to compact a deposit rather than the grain-to-grain stress at any depth within a compacting deposit. Quantitatively, the stress applied to the top of a saturated stratum differs from the effective stress at any depth within the stratum by the submerged weight (per unit area) of the intervening sediments, plus or minus the seepage stress due to vertical flow within the intervening sediments.

Human-made changes in applied stress are of greater practical significance than the absolute value of applied stress, inasmuch as the sediments, before disturbance, are in a state of strength equilibrium with preexisting natural stresses. Change in applied stress within an aquifer system results from either a change in load at the land surface or a change in the position of the potentiometric surface (confined or unconfined), or both. Change in applied stress is uniform throughout a depth interval in which head change is uniform.

The change in applied stress within a confined aquifer system due to changes in the potentiometric surfaces (i.e., in groundwater, a synonym of piezometric surface) can be expressed as

$$\Delta p_a = -(\Delta h_c - \Delta h_u Y_s)$$

where p_a is the applied stress expressed in feet of water, h_c is the head (assumed uniform) in the confined aquifer system, h_u is the head in the overlying unconfined aquifer, and Y_s is the average specific yield (expressed as a decimal fraction) in the interval of water-table fluctuation. Change in stress applied to a fine-grained bed becomes effective in changing the thickness of the bed only as rapidly as the diffusivity of the medium permits a decrease of excess pore pressures, and thus allows the internal grain-to-grain stress (effective stress) to change.

Water only rises to the water level of the main body of water when it is at rest (static or standing water). The situation is quite different when water is flowing. Consider, for example, an elevated storage tank feeding a distribution system pipeline. When the system is at rest, with all of the valves closed, all of the piezometric surfaces are the same height as the free water surface in storage. On the other hand, when the valves are opened and the water begins to flow, the piezometric surface changes. This is an important point because, as water continues to flow down a pipeline, less and less pressure is exerted. This happens because some pressure is lost (used up) to keep the water moving over the interior surface of the pipe (friction). The pressure that is lost is called *head loss*.

Stress, Effective

Effective stress (FL^{-2}; L) is stress (pressure) that is borne by and transmitted through the grain-to-grain contacts of a deposit and thus affects its porosity or void ratio and other physical properties. In one-dimensional compression, effective stress is the average grain-to-grain load per unit area in a plane normal to the applied stress. At any given depth, the effective stress is the weight (per unit area) of sediments and moisture above the water table, plus the submerged width (per unit area) of sediments between the water table and the specified depth, plus or minus the seepage stress (hydrodynamic drag) produced by downward or upward components, respectively, of water movement through the saturated sediments above the specified depth. Thus, effective stress may be defined as the algebraic sum of the two body stresses, gravitational stress and seepage stress. Effective stress may also be defined as the difference between geostatic and neutral stress. In an aquifer system, a given change in applied stress results in an immediate equivalent change in effective stress within the aquifers (coarse-grained beds). The increase in stress applied to an interbedded aquitard, however, becomes an increased effective stress within the aquitard only as rapidly as excess pore pressures can decrease. Because of the low diffusivity of the aquitards, months or years may be required to reach equilibrium—that is, for the change in applied stress to become fully effective.

Stress, Geostatic

Geostatic stress (FL^{-2}; L) is the total load per unit area of sediments and water above some plane of reference.

Stress, Gravitational

Gravitational stress (FL^{-2}; L) is the downward stress within a body of sediments produced by the weight per unit area of sediments and moisture above the water table plus the submerged (buoyed up) weight per unit area of sediments below the water table. Gravitational stress differs from *geostatic (total) stress* in that, below the water table, it includes only the submerged weight of the deposits, whereas the geostatic stress includes the full weight of the saturated deposits (solids plus contained water).

Stress, Neutral

Neutral stress (FL^{-2}; L) is the fluid pressure exerted equally in all directions at a point in a saturated deposit by the head of water. The neutral stress (pressure) is equal to the pressure head multiplied by the unit weight of water, or

$$u_w = \gamma_w \bullet h_p$$

where u_w is the neutral pressure, γ_w is the unit weight of water, and h_p is the pressure head (Terzaghi and Peck, 1948). Neutral pressure is transmitted to the base of the deposit through the pore water and does not have a measurable influence on the void ratio or on any other mechanical property of the deposits. The total load per unit area (geostatic stress), p, normal to any horizontal plane of reference in a saturated deposit, is comprised of two components: a neutral stress (u_w) and an effective stress (p'). Therefore, $p = p' + u_w$.

Injection of Produced Wastewater

Stress, Preconsolidation

Preconsolidation stress (FL^{-2}; L) is the maximum antecedent effective stress to which a deposit has been subjected and which it can withstand without undergoing additional permanent deformation. Stress changes in the range less than the preconsolidation stress produce elastic deformations of small magnitude. In fine-grained materials, stress increases beyond the preconsolidation stress produce much larger deformations that are principally inelastic (nonrecoverable).

Stress, Seepage

Seepage stress (FL^{-2}; L) occurs when water flows through a porous medium and force is transferred from the water to the medium by viscous friction. The force transferred to the medium is equal to the loss of hydraulic head. This force, called the *seepage force*, is exerted in the direction of flow. The vertical seepage force (F) at the base of a stratum across which a hydraulic head differential exists can be expressed as

$$F = (h_t - h_b)\gamma_w \bullet A$$

where h_t and h_b are the heads at the top and bottom, respectively, of the stratum; γ_w is the unit weight of water; and A is the cross-sectional area normal to the direction of seepage. Under conditions of steady vertical flow, the seepage force is distributed through the body of the medium in the same way as a gravitational force. The average vertical seepage force per unit volume ($\underline{J}$), analogous to average unit weight, is

$$\underline{J} = F/(A \bullet m) = [(h_t - h_b)\gamma_w]/m$$

where m is the thickness of the stratum. The seepage force per unit area, referred to in this book as the *seepage* (J), is

$$J = \underline{J} \bullet m = (h_t - h_b)\gamma_w$$

This vertical seepage stress is algebraically additive with the gravitational stress at the base of the stratum in question, and the sum is transmitted downward through the granular structure of the aquifer system. If the seepage stress, or pressure, is expressed as an equivalent head of water, then γ_w is not required and the expression is simply

$$J = h_t - h_b$$

Subsidence

Subsidence is the sinking or settlement of the land surface due to any of several processes. As commonly used, the term relates to the vertical downward movement of natural surfaces although small-scale horizontal components may be present. Land subsidence occurs when large amounts of fluids have been withdrawn from certain types of rocks, such as fine-grained sediments. The rock compacts because the fluid is partly responsible for holding the ground (overlayment or overburden). The term does not include landslides, which have large-scale horizontal displacements (mass movement), or settlement of artificial fills.

Subsidence/Head Decline Ratio

The subsidence/head decline ratio is the ratio between subsidence and the hydraulic head decline in the coarse-grained beds of the compacting aquifer system.

Unit Compaction/Head Decline Ratio

The unit compaction/head decline ratio (L^{-1}) is the ratio between the compaction per unit thickness of the compacting deposits and the head decline in the coarse-grained beds of the compacting aquifer system; it equals specific unit compaction if the observed head decline is a direct measure of increase in applied stress.

WELLS

It is well known that humans have been digging wells for centuries and even longer. In the early days of human occupancy of the globe, many ensured that their caves or abodes were as near to drinking water sources as possible. On occasion, however, it became necessary for nomadic types to leave the convenience of readily available drinking water sources and to search for life-sustaining water elsewhere. Eventually, through trial and error and much searching and digging, water was found at various subsurface ground levels in varying locations. Digging a well to find water became commonplace and is practiced extensively to this very day. Later in humankind history, when the importance of other subsurface fluids grew, oil and gas wells were dug (drilled) in record numbers. With the advent of horizontal drilling and fracking technology, new wells and old ones were and are commonly drilled.

As time continued to pass, other wells were drilled, but instead of removing subsurface contents for human use, many of the new wells were drilled to accommodate the injection and storage of petroleum products for future and emergency use. Another relatively recent subsurface injection practice is the treatment of domestic wastewater at conventional treatment plants and then conveying the treated effluent to additional advanced treatment processes. These advanced treatment processes (e.g., reverse osmosis, nanofiltration) treat the already treated wastewater to drinking water standards. The drinking water quality of advanced treated wastewater is verified through laboratory testing; moreover, the water must also be approved by regulators who enforce the tenets of the Safe Drinking Water Act.

The reader is probably experiencing the so-called yuck factor while considering if this advanced treated wastewater could be transported to the kitchen tap for household uses. It might be, but that is fodder for a different extensive and ongoing discussion elsewhere. Here, we are talking about treating wastewater to drinking water standards for its legal injection into the subsurface. But, why would the treated wastewater be injected into the subsurface? Is the intention to store the water for some future use? Is the treated wastewater injected into the subsurface to replace water that has been withdrawn from the subsurface?

The last question brings us close to the direction where this discussion is headed; that is, treating wastewater to drinking water standards in order to replace groundwater withdrawal is the current plan of the Hampton Roads Sanitation District (HRSD) in the Tidewater region of southeastern Virginia (which includes the communities of Norfolk, Virginia Beach, Suffolk, Chesapeake, Portsmouth, Newport News, Hampton, and Williamsburg). HRSD is renowned globally as the premier wastewater treatment operation in the world. HRSD's treated water injection program into the Tidewater subsurface is targeted to replace and store groundwater that has been withdrawn over the years. In addition, the focus of HRSD's Sustainable Water Initiative for Tomorrow (SWIFT) has another dual goal: alleviating the serious problem with land subsidence in the region and lowering discharge of nutrients (nitrogen and phosphorus) into local rivers by sequestering them underground. A brief description of HRSD's SWIFT program is provided in Sidebar 10.1.

SIDEBAR 10.1   HAMPTON ROADS SANITATION DISTRICT'S SWIFT PROGRAM[*]

HRSD proposes to add advanced treatment processes to several of its facilities to produce water that exceeds drinking water standards and to pump this clean water into the ground. This will ensure a sustainable source of water to meet current and future groundwater needs throughout eastern Virginia while improving water quality in local rivers and the Chesapeake Bay. Projected benefits from HRSD's SWIFT program include the following:

---

[*] The information was provided by a personal communication with Ted Henifin, P.E., General Manager of HRSD, and on research by the author at the HRSD SWIFT sites for a forthcoming book describing wastewater injection to reduce or stabilize land subsidence.

Injection of Produced Wastewater

- Eliminate HRSD discharge to the James, York, and Elizabeth rivers except during significant storms.
- Restore rapidly dwindling groundwater supplies in eastern Virginia upon which hundreds of thousands of Virginia residents and businesses depend.
- Create huge reductions in the discharge of nutrients, suspended solids, and other pollutants to the Chesapeake Bay.
- Make available significant allocations of nitrogen and phosphorous to support regional needs.
- Protect groundwater from saltwater contamination/intrusion.
- Reduce the rate of land subsidence, effectively slowing the rate of sea level rise by up to 25%.
- Extend the life of protective wetlands and valuable developed low-lying lands.

HRSD is working to obtain approvals to begin construction in 2020 (many of the approvals have now been obtained), with full-scale operations replenishing the aquifer (120 million gallons per day) projected by 2030. With regard to costs, the construction cost of this project, which will include multiple new advanced treatment facilities, is estimated at $1 billion. The effort has the potential to be integrated into and funded by HRSD as part of the U.S. Environmental Protection Agency-mandated wet weather management plan with minimal to no impact on HRSD's existing 20-year forecast. Operating costs are estimated to be between $20 million and $40 million annually, which could be recovered with a modest groundwater withdrawal fee. Will HRSD's wastewater injection program accomplish the goals it has set for itself? The jury is out on this one, but it is difficult to argue against this innovative approach until it has been proven to be non-workable or detrimental to the environment. Simply, time will tell.

As technology continued to grow, so did the generation of liquid waste products, including hazardous wastes. Because many hazardous wastes are in liquid form they are difficult to store and treat. Thus, it has become convenient to inject many different types of waste products into the subsurface. One of these liquid wastes is produced wastewater.

Injecting liquid wastes into the subsurface raises many environmental concerns. Probably the main concern with subsurface injection of liquid wastes is the possible contamination of groundwater. In this regard, the injection of produced wastewater is a major concern. Properly constructed injection wells are designed to protect drinking water supplies. The author has encountered a great deal of skepticism regarding the integrity and safety of injection wells. To help the reader understand injection wells and how they are constructed to ensure that drinking water sources are not contaminated, it is important to highlight the difference between conventional groundwater withdrawal wells and injection wells by first discussing conventional drinking water wells and then explaining injection wells. Again, comparing the two types of wells will help readers understand their similarities and differences.

CONVENTIONAL WATER WELLS

The most common method for withdrawing groundwater is to penetrate the aquifer with a vertical well and then pump the water up to the surface. In the past, when someone wanted a well, they simply dug (or hired someone to dig) and hoped that they would find water in a quantity and quality suitable for their needs. Today, in most locations in the United States, developing a well water supply usually involves a more complicated step-by-step process. Local, state, and federal requirements specify the actual requirements for development of a well supply in this country. The standard sequence for developing a well supply generally involves a seven-step process:

1. *Application*—Depending on the location, filling out and submitting an application (to the applicable authorities) to develop a well supply is standard procedure.
2. *Well site approval*—Once the application has been made, local authorities check various local geological and other records to ensure that the siting of the proposed well coincides with mandated guidelines for approval.
3. *Well drilling*—The well is drilled.
4. *Preliminary engineering report*—After the well is drilled and the results documented, a preliminary engineering report is made on the suitability of the site to serve as a water source. This procedure involves performing a pump test to determine if the well can supply the required amount of water. The well is generally pumped for at least 6 hours at a rate equal to or greater than the desired yield. A stabilized drawdown should be obtained at that rate and the original static level should be recovered within 24 hours after pumping stops. During this test period, samples are taken and tested for bacteriological and chemical quality.
5. *Submission of documents for review and approval*—The application and test results are submitted to an authorized reviewing authority that determines if the well site meets approval criteria.
6. *Construction permit*—If the site is approved, a construction permit is issued.
7. *Operation permit*—When the well is ready for use, an operation permit is issued.

Note: Check with local regulatory authorities to determine specific well site requirements.

1. Minimum well lot requirements
    * 50 feet from well to all property lines
    * All-weather access road provided
    * Lot graded to divert surface runoff
    * Recorded well plat and dedication document
2. Minimum well location requirements
    * At least 50 feet horizontal distance from any actual or potential sources of contamination involving sewage
    * At least 50 feet horizontal distance from any petroleum or chemical storage tank or pipeline or similar source of contamination, except where plastic-type well casing is used the separation distance must be at least 100 feet
3. Vulnerability assessment
    * Is the wellhead area 1000 ft radius from the well?
    * What is the general land use of the area (residential, industrial, livestock, crops, undeveloped, other)?
    * What are the geologic conditions (sinkholes, surface, subsurface)?

Water supply wells may be characterized as shallow or deep. In addition, wells are classified as follows:

1. Class I—Cased and grouted to 100 ft
2. Class II A—Cased to a minimum of 100 ft and grouted to 20 ft
3. Class II B—Cased and grouted to 50 ft

Note: During the well development process, mud and silt forced into the aquifer during the drilling process are removed, allowing the well to produce the best-quality water at the highest rate from the aquifer.

Injection of Produced Wastewater

SHALLOW WELLS

Shallow wells are those that are less than 100 ft deep. Such wells are not particularly desirable for municipal supplies because the aquifers they tap are likely to fluctuate considerably in depth, making the yield somewhat uncertain. Municipal wells in such aquifers cause a reduction in the water table (or phreatic surface) that affects nearby private wells, which are more likely to utilize shallow strata. Such interference with private wells may result in damage suits against the community. Shallow wells may be dug, bored, or driven:

- *Dug wells*—Dug wells are the oldest type of well and date back many centuries; they are dug by hand or by a variety of unspecialized equipment. They range in size from approximately 4 to 15 ft in diameter and are usually about 20 to 40 ft deep. Such wells are usually lined or cased with concrete or brick. Dug wells are prone to failure from drought or heavy pumpage. They are vulnerable to contamination and are not acceptable as a public water supply in many locations.
- *Driven wells*—Driven wells consist of a pipe casing terminating in a point slightly greater in diameter than the casing. The pointed well screen and the lengths of pipe attached to it are pounded down or driven in the same manner as a pile, usually with a drop hammer, to the water-bearing strata. Driven wells are usually 2 to 3 inches in diameter and are used only in unconsolidated materials. This type of shallow well is not acceptable as a public water supply.
- *Bored wells*—Bored wells range from 1 to 36 inches in diameter and are constructed in unconsolidated materials. The boring is accomplished with augers (either hand or machine driven) that fill with soil and then are drawn to the surface to be emptied. The casing may be placed after the well is completed (in relatively cohesive materials), but must advance with the well in noncohesive strata. Bored wells are not acceptable as a public water supply.

DEEP WELLS

Deep wells are the usual source of groundwater for municipalities. Deep wells tap thick and extensive aquifers that are not subject to rapid fluctuations in water level (remember that the *piezometric surface* is the height to which water will rise in a tube penetrating a confined aquifer) and that provide a large and uniform yield. Deep wells typically yield water of more consistent quality than shallow wells, although the quality is not necessarily better. Deep wells are constructed by a variety of techniques; we discuss two of these techniques below:

- *Jetted wells*—Jetted well construction commonly employs a jetting pipe with a cutting tool. This type of well cannot be constructed in clay or hardpan or where boulders are present. Jetted wells are not acceptable as a public water supply.
- *Drilled wells*—Drilled wells are usually the only type of well allowed for use in most public water supply systems. Several different methods of drilling are available, all of which are capable of drilling wells of extreme depth and diameter. Drilled wells are constructed using a drilling rig that creates a hole into which the casing is placed. Screens are installed at one or more levels when water-bearing formations are encountered.

COMPONENTS OF A WELL

The components that make up a well system include the well itself, the building and the pump, and related piping system. In this section, we focus on the components that make up the well itself. Many of these components are shown in Figure 10.5.

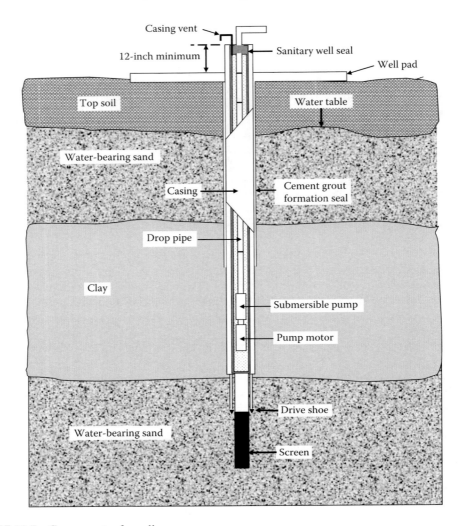

FIGURE 10.5 Components of a well.

Well Casing

A well is a hole in the ground called the *borehole*. A well casing placed inside the borehole prevents the walls of the hole from collapsing and prevents contaminants (either surface or subsurface) from entering the water source. The casing also provides a column of stored water and housing for the pump mechanisms and pipes. Well casings constructed of steel or plastic material are acceptable. The well casing must extend a minimum of 12 inches above grade.

Grout

To protect the aquifer from contamination, the casing is sealed to the borehole near the surface and near the bottom where it passes into the impermeable layer with grout. This sealing process keeps the well from being polluted by surface water and seals out water from water-bearing strata that have undesirable water quality. Sealing also protects the casing from external corrosion and restrains unstable soil and rock formations. Grout consists of near cement that is pumped into the annular space (it is completed within 48 hours of well construction); it is pumped under continuous pressure starting at the bottom and progressing upward in one continuous operation.

Well Pad

The well pad provides a ground seal around the casing. The pad is constructed of reinforced concrete 6 ft by 6 ft (6 inches thick) with the well head located in the middle. The well pad prevents contaminants from collecting around the well and seeping down into the ground along the casing.

Sanitary Seal

To prevent contamination of the well, a sanitary seal is placed at the top of the casing. The type of seal varies depending on the type of pump used. The sanitary seal contains openings for power and control wires, pump support cables, a drawdown gauge, discharge piping, pump shaft, and air vent, while providing a tight seal around them.

Well Screen

Screens can be installed at the intake points on the end of a well casing or on the end of the inner casing on gravel-packed wells. These screens perform two functions: (1) supporting the borehole, and (2) reducing the amount of sand that enters the casing and the pump. They are sized to allow passage of the maximum amount of water while preventing passage of sand, sediment, or gravel.

Well Casing Vent

The well casing must have a vent to allow air into the casing as the water level drops. The vent terminates 18 inches above the floor with a return bend pointing downward. The opening of the vent must be screened with No. 24 mesh stainless steel to prevent entry of vermin and dust.

Drop Pipe

The drop pipe or riser is the line leading from the pump to the well head. It ensures adequate support so that an aboveground pump does not move and so that a submersible pump is not lost down the well. This pipe is either steel or PVC. Steel is the most desirable.

Miscellaneous Well Components

Miscellaneous well components include the following:

- *Gauges and air lines* measure the water level of the well.
- *Check valve* is located immediately after the well to prevent system water from returning to the well. It must be located above ground and protected from freezing.
- *Flowmeter* is required to monitor the total amount of water withdrawn from the well, including any water blown off.
- *Control switches* control well pump operation.
- *Blowoff valve* is located between the well and storage tank and is used to flush the well of sediment or turbid or superchlorinated water.
- *Sample taps* include (1) raw water sample taps, which are located before any storage or treatment to permit sampling of the water directly from the well, and (2) entry-point sample taps located after treatment.
- *Control valves* isolate the well for testing or maintenance or are used to control water flow.

WELL EVALUATION

After a well is developed, conducting a pump test determines if it can supply the required amount of water. The well is generally pumped for at least 6 hours (many states require a 48-hour yield and drawdown test) at a rate equal to or greater than the desired yield. *Yield* is the volume or quantity of water discharged from a well per unit of time (e.g., gpm, ft³/sec). Regulations usually require that a well produce a minimum of 0.5 gpm per residential connection. *Drawdown* is the difference between the static water level (level of the water in the well when it has not been used for some time and has stabilized) and the pumping water level in a well. Drawdown is measured by using an air line and pressure gauge to monitor the water level during the 48 hours of pumping.

The procedure calls for the air line to be suspended inside of the casing down into the water. At the other end are the pressure gauge and a small pump. Air is pumped into the line (displacing the water) until the pressure stops increasing. The highest pressure reading on the gauge is recorded. During the 48 hours of pumping, the yield and drawdown are monitored more frequently during the beginning of the testing period, because the most dramatic changes in flow and water level usually occur then. The original static level should be recovered within 24 hours after the pumping stops.

Testing is performed on a bacteriological sample for analysis by the most probable number (MPN) method every half hour during the last 10 hours of testing. The results are used to determine if chlorination is required or if chlorination alone will be sufficient to treat the water. Chemical, physical, and radiological samples are collected for analysis at the end of the test period to determine if treatment other than chlorination may be required.

> **Note:** Recovery from the well should be monitored at the same frequency as during the yield and drawdown testing and for at least the first 8 hours, or until 90% of the observed drawdown is obtained.

Specific capacity (often referred to as the *productivity index*) is a test method for determining the relative adequacy of a well; over a period of time, it is a valuable tool for evaluating well production. Specific capacity is expressed as a measure of well yield per unit of drawdown (yield divided by drawdown). When conducting this test, if possible always run the pump for the same length of time and at the same pump rate.

WELL PUMPS

Pumps are used to move the water out of the well and deliver it to the storage tank or distribution system. The type of pump chosen should provide optimum performance based on the location and operating conditions, required capacity, and total head. Two types of pumps commonly installed in groundwater systems are *lineshaft turbines* and *submersible turbines*. Whichever type of pump is used, they are rated on the basis of pumping capacity expressed in gpm (e.g., 40 gpm), not on horsepower.

ROUTINE OPERATION AND RECORDKEEPING REQUIREMENTS

Ensuring the proper operation of a well requires close monitoring; wells should be visited regularly. During routine monitoring visits, check for any unusual sounds in the pump, line, or valves and for any leaks. In addition, as a routine, cycle valves to ensure good working condition. Check motors to make sure they are not overheating. Check the well pump to guard against short cycling. Collect a water sample for a visual check for sediment. Also, check chlorine residual and treatment equipment. Measure gallons on the installed meter for one minute to obtain the pump rate in gallons per minute (look for gradual trends or big changes). Check water level in the well at least monthly (perhaps more often in summer or during periods of low rainfall). Finally, from recorded meter readings, determine gallons used and compare with water consumed to determine possible distribution system leaks. Along with meter readings, other records must be accurately and consistently maintained for water supply wells. Such recordkeeping is absolutely imperative. The records (an important resource for troubleshooting) can be useful when problems develop or can be helpful in identifying potential problems. A properly operated and managed waterworks facility keeps the following records of well operation.

Injection of Produced Wastewater

The well log provides documentation of what materials were found in the borehole and at what depth. The well log includes the depths at which water was found, the casing length and type, the depth at which various types of soils were found, testing procedures, well development techniques, and well production. In general, the following items should be included in the well log:

1. Well location
2. Who drilled the well
3. When the well was completed
4. Well class
5. Total depth to bedrock
6. Hole and casing size
7. Casing material and thickness
8. Screen size and locations
9. Grout depth and type
10. Yield and drawdown (test results)
11. Pump information (type, horsepower, capacity, intake depth, and model number)
12. Geology of the hole
13. A record of yield and drawdown data

Pump data that should be collected and maintained include the following:

1. Pump brand and model number
2. Rate capacity
3. Date of installation
4. Maintenance performed
5. Date replaced
6. Pressure reading or water level when the pump is set to cut on and off
7. Pumping time (hours per day the pump is running)
8. Output in gallons per minute

A record of water quality should also be maintained, including bacteriological, chemical, physical (inorganic, metals, nitrate/nitrite, VOCs), and radiological reports.

System-specific monthly operation reports should contain information and data from meter readings (total gallons per day and month), chlorine residuals, amount and type of chemicals used, turbidity readings, physical parameters (pH, temperature), pumping rate, total population served, and total number of connections.

A record of water level (static and dynamic levels) should be maintained, as well as a record of any changes in conditions (such as heavy rainfall, high consumption, leaks, and earthquakes) and a record of specific capacity.

Well Maintenance

Wells do not have an infinite life, and their output is likely to reduce with time as a result of hydrological and/or mechanical factors. Protecting the well from possible contamination is an important consideration. Potential problems can be minimized if a well is properly located (based on knowledge of the local geological conditions and a vulnerability assessment of the area).

During the initial assessment, ensuring that the well is not located in a sinkhole area is important. Locations where unconsolidated or bedrock aquifers could be subject to contamination must be identified. Several other important determinations must also be made: Is the well located on a floodplain? Is it located next to a drainfield for septic systems or near a landfill? Are petroleum or gasoline storage tanks nearby? Is any pesticide or plastics manufacturing conducted near the well site?

Along with proper well location, proper well design and construction prevent wells from acting as conduits for the vertical migration of contaminants into the groundwater. Basically, the pollution potential of a well equals how well it was constructed. Contamination can occur during the drilling process, and an unsealed or unfinished well is an avenue for contamination. Any opening in the sanitary seal or break in the casing may cause contamination, as can a reversal of water flow. In routine well maintenance operations, corroded casing or screens are sometimes withdrawn and replaced, but this is difficult and not always successful. Simply constructing a new well may be less expensive.

Troubleshooting Well Problems

During operation, various problems may develop; for example, the well may pump sand or mud. When this occurs, the well screen may have collapsed or corroded, causing the slot openings of the screen to become enlarged (allowing debris, sand, and mud to enter). If the well screen is not the problem, the pumping rate should be checked, as it may be too high. In the following, we provide a few other well problems, their probable causes, and the remediation required:

1. If the water is white, the pump might be sucking air; reduce the pump rate.
2. If water rushes backwards when the pump shuts off, check the valve, as it may be leaking.
3. If the well yield has decreased, check the static water level. A downward trend in static water level suggests that the aquifer is becoming depleted, which could be the result of the following:
   - Local overdraft (well spacings are too close)
   - General overdraft (pumpage exceeds recharge)
   - Temporary decrease in recharge (dry cycles)
   - Permanent decrease in recharge (less flow in rivers)
   - Decreased specific capacity (if it has dropped 10 to 15%, determine the cause; it may be a result of incrustation)

Note: Incrustation occurs when clogging, cementation, or stoppage of a well screen and water-bearing formation occurs. Incrustations on screens and adjacent aquifer materials result from chemical or biological reactions at the air–water interface in the well. The chief encrusting agent is calcium carbonate, which cements the gravel and sand grains together. Incrustation could also be a result of carbonates of magnesium, clays and silts, or iron bacteria. Treatment involves pulling the screen and removing incrusted material, replacing the screen, or treating the screen and water-bearing formation with acids. If severe, treatment may involve rehabilitating the well.

   - Pump rate is dropping, but water level is not—probable cause is pump impairment
   - Worn impellers
   - Change in hydraulic head against which the pump is working (head may change as a result of corrosion in the pipelines, higher pressure setting, or newly elevated tank)

WELL ABANDONMENT

In the past, common practice was simply to walk away and forget about a well when it ran dry. Today, while dry or failing wells are still abandoned, we know that they must be abandoned with care (and not completely forgotten). An abandoned well can become a convenient (and dangerous) receptacle for wastes, thus contaminating the aquifer. An improperly abandoned well could also become a haven for vermin or, worse, a hazard for children. A temporarily abandoned well must be sealed with a watertight cap or wellhead seal. The well must be maintained so it does not become a source or channel of contamination during temporary abandonment.

When a well is permanently abandoned, all casing and screen materials may be salvaged. The well should be checked from top to bottom to ensure that no obstructions interfere with plugging and sealing operations. Prior to plugging, the well should be thoroughly chlorinated. Bored wells

Injection of Produced Wastewater

should be completely filled with cement grout. If the well was constructed in an unconsolidated formation, it should be completely filled with cement grout or clay slurry introduced through a pipe that initially extends to the bottom of the well. As the pipe is raised, it should remain submerged in the top layers of grout as the well is filled.

Wells constructed in consolidated rock or that penetrate zones of consolidated rock can be filled with sand or gravel opposite zones of consolidated rock. The sand or gravel fill is terminated 5 feet below the top of the consolidated rock. The remainder of the well is filled with sand–cement grout.

INJECTION WELLS

When other methods of managing liquid wastewater are either not possible or too costly, subsurface injection of liquid waste is used as a disposal method in many parts of the country. The petroleum industry, since the 1930s, has used subsurface injection to dispose of brine wastewater that is produced with oil and gas. More recently, chemical and manufacturing industries have begun to dispose of liquid wastes into the subsurface in a number of states. In many locations in the United States, several municipalities have adopted subsurface injection for the disposal of effluent from sewage treatment plants because stringent water quality regulations make surface disposal costly. As mentioned earlier, the Hampton Roads Sanitation District (HRSD), through its SWIFT program, is planning to inject treated effluent into the subsurface to reduce the outfalling of nutrients into local rivers and also in an attempt to reduce or at least to put on hold local land subsidence in the Tidewater region.

Interest in subsurface injection for waste disposal stems partly from the recognition that surface disposal of liquid waste may create a potential for degrading freshwater resources. One aspect of the protection of drinking water supplies that impacts oil and gas drilling, production, and processing is the Safe Drinking Water Act (SDWA), which establishes a framework for the Underground Injection Control (UIC) Program to prevent the injection of liquid wastes into an underground source of drinking water (USDW). The USEPA and states implement the UIC Program, which sets standards for safe waste injection practices and bans certain types of injection altogether. The UIC Program provides these safeguards so that injection wells do not endanger USDWs. The first federal UIC regulations were issued in 1980.

The USEPA currently groups underground injection wells into five classes for regulatory control purposes and has a sixth class under consideration. Each class includes wells with similar functions, construction, and operating features so that technical requirements can be applied consistently to the class.

Class I Wells

Class I wells operate under permit (valid for up to 10 years) and are used to inject hazardous and nonhazardous fluids (industrial and municipal wastes) into isolated formations beneath the lowermost underground source of drinking water (USDW), ranging from 1700 to more than 10,000 feet in depth (Figure 10.6) (USEPA, 2015a). The injection zone is below and separated from USDWs by an impermeable cap rock called the *confining layer*. Class I wells inject wastes produced in petroleum refining, metal production, chemical production, pharmaceutical production, commercial disposal food production, and municipal wastewater treatment. Based on the characteristics of

---

DID YOU KNOW?

At the time of this writing, there are approximately 480 Class I wells in the United States, of which 120+ are hazardous and 350+ are nonhazardous or municipal wells. Texas has the greatest number of Class I hazardous wells (65), followed by Louisiana (18). Florida has the greatest number of nonhazardous wells (the majority of which are municipal wells) (USEPA, 2001).

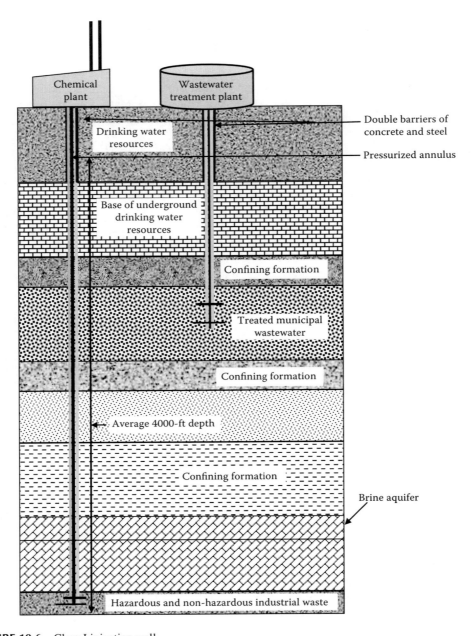

FIGURE 10.6 Class I injection well.

the fluids injected, Class I wells fall into one of four subcategories: (1) hazardous waste disposal wells (~17% of Class I wells), (2) non-hazardous industrial waste disposal wells (~53%), (3) municipal wastewater disposal wells (~30%), and (4) radioactive waste disposal wells (none currently in operation in the United States) (USEPA, 2015a).

Ensuring that injected fluids in Class I wells travel only to their intended location and safely away from USDWs, and that they remain there for as long as they pose a risk to human health or the environment, can be accomplished by injection engineering technology. Moreover, regional and local geologic characterization and compliance with findings via site-specific mathematical models also help to ensure that Class I well contents are properly injected and stored.

Class I wells are designed and constructed to prevent the movement of injected wastewaters into USDWs. Figure 10.6 shows the key construction elements of a typical Class I well. Wells typically consist of three or more concentric layers of pipe: surface casing, long string casing, and injection tubing (all three layers are required of Class I wells per 40 CFR 146.65(c)). The well casing is important because it prevents the borehole from caving in and contains the tubing. It typically is constructed of a corrosion-resistant material such as steel or fiberglass-reinforced plastic. The surface casing is the outermost of the three protective layers; it extends from the surface to below the lowermost USDW. The long string casing extends from the surface to or through the injection zone. The long string casing terminates in the injection zone with a screened, perforated, or open-hole completion, where injected fluids exit the tubing and enter the receiving formation. The well casing design and materials vary based on the physical and chemical nature of injected and naturally occurring fluids in the rock formation, as well as the formation's characteristics. The wastewater must be compatible with the well materials that come into contact with it. Cement made of latex, mineral blends, or epoxy is used to seal and support the casing.

Class I wells are the most strictly regulated and are further regulated under the Resource Conservation and Recovery Act (RCRA) and the Safe Drinking Water Act (SDWA). The Hazardous and Solid Waste Amendments (HSWA) to the RCRA added significant restrictions on the disposal of hazardous waste. Under these amendments, land disposal of hazardous waste, which includes Class I hazardous waste injection wells, is prohibited unless the waste has been treated to become non-hazardous or the disposer can demonstrate that the waste will remain where it has been placed for as long as it remains hazardous, which has been defined as 10,000 years by regulation. The USEPA publication *Class I Underground Injection Control Program: Study of the Risks Associated with Class I Underground Injection Wells* (USEPA, 2001) synthesizes existing information on the Class I program and documents studies of the risks to human health or the environment posed by Class I injection wells.

Class II Wells

Class II wells are used to inject brines and other fluids associated with oil and gas production (USEPA, 2015b). Class II fluids are primarily brines (saltwater) that are brought to the surface while producing oil and gas. It is estimated that over 2 billion gallons of brine are injected in the United States every day. The number of Class II wells varies from year to year based on fluctuations in oil and gas demand and production. In 2015, approximately 180,000 Class II wells were in operation in the United States. Most oil and gas injection wells are in Texas, California, Oklahoma, and Kansas. Class II wells fall into one of three categories: disposal wells, enhanced recovery wells, and hydrocarbon storage wells:

- *Disposal wells*—During extraction of oil and gas, brines are also brought to the surface. This brine water is naturally occurring, deep-basin formation water. Formation brines not associated with oil and gas extraction are usually separated from fresh groundwater by a transition zone of slightly to very saline water, which normally lessens the degree of natural or induced salinization. Human contamination of freshwater by oil- or gas-field brine, in contrast, is not associated with a transition zone but instead brings concentrated brine into direct contact with freshwater. Therefore, salinization of fresh groundwater by oil- or gas-field brine is often very abrupt and characterized by large increases in dissolved solids within relatively short time periods and short distances (Richter and Kreitler, 1993). When these brines are brought to the surface, they are separated from hydrocarbons at the surface and reinjected into the same or similar underground formations for disposal. Wastewater from hydraulic fracturing activities can also be injected into Class II wells. Class II disposal wells make up about 20% of the total number of Class II wells.
- *Enhanced recovery wells*—Various fluids consisting of brine, steam, freshwater, carbon dioxide, or polymers are injected into oil-bearing formations to recover residual oil and, in limited applications, natural gas. The injected fluids thin (decrease the viscosity) or displace

DID YOU KNOW?

In the Energy Policy Act of 2005, Congress created a broad exemption for hydraulic fracturing under the SDWA. Specifically, hydraulic fracturing—except when using diesel fuel—is excluded from the definition of underground injection and is not subject to regulation under the UIC program (SDWA Section 1421(d)(1)(B)).

small amounts of extractable oil and gas. The oil and gas are then available for recovery. In a typical configuration, a single injection well is surrounded by multiple production wells that bring oil and gas to the surface (see Figure 10.7). The USEPA's Underground Injection Control (UIC) Program does not regulate wells that are used solely for production; however, whenever diesel fuels are used in fluids or propping agents, the USEPA does have authority to regulate hydraulic fracturing. During fracking operations, another enhanced recovery process, a viscous fluid is injected under high pressure until the desired fracturing is achieved, followed by a proppant such as sand. The pressure is then released and the proppant holds the fractures open to allow fluid to return to the wall. Enhanced recovery wells are the most numerous type of Class II wells; they represent as much as 80% of the total number of Class II wells.

- *Hydrocarbon storage wells*—Underground formations (such as salt caverns) are commonly used as injection sites for storing the U.S. Strategic Petroleum Reserve. At the present time, over 100 liquid hydrocarbon storage wells operate in the United States.

Most injection wells associated with oil and gas production are Class II wells (Figure 10.7). These wells may be used to inject water and other fluids (e.g., liquid CO_2) into oil- and gas-bearing zones to enhance recovery, or they may be used to dispose of produced water. The regulation specifically prevents the disposal of waste fluids into USDWs by limiting injection only to formations that are not underground sources of drinking water. The UIC Program is designed to prevent contamination of water supplies by setting minimum requirements for state UIC programs. The basic purpose of the UIC programs is to prevent contamination of USDWs by keeping injected fluids within the intended injection zone. The injected fluids must not endanger a current or future public water supply. The UIC requirements that affect the siting, construction, operation, maintenance, monitoring, testing, and, finally, closure of injection wells have been established to address these concepts. All injection wells require authorization under general rules or specific permits.

The law was written with the understanding that states are best suited to have primary enforcement authority (primacy) for UIC programs. In the SDWA, Congress cautioned the USEPA against a "one-size-fits-all" regulatory scheme and mandated consideration of local conditions and practices. Section 1421(b)(3)(A) requires that UIC regulations permit or provide consideration of varying geological, hydrological, or historical conditions in different states and in different areas within a state. Section 1425 allows a state to obtain primacy from USEPA for oil- and gas-related injection wells, without being required to adopt the complete set of applicable federal UIC regulations. The state must be able to demonstrate that its existing regulatory program is protecting USDWs as effectively as the federal requirements. To date, 40 states have obtained primacy for oil and gas injection wells (Class II), although not all of these states have oil and gas production. The USEPA administers UIC programs for ten states, seven of which are oil and gas states, and all other federal jurisdictions and Indian Lands (USEPA, 2015c).

Class III Wells

Class III wells may inject fluids associated with solution mining minerals; these injected fluids dissolve and extract minerals such as uranium, salt, copper, and sulfur (USEPA, 2015d). Class III production wells, which bring mining fluids to the surface, are not regulated under the UIC Program.

Injection of Produced Wastewater

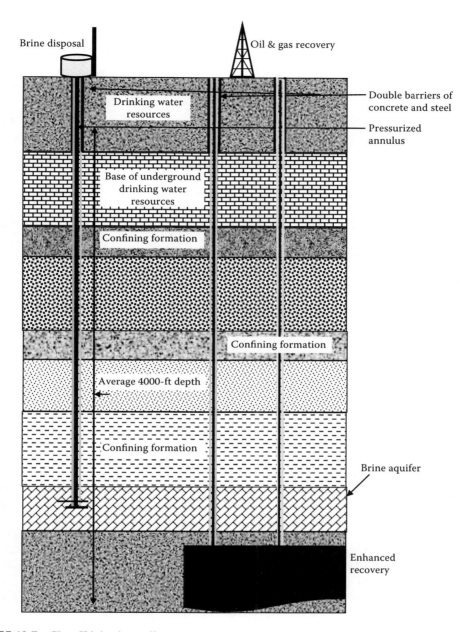

FIGURE 10.7 Class II injection well.

About 170 mining sites with approximately 18,500 Class III wells operate across the nation. More than 50% of the salt and 80% of the uranium extraction in the United States involve the use of Class III injections wells.

The most common method for extracting uranium in the United States is by *in situ* leaching (ISL). The majority of Class III wells in the United States are uranium ISL mines. Typically, a uranium mining operation requires injection, extraction, and monitoring wells. The process includes drilling into the formation containing the uranium and injecting a solution known as a *lixiviant* (meaning to leach out) into the mineral bearing rocks, where the solution is allowed to remain in

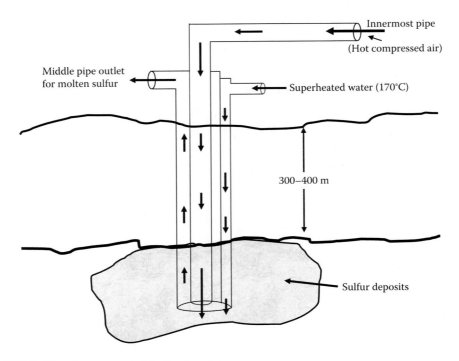

FIGURE 10.8 Frasch process used for *in situ* mining.

contact with the rocks long enough to dissolve the uranium ore. When the lixiviant is almost saturated with uranium, the fluid is brought to the surface via a production well where the uranium is separated from the lixiviant. Finally, the lixiviant is injected again to extract more uranium.

The mining process for salt, copper, and sulfur differs from uranium ISL mining. Salt solution mining wells inject clean water to dissolve the salt. The resulting brine is pumped to the surface where the salt is extracted. Two extraction methods are used. In one method, under normal flow, waste is injected into the well tubing. The saturated fluid is produced through the annulus between the tubing and the casing. Another method is used if the salt is contained in a dome; in that case, a single well is typically used. If the salt is contained in multiple, bedded layers, multiple injection wells are used. Salt solution mining wells make up 5% of the Class III wells (USEPA, 2015d). Copper is mined using injection wells in only in a few states. A sulfuric acid solution is used to dissolve the copper ore. Sulfur may be mined via the Frasch process, where super-heated steam is injected into the mineral-bearing formation to generate a sulfur solution (via hot water melting) that can be recovered (see Figure 10.8). Injection wells are not being used to extract sulfur at this time.

Class IV Wells

Class IV wells may inject hazardous or radioactive wastes into or above an underground source of drinking water and are banned unless specifically authorized under other statues for groundwater remediation (USEPA, 2015e). These wells were banned by the USEPA in 1984. The only time these wells can be operated is if and when they are used as part of a USEPA- or state-authorized groundwater clean-up action—that is, the clean-up of groundwater contaminated by hazardous chemicals. A common method for cleaning contaminated groundwater is to use the *pump and treat* process. In this process, contaminated water is brought to the surface and treated to remove as much of the contaminant as possible. Then, the treated water is injected, through a well, back into the same formation. This process is continued until contaminant concentrations are reduced and additional removal is impossible. Fewer than 32 waste clean-up sites with Class IV wells exist in the United States.

Class V Wells

Class V includes all underground injection not included in Classes I to IV (USEPA, 2015f). Generally, most Class V wells inject nonhazardous fluids into or above a USDW and are onsite disposal systems, such as floor and sink drains that discharge to dry wells, septic systems, leach fields, and drainage wells. This disposal can pose a threat to groundwater quality if not managed properly. Injection practices or wells that are not covered by the UIC Program include single-family septic systems and cesspools, as well as non-residential septic systems and cesspools serving fewer than 20 persons that inject *only* sanitary wastewater. Most Class V wells are unsophisticated shallow disposal systems. Examples include stormwater drainage wells, septic system leach fields, and agricultural drainage wells.

As noted earlier, the Hampton Roads Sanitation District (HRSD) in Tidewater, Virginia, is in the process of testing the feasibility of aquifer replenishment by recharged clean water, purified by the advanced treatment of wastewater treatment plant (WWTP) effluent. The process includes evaluating all of the essential elements of recharging clean water into the Potomac Aquifer Systems (PAS) at seven HRSD WWTPs. This evaluation includes determining the capacity of individual injection wells at the seven WWTPs, projecting the injection capacity with the existing site area of the seven WWTPs, and characterizing the regional beneficial hydraulic response of the PAS to clean water injection.

The HRSD's clean water injection plan is intended to be as innovative, inventive, far-reaching, and far-thinking as possible to meet future goals the organization has set. Challenges the project faces involve a combination of technical, financial, and institutional complexities related to the management of the region's water supply and receiving water resources. HRSD is exploring possibilities using nontraditional approaches that provide benefits on a larger scale beyond what the current wastewater treatment and disposal model can achieve. The process of treating wastewater to drinking water quality and injecting it into the aquifer beneath the lower Chesapeake Bay is a daunting challenge that could reduce the potential damage caused by discharge to the lower James River and the Chesapeake Bay. Additionally, HRSD's SWIFT program may halt, delay, or even mitigate land subsidence in the region. This last item is the ultimate goal. Is it achievable? Time will tell, literally. The SWIFT project is scheduled to be fully implemented and operational by 2030. Readers will be kept informed on the progress and results of SWIFT, because the author has been given *carte blanche* access to the operation and results and contracted to report future results.

Class VI Wells

Class VI wells are used to inject carbon dioxide (CO_2) for the purpose of sequestration in deep rock formations (see Figure 10.9) (USEPA, 2015g). Officially, this long-term underground storage is called *geological sequestration* (GS), which is designed to reduce CO_2 emissions to the atmosphere and mitigate climate change.

DID YOU KNOW?

It should be clear by now that subsurface injection into wells is not a natural process whereby Mother Nature automatically injects disposal materials into the subsurface. Instead, subsurface injection is the forcing of liquid through a well into underground rock openings that generally are filled with water. Sometimes the weight of the liquid column in a well provides sufficient force for injection. In this application, the well is called a *gravity injection well*. Commonly, another force is added to the weight of the liquid to cause injection. Pumps add this force by increasing the pressure on the liquid until its pressure, at the point of injection, exceeds the pressure of the water in the underground rock openings. Where a pump is employed, the well is called a *pressure injection well*.

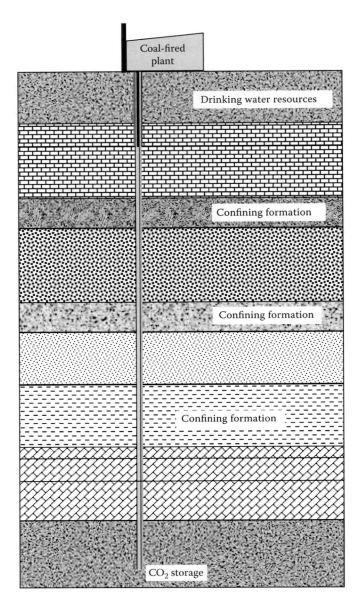

FIGURE 10.9 Class VI injection well.

THOUGHT-PROVOKING QUESTIONS

1. Do you think that using injection wells to dispose of produced wastewater is a good idea?
2. Do you think injecting produced wastewater will reduce subsidence in certain areas?

REFERENCES AND RECOMMENDED READING

AGI. (1957). *Glossary of Geology and Related Sciences.* American Geological Institute, Washington, DC.
AOGC. (2008). *Mission Statement.* Arkansas Oil and Gas Commission, Little Rock (http://www.aogc.state.ar.us/mission.pdf).
ASCE. (1962). *Nomenclature for Hydraulics*, Manuals and Reports on Engineering Practice No. 43. American Society of Civil Engineers, Reston, VA.

Bellabarba, M. et al. (2008). Ensuring zonal isolation beyond the life of the well. *Oilfield Review*, 20(1):18–31.

BLM. (2006). *Scientific Inventory of Onshore Federal Lands' Oil and Gas Resources and the Extent and Nature of Restrictions or Impediments to their Development*. Bureau of Land Management, U.S. Department of the Interior, Washington, DC.

Bonner, T. and Willer, L. (2005). Royalty management 101: basics for beginners and refresher course for everyone, in *Proceedings of the 25th Anniversary Convention of National Association of Royalty Owners (NARO)*, Oklahoma City, November 3–5.

Bromley, M. (1985). *Wildlife Management Implications of Petroleum Exploration and Development in Wildland Environments*, General Technical Report INT-199. U.S. Department of Agriculture, Intermountain Research Station, Ogden, UT.

Bull, W.B. (1964). *Alluvial Fans and Near-Surface Subsidence in Western Fresno County, California*, USGS Professional Paper 437-A. U.S. Geological Survey, Reston, VA.

Catskill Mountainkeeper. (2008). *The Marcellus Shale—America's Next Super Giant.* Catskill Mountainkeeper, Youngsville, NY (www.catskillmountainkeeper.org/our-programs/fracking/marcellus-shale/).

Chesapeake Energy Corporation. (2008). Drilling 101, paper presented to New York Department of Environmental Conservation, Albany.

Douglas, D. (2008). *Anger over Road Damage Caused by Barnett Shale Development*. WFAA-TV, Dallas–Fort Worth, TX (www.wfaa.com/news/local/64791902.html).

Franz, Jr., J.H. and Jochen, V. (2005). *When Your Gas Reservoir Is Unconventional, So Is Our Solution*, Shale Gas White Paper 05-OF-299. Schlumberger, Houston, TX.

Halliburton. (2008). *U.S. Shale Gas: An Unconventional Resource. Unconventional Challenges*. Halliburton, Houston, TX, 8 pp.

Harper, J. (2008). The Marcellus Shale—an old "new" gas reservoir in Pennsylvania. *Pennsylvania Geology*, 28(1):2–13.

Harrill, J.R. (1986). *Ground-Water Storage Depletion in Pahrump Valley, Nevada–California, 1962–75*, USGS Water-Supply Paper 2279. U.S. Geological Survey, Denver, CO.

Inter-Agency Committee on Land Subsidence in the San Joaquin Valley. (1958). *Progress Report on Land-Subsidence Investigations in the San Joaquin Valley, California, Through 1957*. Inter-Agency Committee on Land Subsidence in the San Joaquin Valley, Sacramento.

Lantz, G. (2008). Drilling green along Trinity Trails. *The Barnett Shale Magazine*, Summer.

Lofgren, B.E. (1960). Near-surface land subsidence in western San Joaquin Valley, California. *Journal of Geophysical Research*, 65(3):1052–1062.

Lofgren, B.E. (1969). Land subsidence due to the application of water. *Reviews in Engineering Geology*, 2:271–304.

Lohman, S.W. et al. (1972). *Definitions of Selected Ground-Water Terms—Revisions and Conceptual Refinements*, USGS Water-Supply Paper 1988. U.S. Geological Survey, Washington, DC.

Marshall Miller & Associates. (2008). Marcellus Shale, paper presented to Fireside Pumpers, Bradford, PA.

Parshall, J. (2008). Barnett Shale showcases tight-gas development. *Journal of Petroleum Technology*, 60(9):43–55.

Perryman Group, The. (2007). *Bounty from Below: The Impact of Developing Natural Gas Resources Associated with the Barnett Shale on Business Activity in Fort Worth and the Surrounding 14-County Area*. The Perryman Group, Waco, TX.

Prokopovich, N.P. (1963). Hydrocompaction of soils along the San Luis Canal alignment, western Fresno County, California, in *Geological Society of America Abstracts for 1962*, GSA Special Paper 73. Geological Society of America, Boulder, CO.

Richter, B.C. and Kreitler, C.W. (1993). *Geochemical Techniques for Identifying Sources of Ground-Water Salinization*. CRC Press, Boca Raton, FL.

Railroad Commission of Texas. (2008). *Newark, East (Barnett Shale) Field*. Oil and Gas Division, Railroad Commission of Texas, Austin.

Satterfield, J., Mantell, M., Kathol, D., Hebert, F., Patterson, K., and Lee, R. (2008). Managing Water Resource's Challenges in Select Natural Gas Shale Plays, paper presented at the Ground Water Protection Council Annual Forum, Cincinnati, OH, September 21–24.

Spellman, F.R. (2015). *The Science of Water: Concepts and Applications*, 3rd ed. CRC Press, Boca Raton, FL.

Terzaghi, K. and Peck, R.B. (1948). *Soil Mechanics in Engineering Practice*. John Wiley & Sons, New York.

USDOI. (2008). *Arkansas: Reasonably Foreseeable Development Scenario for Fluid Minerals*. Eastern States Field Office, Bureau of Land Management, U.S. Department of the Interior, Jackson, MS.

USDOI and USDA. (2007). *Surface Operating Standards and Guidelines for Oil and Gas Exploration and Development: The Gold Book*, 4th ed., BLM/WO/ST-06/021+3071/REV07. Bureau of Land Management, U.S. Department of the Interior, Denver, CO.

USEPA (2001). *Class I Underground Injection Control Program: Study of the Risks Associated with Class I Underground Injection Wells.* U.S. Environmental Protection Agency, Washington, DC (https://www.epa.gov/uic/class-i-underground-injection-control-program-study-risks-associated-class-i-underground).

USEPA (2015a). *Class I Industrial and Municipal Waste Disposal Wells.* U.S. Environmental Protection Agency, Washington, DC (https://www.epa.gov/uic/class-i-industrial-and-municipal-waste-disposal-wells).

USEPA (2015b). *Class II Oil and Gas Related Injection Wells.* U.S. Environmental Protection Agency, Washington, DC (https://www.epa.gov/uic/class-ii-oil-and-gas-related-injection-wells).

USEPA (2015c). *General Information about Injection Wells.* U.S. Environmental Protection Agency, Washington, DC (https://www.epa.gov/uic/general-information-about-injection-wells).

USEPA (2015d). *Class III Injection Wells for Solution Mining.* U.S. Environmental Protection Agency, Washington, DC (https://www.epa.gov/uic/class-iii-injection-wells-solution-mining).

USEPA (2015e). *Class IV Shallow Hazardous and Radioactive Injection Wells.* U.S. Environmental Protection Agency, Washington, DC (https://www.epa.gov/uic/class-iv-shallow-hazardous-and-radioactive-injection-wells).

USEPA (2015f). *Basic Information About Class V Injection Wells.* U.S. Environmental Protection Agency, Washington, DC (https://www.epa.gov/uic/basic-information-about-class-v-injection-wells).

USEPA (2015g). *Class VI—Wells Used for Geologic Sequestration of CO_2.* U.S. Environmental Protection Agency, Washington, DC (https://www.epa.gov/uic/class-vi-wells-used-geologic-sequestration-co2).

Venesky, T. (2008). State-owned parcels eyed for gas deposits. *Times Leader,* March 4 (www.timesleader.com/stories/State-owned-parcels-eyed-for-gas-deposits,110359).

Warner, D.L. (1984). Foreword, in *Subsurface Injection of Liquid Waste with Emphasis on Injection Practices in Florida,* USGS Water-Supply Paper 2281. U.S. Geological Survey, Washington, DC.

11 Offsite Treatment/Disposal

The *rule of capture* generally permits a landowner to drain or "capture" oil and natural gas from a neighboring property without liability or recourse.

OFFSITE COMMERCIAL DISPOSAL

When onsite management is not practical, operators may send their produced wastewater offsite to a commercial disposal facility. Typically, produced wastewater is removed from remote well locations periodically and transported via truck to an offsite facility. Offsite commercial disposal is usually a necessity because of the remote location of the fracking operation. Many fracking operations are located and constructed in remote rural, wilderness regions. Often, these operational locations are constructed on hilltops, hill sides, woodland areas, or areas where access can be challenging. This is particularly an issue with regard to the fracking and local environment interface, such as occurs during the construction of access roads and well pads. Because fracking operation sites are often remote, the surface disturbance required to construct access roads and well pads has been and continues to be an issue. Also, when utilizing offsite disposal, the produced wastewater must be transported on roads that are often no more than bulldozed paths through a forest or otherwise over rough terrain. These makeshift dirt roads can in no way be compared to paved roads that are standard in cities and counties or connection roads to interstates and turnpikes. These makeshift roads are usually narrow and subject to serious degradation during adverse weather conditions. In addition, many of these rough but pristine areas are not normally subjected to air quality threats; however, heavy equipment and tanker truck traffic on the makeshift roads may make air quality in the area become an issue.

Other considerations associated with traditional oil and gas development include conflicts that arise from split estates. In some instances, mineral rights and surface rights are not owned by the same party, a situation that is referred to as a *split estate* or *severed mineral rights*. The condition of split estate is more prevalent in western states, where the federal government owns much of the mineral rights (All Consulting, 2004). In the Midwestern and eastern states, where shale gas development resources are more prevalent, only 4% of the lands are associated with a federal split estate (BLM, 2006). However, these same areas frequently have private–private split estate scenarios where the surface owner differs from the mineral estate owner. In these cases, the mineral owner may be another individual or a business enterprise such as a coal company.

No matter who owns the property and mineral rights, contaminated well pad areas can be an issue. An accidental fracking wastewater spill is not that uncommon, and spillage caused during truck loading is also common. Spilled produced wastewater can eventually make its way to local lakes and streams and even to groundwater. Spilled produced wastewater with harmful chemicals mixed in can kill off surrounding wildlife and livestock in nearby farms. Many property owners who are in a split estate situation, regardless of its nature, can become embroiled in conflicts—especially in areas where active mineral resource development is not commonplace. Landowners can be surprised to find that the mineral leaseholder is entitled to reasonable use of the land surface even though they do not own the surface. However, it is important to understand that surface owners who do not own mineral rights are still afforded certain protections. If the mineral owner does not own the surface where fracking operations will occur, a separate agreement may be negotiated (in some states it is required) with the landowner to ensure that they are compensated for the use of the land and to set requirements for reclaiming the land when operations are complete (Bonner and Willer, 2005).

Shale gas development from fracking operations within or near existing communities has created challenges for production companies. New technologies have generally allowed these challenges to be met successfully. In some cases, a combination of modern shale gas technologies and the innovative use of best management practices has been required to allow development to continue without compromising highly valued community resources. In one instance, Chesapeake Energy Corporation constructed a well pad to develop natural gas from the Barnett Shale play near a popular Fort Worth area known as Trinity Trails. Located on private land, Trinity Trails consists of a 35-mile network of paved and natural surface pathways. The drilling pad was constructed approximately 200 feet from one portion of the trail. During the initial planning stages, proposed use of this land for development of natural gas was met with significant opposition by the public. Maintaining healthy populations of upland hardwood forest habitat was important to the community because such woodlots are rare in urban settings. To address the concerns of the community, the company sponsored public meetings and opinion surveys, provided landscape plans, planted trees and shrubs, and enhanced the general area by improving irrigation and lowering maintenance requirements. The well pad was specifically designed to be as small as possible in order to reduce the footprint of the well. Special construction practices were used to help preserve many of the existing trees. The construction zone was isolated from view using a 16-foot barrier fence with sound baffling. This approach benefited both partners: The company was able to produce the shale gas, important community resources were protected, and at no point in the process was a portion of the trail closed (Lantz, 2008).

THOUGHT-PROVOKING QUESTIONS

1. Are the casing and cementing procedures used to seal the well bore and prevent contamination of groundwater adequate?
2. Was Vice President Dick Cheney's exclusion of hydraulic fracturing from the 2005 Safe Drinking Water Act a correct move?
3. The U.S. Environmental Protection Agency (USEPA) has several case studies ongoing to determine the environmental impact of hydraulic fracturing. What do you think they will find?
4. If the USEPA case studies determine that hydraulic fracturing is causing harm to groundwater sources, what actions should be taken?
5. Do you think protecting wildlife around or close to fracking sites is more important than mining the gas or oil?

REFERENCES AND RECOMMENDED READING

ALL Consulting and Montana Board of Oil and Gas Conservation (MBOGC). (2002). *Handbook on Best Management Practices and Mitigation Strategies for Coal Bed Methane in the Montana Portion of the Powder River Basin*, prepared for National Energy Technology Laboratory, National Petroleum Technology Office, U.S. Department of Energy, Tulsa, OK, 44 pp.

ALL Consulting and Montana Board of Oil and Gas Conservation (MBOGC). (2004). *Coal Bed Natural Gas Handbook: Resources for the Preparation and Review of Project Planning Elements and Environmental Documents*, prepared for National Petroleum Technology Office, U.S. Department of Energy, Washington, DC, 182 pp.

Bellabarba, M. et al. (2008). Ensuring zonal isolation beyond the life of the well. *Oilfield Review*, 20(1):18–31.

BLM. (2006). *Scientific Inventory of Onshore Federal Lands' Oil and Gas Resources and the Extent and Nature of Restrictions or Impediments to their Development*. Bureau of Land Management, U.S. Department of the Interior, Washington, DC.

Bonner, T. and Willer, L. (2005). Royalty management 101: basics for beginners and refresher course for everyone, in *Proceedings of the 25th Anniversary Convention of National Association of Royalty Owners (NARO)*, Oklahoma City, November 3–5.

Bromley, M. (1985). *Wildlife Management Implications of Petroleum Exploration and Development in Wildland Environments*, General Technical Report INT-199. U.S. Department of Agriculture, Intermountain Research Station, Ogden, UT.

Catskill Mountainkeeper. (2008). *The Marcellus Shale—America's Next Super Giant.* Catskill Mountainkeeper, Youngsville, NY (www.catskillmountainkeeper.org/our-programs/fracking/marcellus-shale/).

Chesapeake Energy Corporation. (2008). Drilling 101, paper presented to New York Department of Environmental Conservation, Albany.

Douglas, D. (2008). *Anger over Road Damage Caused by Barnett Shale Development.* WFAA-TV, Dallas–Fort Worth, TX (www.wfaa.com/news/local/64791902.html).

Franz, Jr., J.H. and Jochen, V. (2005). *When Your Gas Reservoir Is Unconventional, So Is Our Solution*, Shale Gas White Paper 05-OF-299. Schlumberger, Houston, TX.

Halliburton. (2008). *U.S. Shale Gas: An Unconventional Resource. Unconventional Challenges.* Halliburton, Houston, TX, 8 pp.

Harper, J. (2008). The Marcellus Shale—an old "new" gas reservoir in Pennsylvania. *Pennsylvania Geology*, 28(1):2–13.

Lantz, G. (2008). Drilling green along Trinity Trails. *The Barnett Shale Magazine*, Summer.

Marshall Miller & Associates. (2008). Marcellus Shale, paper presented to Fireside Pumpers, Bradford, PA.

Parshall, J. (2008). Barnett Shale showcases tight-gas development. *Journal of Petroleum Technology*, 60(9):48–55.

Perryman Group, The. (2007). *Bounty from Below: The Impact of Developing Natural Gas Resources Associated with the Barnett Shale on Business Activity in Fort Worth and the Surrounding 14-County Area.* The Perryman Group, Waco, TX.

Railroad Commission of Texas. (2008). *Newark, East (Barnett Shale) Field.* Oil and Gas Division, Railroad Commission of Texas, Austin.

Satterfield, J., Mantell, M., Kathol, D., Hebert, F., Patterson, K., and Lee, R. (2008). Managing Water Resource's Challenges in Select Natural Gas Shale Plays, paper presented at the Ground Water Protection Council Annual Forum, Cincinnati, OH, September 21–24.

USDOI. (2008). *Arkansas: Reasonably Foreseeable Development Scenario for Fluid Minerals.* Eastern States Field Office, Bureau of Land Management, U.S. Department of the Interior, Jackson, MS.

USDOI and USDA. (2007). *Surface Operating Standards and Guidelines for Oil and Gas Exploration and Development: The Gold Book*, 4th ed., BLM/WO/ST-06/021+3071/REV07. Bureau of Land Management, U.S. Department of the Interior, Denver, CO.

Venesky, T. (2008). State-owned parcels eyed for gas deposits. *Times Leader*, March 4 (www.timesleader.com/stories/State-owned-parcels-eyed-for-gas-deposits,110359).

Section V

*Safety and Health
Impacts of Working with
Produced Wastewater*

12 Safety and Health Considerations

Date of Incident:	02/07/12
Location of Incident:	Garden City, Texas
Description of Incident:	Worker performing pre-fracking operations was struck and killed when a pipe exploded under pressure.

OSHA (2012)

INTRODUCTION*

Over the years, in several of my books, articles, and classroom lectures, I have made the point, over and over again, that there is one word in our current vernacular that is responsible for a high level of consternation and trepidation among many industrial organizations (this is especially true for the oil and gas drilling industry—the so-called privileged upper 1%) while at the same time representing a blessing, a God-sent protective device against those bad-boy industrialists who dare to foul the air we breathe and the water we drink (shame on them!). That Earth-shaking word? *Regulation*, of course

Awhile back, Ferry (1990) voiced a familiar refrain: "We do not like rules and regulations. We are free, we have choices, and we intend to make them—we don't like others telling us what we can or cannot do." We can extend this idea further: "Yes, nobody should be able to tell us what to do. We have choices, and we should be able to make them." The problem is that we often give little thought to their repercussions when we make choices that satisfy our immediate wants and desires. Sound familiar? Most of us know we need rules and regulations to live with other people in our society, but the fact is we don't like rules—and we often don't abide by them. (Been on an interstate lately? The speed limit may be 65 mph, but you'll find that most people are going 70, 75, 85, or faster. The statistics say we are far less safe on the road at that speed, but rules are made to be broken—right?)

From management's point of view, there are good rules and bad rules. Management likes rules requiring workers to show up on time, to put in an honest day's work, to maintain good order and discipline in the workplace, to focus on company goals. Rules that are good for the company are obviously good rules—right?

So what are the bad rules? Typically, a business manager views any rule, law, regulation, or other requirement placed upon his or her company by an outside regulatory agency as a bad rule. Why? Primarily for the "headache-making" problems pointed out by Ferry earlier. For the purposes of illustration, let's consider the federal Occupational Safety and Health (OSH) Act and break these problems down one by one. After reading the following sections, you can make your own judgment as to what the real problem is.

A NETWORK OF CONFUSING AND CONSTRAINING RULES AND STANDARDS

Anyone who has attempted to read and then to comply with 29 CFR 1910, Occupational Safety and Health Standards for General Industry, will probably agree with Terry's statement. Much of the material contained within this Occupational Safety and Health Administration (OSHA) bible is indeed difficult to comprehend, a problem that is greatly exacerbated for those who have very

---

* Adapted from Spellman, F.R. and Whiting, N., *Safety Engineering: Principles and Practices*, 2nd ed. Government Institutes Press, Latham, MD, 2005.

limited safety experience. The problem is compounded by the ambiguity and vagary that contribute to the warp and woof of the twisted fabric whose tightness depends almost entirely on how the material is interpreted by the reader. More importantly, how might this material that has been interpreted and in use by a company be interpreted by OSHA auditors? Note that the OSHA auditor usually has the final word on interpretation. I know this from experience. I have been there, and I did win all of my cases in court. Always contest an OHSA citation, unless you are guilty of noncompliance; if guilty, look for another occupation.

COSTLY MODIFICATIONS OF EXISTING INSTALLATIONS TO MEET NEW LEGAL DEMANDS

This item not only is a headache-generator for any site manager but can also be very costly to the company in terms of both money and workers' time. Companies are in business, obviously, to make money—to hold or improve their bottom line. The last thing any manager who is fighting competition and other costly impediments to making his or her company profitable wants is to have "those briefcase-carrying so and so's coming into my site and telling *me* I have to have this and I have to have that. Not only do they waste my time, but they also make me spend money on things that cut profits—things that don't contribute to the bottom line." Have you heard anything similar before?

INSPECTIONS, FINES, OR TIME-CONSUMING LEGAL HEARINGS

A typical OSHA workplace inspection (commonly called an *audit*) and to a degree the citations that can lead to fines being assessed on the employer for noncompliance are, for the manager, a trying experience. How trying (Spellman, 1998)?

> You're in a business under OSHA's regulatory supervision. You've heard the industry horror stories about auditors called in on employee complaints, who ask if you mind if they have a "look around"—and later, thousands of dollars worth of fines and hours of fear, pain, and/or aggravation later, they leave again (hopefully not to return). But maybe you don't think that can happen to you. Believe me, it can. If you are at all casual, careless, or haphazard about required compliance, you're running an OSHA risk. Even facilities that make the strongest possible attempts to comply—that are, in fact, in compliance down to the dotting of the i's and crossing of the t's—will be cited by OSHA for an interpretation of the smallest detail on 3rd or 4th level instructions. Better that than having OSHA come down heavy—but a headache generator at the very least.

This rather pointed assessment of OSHA and its auditing process may seem silly or ridiculous to many. But have you been there? If not, then the reality is beyond imagination. If you have been there, then you may feel that this description is a rather mild one, an understatement.

The authority of OSHA is nothing to ignore. Formal regulatory inspections and fines that result from any noncompliance findings are not only costly but also a major contributor to every manager's headache—the one that begins at the base of your skull and makes even your eyebrows ache, the one that results from dealing with regulatory requirements. But, much more contributes to management's dilemma in dealing with regulatory requirements; for example, consider the legal ramifications of regulatory noncompliance. In addition to the civil or criminal penalties that might result from employer violations of or noncompliance with OSHA regulations, possible legal actions may result from noncompliance (not uncommon in this age of "let me sue you before you sue me"). Employees can sue an employer for making them work in an unsafe workplace, for making them perform unsafe work actions, or for injuries incurred while working on the job.

In addition to regulatory penalties, an employer may be exposed to workers' compensation liability for employee injuries. Another potential headache generator can be product liability. Most managers need not be told that the company can be held liable if a product it manufactures or sells causes personal injury or property damage to buyers or third parties; however, it does sometimes

Safety and Health Considerations

surprise managers (but usually not for long) that this liability is not lessened if the firm produces finished products from components manufactured by someone else. The simple fact is that if a component causes injury, the firm that assembled and sold the product can be held liable. This is, of course, a major concern and consideration for any company that produces a product to be sold to any consumer. One of an employer's worst nightmares (and sometimes its most significant headache generator) can be employee complaints—especially when these complaints are made to legal counsel and eventually in a court of law. Actually, whenever an employee decides (for whatever reason) that he or she is going to take legal action against an employer (whether the employer is in the right or wrong) in a court of law, the headache soon turns into the migraine variety.

In some instances, of course, employees should take legal action against their employers. Some employers have absolutely no regard for the safety, health, and wellbeing of their employees. In these cases, the court house door is wide open for litigants and their lawyers. How bad can this situation be in the real world? Just check your local newspaper articles and television news stories to get an answer to this question. Almost daily, an employee or group of employees sues their employer or employers for some infraction of safety and health regulations. To be fair to both sides, many of the suits brought by employees against their employer are frivolous. Some employees make false claims against their employers because they don't like the employer, a supervisor, company policies, or the results of last night's football game, or they feel they have been improperly disciplined or terminated—for whatever reason. Lawsuits generated by disgruntled employees are fairly common, with or without cause, but one cannot overlook those cases when employees have just cause to sue their employers.

ABOVE ALL, AN INCREASINGLY BURDENSOME TASK OF RECORDKEEPING AND PAPERWORK

Shortly after being hired, new safety officials discover that the training and recordkeeping function is vital, necessary, costly—and frequently overwhelming. It cannot be avoided. The new safety official also soon discovers that drawing the line between what is required by regulation and what is simply required for efficiency is difficult. Literally hundreds of records are required, for a variety of reasons and purposes. The importance of keeping and maintaining up-to-date and accurate records cannot be stressed enough. Safety officials soon discover that, although they may be tedious and time consuming to prepare, written records are their first, second, and sometimes third line of defense. Line of defense? Absolutely. Remember, safety officials hold a precarious position, constantly walking a very fine line. The seasoned safety professional will instantly understand this last statement. Accurate and complete recordkeeping is essential to the safety official's job, professional standing, and personal wellbeing.

Exactly what kind of records is the safety official responsible for? Good question. The safety official is primarily responsible for complying with recordkeeping under the OSH Act. Also, such recordkeeping is concerned with workplace safety, health, the workplace environment, and other administrative functions. Here is an important point that you should remember—one that you should learn to live by if you are going to become a safety official, if you are going to survive as a safety official (Ferry, 1990):

> Wisdom in record keeping is too often a matter of hindsight. As company safety engineer you must be able to anticipate what records will be needed by knowing what is required. The excuse of not recognizing a needed record-keeping function is unacceptable not only to OSHA but also to a court of law.

Simply put, whatever the safety official does and says as part of his or her job should be covered by a piece of paper. Notwithstanding the trend toward a paperless working environment, do not get caught in the trap of not having a piece of paper that is acceptable in a court of law, a stockholders' meeting, or for whatever other purpose an auditor, for example, can come up with. "In this CYA world, I want it on paper, please!"

SAFETY AND HEALTH REGULATIONS AND STANDARDS

Again, why do we need regulations? Actually, based on personal experience, a better question would also seek to answer when we need regulations and how they are enforced. In dealing with the implications of various regulations for many years, I have come to realize that whether a regulation is a good regulation or a bad regulation is nothing more than a judgment call—a personal judgment call. This is especially the case with regulations that affect the environment and are designed to reduce, limit, or do away with pollution or contamination of the three environmental media of air, water, and soil. Preventing pollution and contamination of Earth's environmental media is, of course, a focus of this book.

Again, on a personal note, in regard to shale gas hydraulic fracturing and its environmental consequences, I am reminded of one of those homemade signs in bold letters and carried by an Occupier (part of the 99%) that stated the following:

Fracking jobs are grave-digging jobs!

There is another sign that I like and have often posted here and there:

Safety is all about fear!

Why are fracking jobs grave-digging jobs? Isn't this statement over the top, scare talk to instill fear? Depends on how unsafe you are. Based on personal experience and published data, fracking is, indeed, dangerous. According to the Institute for Southern Studies, approximately 453,000 workers are employed by the U.S. oil and gas extraction industry; approximately 50% of these people work for well services companies that conduct hydraulic fracking. Occupational deaths in the oil and gas extraction industry from 2003 to 2009 were 27.5 per 100,000 workers. To date, the number of gas frackers killed in highway crashes stands at more than 300 and counting. Fracking workers are more than seven times more likely to die on the job than other types of workers. The Institute also pointed out that oil and gas field workers work an average 20-hour shift (Institute for Southern Studies, 2012).

Getting back to the fear mongering deliberately injected into this discussion, undoubtedly fear is an important component of acting safely; it enters the picture (or should) no matter what the job or activity. With regard to produced wastewater operations, amen to fear if the thought of receiving crippling injuries to one's neck, back, knees, or shoulders; of being paralyzed or dying as a result of an explosion, seismic activity, or a sinkhole opening up with little or no warning; of falling asleep while working long shifts and being struck by moving equipment or high-pressure lines; of working in extreme temperatures and confined spaces; or of getting a leg caught in a cable and being pulled 30 to 40 feet into the air and dropped on the hard ground makes a worker fearful. I know of few workers who do not fear losing a finger, hand, arm, foot, leg, or their eyesight from their work activities.

Safety is all about fear!

In addition to the fear of being injured or worse on the job, workers may also fear violating company safety rules because it could cause them to lose their jobs. Also, workers may fear injuring one or more of their co-workers if they are unsafe.

Safety is all about fear!

Fear is good. Working without being injured is good.

OSHA STANDARDS APPLICABLE TO FRACKING OPERATIONS

OCCUPATIONAL SAFETY AND HEALTH ACT OF 1970 (OSH ACT)

Although some federal safety legislation was passed prior to 1970, this legislation affected only a small fraction of the American workforce. At the end of the 1960s, two shortcomings became blatantly obvious: (1) a new national policy needed to be established that would encompass the majority of industries, and (2) states had generally failed to meet their voluntary obligations for health and safety in the workplace. To solve these shortcomings, the OSH Act (designed to "ensure so far as possible every working man and woman in the Nation safe and healthful working conditions and to preserve our human resources") was signed by President Nixon on December 29, 1970. Since its effective date on April 28, 1971, this single act has had enormous impact on the safety and health movement within the United States, more than any other legislation. The law affects approximately 60 million employees in over 4 million establishments but excludes employees of state and federal government, who are protected under regulations similar to those within the Occupational Safety and Health Administration (OSHA). Under the Secretary of Labor, the Assistant Secretary of Labor for Occupational Safety and Health has the responsibility to guide and administer OSHA. Under the provisions of the Act, each employer covered by the Act has the following duties:

1. The general duty to furnish each of his/her employees employment and places of employment that are free from recognized hazards that are causing or likely to cause death or serious physical harm (which means that even if a hazard in the workplace is not specifically covered by a regulation, the employer must protect the employee anyway). This is commonly referred to as the Act's *General Duty Clause*; safety professionals and in particular OSHA professionals view it as a "safety net."
2. The specific duty of complying with safety and health standards promulgated under the act.

Each employee has the duty to comply with the safety and health standards, as well as all rules, regulations, and orders applicable to his own actions and conduct on the job. Experience has shown that when employees are informed of this requirement under OSHA they are surprised. They often view the Act as applying only to the employer.

The OSHA regulations (compliance with these regulations being the major concern of this text) take two basic forms: They are either *specific standards* or *performance standards*. Specific standards explain exactly how to comply; for example, the OSHA regulation covering means of egress from buildings very specifically lists requirements for means of egress, exit access, exit discharge, and so forth. A performance standard lists the ultimate goal of compliance but does not explain exactly how to accomplish it. A good example of a performance standard is the General Duty Clause, which states that the employer must protect the health and safety of the employee even if no OSHA regulation currently covers the work activity in question. These standards do not explain how to accomplish this—that is left up to the employer.

In general, safety and health regulations governing labor practices are listed under Title 29 of the Code of Federal Regulations (CFR). The occupational safety and health regulations are found in Parts 1900 to 1999. The actual workplace regulations we are concerned with in this text are 29 CFR Part 1910 (General Industry Standards) and Part 1926 (Construction Standards). To gain an understanding of what is contained in one of these parts, let's take a look at Part 1910, which is divided into subparts A to Z, as shown in Table 12.1. OSHA determines how well a program is working by reviewing the company's injury data and insurance costs. If the program is effective, the injury and insurance costs will reflect that.

TABLE 12.1
Occupational Safety and Health Administration Standards (29 CFR 1910 Subparts A–Z)

Subpart	Description
Subpart A. *General*	Provides provisions for OSHA's initial implementation of regulations
Subpart B. *Adoption and Extension of Established Federal Standards*	Explains which businesses are covered by OSHA regulations
Subpart C. *General Safety and Health Provisions*	Provides the right for an employee to gain access to exposure and medical records
Subpart D. *Walking and Working Surfaces*	Establishes requirements for fixed and portable ladders, scaffolding, manually propelled ladder stands, and general walking surfaces
Subpart E. *Means of Egress*	Establishes general requirements for employee emergency plans and fire prevention plans
Subpart F. *Powered Platforms, Man Lifts, and Vehicle-Mounted Work Platforms*	Mandates the minimum requirements for an elevated safe work platform
Subpart G. *Occupational Health and Environmental Control*	Mandates engineering controls of physical hazards such as ventilation for dusts, control of noise, and control of ionizing and non-ionizing radiation
Subpart H. *Hazardous Materials*	Provides requirements for the use, handling, and storage of hazardous materials
Subpart I. *Personal Protective Equipment*	Provides general requirements for personal protective equipment
Subpart J. *General Environmental Controls*	Mandates the requirements for sanitation, accident prevention signs and tags, confined space entry, and hazardous energy lockout/tagout
Subpart K. *Medical and First Aid*	Requires that an employer provide first-aid facilities or personnel trained in first aid to be at the facility
Subpart L. *Fire Protection*	Mandates portable or fixed fire suppression systems for work places
Subpart M. *Compressed Gas and Compressed Air Equipment*	Provides requirements for air receivers
Subpart N. *Materials Handling and Storage*	Covers the use of mechanical lifting devices, changing a flat tire, and forklift and helicopter operations
Subpart O. *Machinery and Machine Guarding*	Provides requirements for guarding rotating machinery
Subpart P. *Hand and Portable Powered Tools and Other Hand-Held Equipment*	Provides requirements for hand-held equipment
Subpart Q. *Welding, Cutting, and Brazing*	Requires the use of eye protection, face shields with lenses, proper handling of oxygen and acetylene tanks
Subpart R. *Special Industries*	Provides special requirements for textiles, bakery equipment, laundry machinery, sawmills, pulpwood logging, grain handling, and telecommunications
Subpart S. *Electrical*	Requires the use of protection mechanisms for electrical installations
Subpart T. *Commercial Diving Operation*	Mandates requirements for dive teams
Subparts U–Y. Not currently assigned	—
Subpart Z. *Toxic and Hazardous Substances*	Requires monitoring and protective methods for controlling hazardous airborne contaminants

Obviously, to get accurate data upon which to base its judgment, OSHA requires extensive recordkeeping for written programs, injuries, illnesses, safety audits, inspections, corrections, and training. Training is a major part of the OSH Act. Almost every regulation requires some sort of transmission of information and training. Why? Simply because injury statistics show that newer employees without adequate training are far more likely to be injured on the job than those with more experience and training.

Enforcement of the OSH Act is carried out through inspections (audits), citations, and levying civil penalties. These three increasingly punitive steps are designed to achieve a safe workplace by requiring the removal of hazards. If hazardous situations are discovered, follow-up inspections

Safety and Health Considerations

217

ensure that the appropriate corrections are made. OSHA investigates and writes citations based on inspections of the work site. An OSHA inspector may visit a site based on the following:

- An employee complaint (don't you just love them?)
- A report that an injury or fatality has occurred
- A random visit to a high-risk business

If the inspection uncovers one or more violations, the OSHA compliance officer provides an explanation on a written inspection report. The types of violations include the following:

- *de minimis*—A condition that has no direct or immediate relationships to job safety and health (e.g., an error in interpretation of a regulation)
- *General*—Inadequate or nonexistent written programs, lack of training, training records, etc.
- *Repeated*—Violations where, upon reinspection, another violation is found of a previously cited section of a standard, rule, order, or condition violating the general duty clause
- *Serious*—A violation that could cause serious harm or permanent injury to the employee and where the employer did not know or could not have known of the violation
- *Willful*—A violation where evidence shows that the employer knew that a hazardous condition existed that violated an OSHA regulation but made no reasonable effort to eliminate it
- *Imminent danger*—A condition where there is reasonable certainty that an existent hazard can be expected to cause death or serious physical harm immediately or before the hazard can be eliminated through regular procedures

When a compliance officer believes an employer has violated a safety or health requirement of the act, or any standard, rule, or order promulgated under it, he or she will issue a citation. Any citation issued for noncompliance must be posted in clear view near the place where the violation occurred for 3 working days or until corrected, whichever is longer. Does the employer have any recourse when cited by OSHA? Actually, the employer can take either of the following courses of action regarding citations:

1. The employer can agree with the citation and correct the problem by the date given on the citation and pay any fines.
2. The employer can contest the citation, proposed penalty, or correction date, as long as it is done within 15 days of the date the citation was issued.

Specific standards have been developed by OSHA to reduce potential safety and health hazards in the oil and gas drilling, servicing, and storage industry. States also have requirements that provide further work and public safety protections. Before discussing each of the pertinent OSHA standards related to safe shale gas fracking operations, it is important to discuss and describe actual process practices, industry profile data, nature of the work, types of recorded on-the-job injuries specific to the industry, and the number and type of citations issued by OHSA during scheduled or unscheduled audits.

INDUSTRY GROUP 138—OIL AND GAS FIELD SERVICES: PROCESS DESCRIPTION*

Oil and gas well drilling and servicing are part of Major Group 13 in the Standard Industrial Classification (SIC). The classification is further defined by three subdivisions within Industry Group 138 (Oil and Gas Field Services): SIC 1381 (Drilling and Gas Wells), SIC 1382 (Oil and

---

* Adapted from USDOL, *Oil and Gas Well Drilling, Servicing and Storage Standards*, U.S. Department of Labor, Washington, DC, 2012 (http://www.osha.gov/SLTC/oilgaswelldrilling/index.html).

Gas Field Exploration Services), and SIC 1389 (Oil and Gas Services, Not Elsewhere Classified). SIC 1381 includes establishments that are primarily engaged in drilling wells for oil or gas field operations for others on a contract or fee basis. This industry includes contractors that specialize in the following:

- Directional drilling of oil and gas wells on a contract basis
- Redrilling oil and gas wells on a contract basis
- Reworking oil and gas wells on a contract basis
- Spudding (starting to drill) oil and gas wells on a contract basis
- Well drilling of gas, oil, and water intake on a contract basis

SIC 1382 includes companies primarily engaged in performing oil and gas geophysical, geological, or other exploration services on a contract or fee basis:

- Aerial geophysical exploration, oil and gas field on a contract basis
- Exploration, oil and gas field on a contract basis
- Geological exploration, oil and gas field on a contract basis
- Geophysical exploration, oil and gas field on a contract basis
- Seismograph surveys, oil and gas field on a contract basis

SIC 1389 includes establishments primarily engaged in performing oil and gas field services, not elsewhere classified, for others on a contract or fee basis. Services included are excavating slush pits and cellars; grading and building foundations at well locations; well surveying; running, cutting, and pulling casings, tubes, and rods; cementing wells; shooting wells; perforating well casings; acidizing and chemically treating wells; and cleaning out, bailing, and swabbing wells. Establishments that have complete responsibility for operating oil and gas wells for others on a contract or fee basis are classified according to the product extracted rather than as oil and gas field services:

- Acidizing wells on a contract basis
- Bailing wells (removing water, sand, mud, drilling cuttings, or oil from cable-tool drilling) on a contract basis
- Building oil and gas well foundations on a contract basis
- Cementing oil and gas well casings on a contract basis
- Chemically treating wells on a contract basis
- Cleaning lease tanks, oil and gas field on a contract basis
- Cleaning wells on a contract basis
- Derrick building, repairing, and dismantling oil and gas on a contract basis
- Dismantling of oil well rigs (oil field service) on a contract basis
- Erecting lease tanks, oil and gas field on a contract basis
- Excavating slush pits and cellars on a contract basis
- Fishing for tools, oil and gas field on a contract basis
- Gas compressing natural gas at the field on a contract basis
- Gas well rig building, repairing, and dismantling on a contract basis
- Grading oil and gas well foundations on a contract basis
- Hard banding service on a contract basis
- Hot oil treating of oil field tanks on a contract basis
- Hot shot service on a contract basis
- Hydraulic fracturing wells on a contract basis
- Impounding and storing saltwater in connection with petroleum

Safety and Health Considerations

- Lease tanks, oil and gas field: erecting, cleaning, and repairing on a contract basis
- Logging wells on a contract basis
- Mud service, oil field drilling on a contract basis
- Oil sampling service for oil companies on a contract basis
- Oil well logging on a contract basis
- Perforating well casings on a contract basis
- Pipe testing service, oil and gas field on a contract basis
- Plugging and abandoning wells on a contract basis
- Pumping of oil and gas wells on a contract basis
- Removal of condensate gasoline from field gathering lines on a contract basis
- Roustabout service on a contract basis
- Running, cutting, and pulling casings, tubes, and rods on a contract basis
- Servicing oil and gas wells on a contract basis
- Shooting wells on a contract basis
- Shot-hole drilling service, oil and gas field on a contract basis
- Surveying wells on a contract basis, except seismographic
- Swabbing wells on a contract basis

Table 12.2 lists the top ten OSHA citations issued during Fiscal Year 2005 for Industry Group 138, and Table 12.3 lists some of the potential hazards and their sources related to Industry Group 138.

OSHA STANDARDS MOST OFTEN CITED FOR LACK OF COMPLIANCE OR WILLFUL VIOLATION

With regard to ensuring employee safety and health, all of OSHA's subparts listed in Table 12.2 could be employed by employers not only to protect workers but also to ensure OSHA compliance. In this section, we specifically list and describe those standards that are most often cited in Industry Group 138 for lack of compliance or willful violation of Occupational Safety and Health Standards (29 CFR 1910) found during OSHA audits and determined to be causal factors during post-accident/fatality investigations.

TABLE 12.2

Top 10 OSHA Violations Cited (FY 2005)

OSHA Standard	Number of Citations	Description
29 CFR 1901.1200	62	Hazard Communication
29 CFR 1910.146	54	Permit-Required Confined Spaces
OSH Act, Section 5(a)(1)	52	General Duty Clause
29 CFR 1910.132	42	Personal Protective Equipment, General
29 CFR 1910.305	42	Wiring Methods, Components, and Equipment
29 CFR 1910.23	40	Guarding Floor and Well Openings and Holes
29 CFR 1910.134	39	Respiratory Protection
29 CFR 1910.151	37	Medical Services and First Aid
29 CFR 1910.141	33	Sanitation
29 CFR 1910.157	30	Portable Fire Extinguishers

Source: OSHA, *Profile: Oil and Gas Well Drilling and Servicing*, U.S. Occupational Safety and Health Administration, Washington, DC, 2006.

TABLE 12.3
Some Potential Hazards and Their Sources

Hazard	Source
Struck by	Falling/moving pipe; tongs and/or spinning chain, Kelly, rotary table, etc.; high-pressure hose connection failure causing employees to be struck by whipping hose; tools or debris dropped from elevated location in rig; vehicles
Caught in/between	Collars and tongs, spinning chain, and pipe; clothing getting caught in rotary table or oil string
Fire/explosion/high-pressure release	Well blowout, drilling/tripping out/swabbing, etc., resulting in release of gas which might be ignited if not controlled at the surface; welding or cutting near combustible materials; uncontrolled ignition sources near the well head (e.g., heater in the doghouse); unapproved or poorly maintained electrical equipment; aboveground detonation of perforating gun
Rig collapse	Overloading beyond the rated capacity of the rig; improper anchoring or guying; improper raising and lowering of the rig; existing maintenance issues with the rig structure that impact its integrity
Falls	Falls from elevated areas of the rig (e.g., stabbing board, monkey board, ladder); falls from rig floor to grade
Hydrogen sulfide (H_2S) exposure	H_2S release during drilling, swabbing, perforating operations, etc., resulting in employee exposures; in production gauging operations, gauges sometimes exposed to H_2S

SUBPART D, WALKING–WORKING SURFACES

1910.22   General Requirements
1910.23   Guarding Floor and Wall Openings and Holes
1910.24   Fixed Industrial Stairs
1910.27   Fixed Ladders

Note that noncompliance with OSHA's requirements for guarding hole and well openings is listed as one of the top ten violations in Table 12.2.

General requirements

Housekeeping
1910.22(a)(1)—Housekeeping in all places of employment, passageways, storerooms, and service rooms shall be kept clean and orderly in a sanitary condition. Good housekeeping includes cleaning up grindings, shavings, and general debris from work areas on a daily basis.

1910.22(a)(2)—Housekeeping includes maintaining the floor of every work space in a clean, dry condition. Where wet processes are used, dry standing places should be provided where practicable.

Aisles and passageways
1910.22(b)(1)—In aisles and passageways, where mechanical handling equipment is used safe clearance shall be provided in aisles, at loading docks, through doorways, and wherever turns or passage must be made.

Covers and guardrails
1910.22(d)(1)—Floor loading protection requires that every structure have floors or mezzanines approved for load bearing when using for storage. These areas shall be marked on plans of approved design.

Safety and Health Considerations

Guarding floor and wall openings and holes

Protection for floor openings

1910.23(a)(9)—Floor holes into which a person can accidentally walk shall be protected by a cover that leaves no opening more than 1 inch wide. The cover must be securely held in place.

Protection of open-sided floors, platforms, and runways

1910.23(c)(1)—Open-sided floors and platforms 4 feet or more above an adjacent floor or ground level shall be guarded by a standard railing. The railing shall be provided with a toeboard where there is a fall hazard.

1910.23(c)(2)—Every runway shall be guarded by a standard railing 4 feet or more above floor or ground level. Wherever tools, machine parts, or materials are likely to be used on the runway, a toeboard shall be provided on each exposed side.

Stairway railings and guards

1910.23(d)(1)—Every flight of stairs having four or more risers shall be equipped with standard railings and handrails.

Fixed industrial stairs

Stair treads

1910.24(f)—Treads on all stairs shall be reasonably slip resistant.

Fixed ladders

Rungs and cleats

1910.27(b)(1)(ii)—The distance between rungs, cleats, and steps shall not exceed 12 inches and shall be uniform throughout the length of the ladder.

1910.27(b)(1)(iii)—The minimum clear length of rungs shall be 16 inches from left to right.

Protection from deterioration

1910.27(b)(6)—Metal ladders shall be painted or treated to resist corrosion and rusting, particularly ladders formed by individual metal rungs embedded in concrete. Rungs shall have a minimum diameter of 1 inch.

Clearance

1910.27(c)(2)—A clear width of at least 15 inches shall be provided on either side from the centerline of the ladder in the climbing space, except when cages or wells are necessary.

SUBPART E, MEANS OF EGRESS

1910	Subpart E Appendix—Exit Routes, Emergency Action Plans, and Fire Prevention Plans
1910.36	Design and Construction Requirements for Exit Routes
1910.37	Maintenance, Safeguards, and Operational Features for Exit Routes

---

DID YOU KNOW?

Choosing the right ladder for the job is an important part of working safely. Whether for maintenance or for operational reasons, ladders are devices that are used extensively in most industrial settings. In order to be used safely, ladders must be sturdy and in good repair. In addition to noting inoperable eye washes and showers in treatment plant safety audits, the audit will usually detect several ladders that are unsafe for use. Ladders are generally ignored until they are needed. A worker who needs a ladder generally grabs the first one available and uses it. Unfortunately, there is no guarantee that the ladder chosen is safe. If the plant or industrial site does not have an effective ladder inspection program, it is likely that unsafe ladders will be available for workers to use.

Emergency Response

Even though no OSHA standards dedicated specifically to the issue of planning for emergencies exist at present, all OSHA standards are written for the purpose of promoting a safe, healthy, accident-free, and hence emergency-free workplace. For this reason, OSHA standards play a significant role in emergency prevention. A first step when developing emergency response plans is to review these OSHA standards. This can help organizations identify and then correct conditions that might exacerbate emergency situations before they occur.

Typically, when we think of emergency response plans for the workplace, we often conjure up thoughts about the obvious. For example, the first workplace emergency that might come to mind is *fire*—a major concern because fire in the workplace is something that can happen, that happens more often than we might want to think, and because fire can be particularly devastating—in ways we know all too well. Most employees do not need to be informed about the dangers of fire, but employers still have the responsibility to do just that—to provide training for employees on fire, fire prevention, and fire protection. Many local codes go beyond this information requirement, insisting that employers develop and implement fire emergency response and evacuation plans. The primary emphasis has been on the latter, but developing an emergency response plan is critical. Employers that equip their workplaces with fire extinguishers and other firefighting equipment and expect their employees to respond aggressively to extinguish workplace fires must have emergency response plans in place. Also, the employer must ensure that all company personnel called upon to fight a fire are completely trained on how to do so safely (29 CFR 1910.156(c); 29 CFR 1910.157(g)).

Medical emergencies are another commonly considered workplace emergency that must be addressed in emergency response plans. Many facilities satisfy this requirement simply by directing employees to call 911 or some other emergency number whenever a medical emergency occurs in the workplace. Other facilities, though, may require employees to provide emergency first aid. When the employer chooses the employee-supplied first aid option, certain requirements must be met before any employee can legally administer first aid. The first aid responder must be trained and certified to administer first aid. This training must also include training on OSHA's bloodborne pathogen standard, which requires that employees be trained on the dangers inherent in handling and being exposed to human body fluids. Employees must be trained on how to protect themselves from contamination. If the first aid responder or anyone else is exposed to and contaminated by body fluids, the employer must make available the hepatitis B vaccine and vaccination series to all employees who have occupational exposure, as well as post-exposure evaluation and follow-up to all employees who have had an exposure incident (29 CFR 1910.1030).

Under 29 CFR 1910.120 (Hazardous Waste Operations and Emergency Response, or HAZWOPER), another type of emergency that must be covered by an emergency response plan is the release of hazardous materials. Unless the facility operator can demonstrate that the operation does not involve employee exposure or the reasonable possibility for employee exposure to safety or health hazards, the following operations are covered:

1. Cleanup operations required by a governmental body involving hazardous substances conducted at uncontrolled hazardous waste sites, state priority site lists, sites recommended by the USEPA, National Priorities List, and initial investigations of government identified sites that are conducted before the presence or absence of hazardous substance has been ascertained
2. Corrective actions involving cleanup operations at sites covered by the Resource Conservation and Recovery Act of 1976 (RCRA)
3. Voluntary cleanup operations at sites recognized by federal, state, local, or other governmental bodies as uncontrolled hazardous waste sites
4. Operations involving hazardous waste conducted at treatment, storage, and disposal (TSD) facilities regulated by the RCRA
5. Emergency response operations for releases of, or substantial threats of releases of, hazardous substances without regard to the location of the hazard

Safety and Health Considerations

The final requirement impacts the largest number of facilities that meet the criteria requiring full compliance with 29 CFR 1910.120 (HAZWOPER), because many such facilities do not normally handle, store, treat, or dispose of hazardous waste but do use or produce hazardous materials in their processes. Because the use of hazardous materials could lead to an emergency from the release or spill of such materials, facilities using these materials must develop and employ an effective site emergency response plan.

Before discussing the basic goals of an effective emergency response plan, we should define *emergency response*. Considering that individual facilities are different, with different dangers and different needs, defining emergency response is not always easy. For our purposes, however, we use the following definition: "Emergency response is defined as a limited response to abnormal conditions expected to result in unacceptable risk requiring rapid corrective action to prevent harm to personnel, property, or system function" (CoVan, 1995). Another important point about emergency response, one critical for the safety engineer, is that "although emergency response and engineering tend toward prevention, emergency response is a skill area that safety engineers must be familiar with because of regulations and good engineering practice" (CoVan, 1995).

Now that we have defined emergency response, let's move on to the basic goals of an effective emergency response plan. Much of the currently available literature on this topic generally lists the goals as twofold:

1. Minimize injury to facility personnel.
2. Minimize damage to the facility and return to normal operation as soon as possible.

Obviously, these goals make a great deal of good sense, but you may be wondering about the language used, particularly "facility personnel" and "damage to the facility." Remember that we are talking about OSHA requirements here. Under OSHA, the primary emphasis is on protecting the worker; protecting the worker's health and safety is OSHA's only focus. What about people who live offsite—the site's neighbors? What about the environment?

Such questions emphasize the fact that OSHA is not normally concerned with the environment, unless contamination of the environment (at the worksite) might adversely impact worker safety and health. What about the neighbors? Again, OSHA's focus is on the worker. One OSHA compliance office explained that, if employers take every step necessary to protect their employees from harm resulting from the use or production of hazardous materials, then the surrounding community should have little to fear.

This statement was puzzling, so the same OSHA compliance officer was queried about incidents beyond the control of the employer—accidents that could put employees in harm's way and endanger the surrounding community. The answer? "Well, that's the EPA's bag—we only worry about the worksite and the worker."

Fortunately, OSHA, in combination with the USEPA, has taken steps to overcome this blatant shortcoming (we like to think of it as an oversight). Under OSHA's Process Safety Management (PSM) and USEPA's Risk Management Planning (RMP) directive, chemical spills and other chemical accidents that could impact both the environment and neighbors have now been properly addressed. What PSM and RMP really accomplish is changing the typical twofold goal of an effective emergency response plan to a threefold goal.

Let us point out that accomplishment of these two- or threefold goals or objectives is essential in any emergency response. Accomplishing these goals or objectives requires an extensive planning effort prior to the emergency (*prior* being the keyword here, because the attempt to develop an emergency response plan when a disaster is occurring or after one has occurred is both futile and stupid). The safety official must never forget that hazards in any facility can be reduced, but risk is an element of everyday existence and therefore cannot be totally eliminated. The safety engineer's goal must be to keep risk to an absolute minimum. To accomplish this advance planning is critical—and essential. We pointed out earlier that most plans address fire, medical emergencies, and the accidental release or spills of hazardous materials; however, the development of emergency response plans should also factor in other possible emergencies, such as natural disasters, floods, or explosions. Site emergency response plans should include the following:

- Assessment of risk
- Chain of command for dealing with emergencies
- Assessment of resources
- Training
- Incident command procedures
- Site security
- Public relations

The Federal Emergency Management Agency (FEMA), the U.S. Army Corps of Engineers, and several other agencies, as well as numerous publications, provide guidance on how to develop a site emergency response plan. Local agencies, such as fire departments, emergency planning commissions and agencies, HazMat teams, and Local Emergency Planning Committees (LEPCs), also provide information on how to design a site plan. All of these agencies typically recommend that a site's plan contain the elements listed in Table 12.4.

TABLE 12.4
Site Emergency Response Plan

Component	Description
Emergency Response Notification	List of who to call and information to pass on when an emergency occurs
Record of Changes	Table of changes and dates for them
Table of Contents/Introduction	Purpose, objective, scope, applications, policies, and assumptions for the plan
Emergency Response Operations	Details regarding what actions must take place
Emergency Assistance Telephone Numbers	Current list of people and agencies that may be needed in an emergency
Legal Authority and Responsibility	Laws and regulations that provide authority for the plan
Chain of Command	Response organization structure and responsibilities
Disaster Assistance and Coordination	Where additional assistance may be obtained when the regular response organizations are over-burdened
Procedures for Changing or Updating the Plan	Who makes changes and how they are made and implemented
Plan Distribution	List of organizations and individuals who have been given a copy of the plan
Spill Cleanup Techniques	Detailed information about how response teams should handle cleanups
Cleanup/Disposal Resources	List of what is available, where it is obtained, and how much is available
Consultant Resources	List of special facilities and personnel who may be valuable in a response
Technical Library/References	List of libraries and other information sources that may be valuable for those preparing, updating, or implementing the plan
Hazard Analysis	Details regarding the kinds of emergencies that may be encountered, where they are likely to occur, what areas of the community may be affected, and the probability of occurrence
Documentation of Spill Events	Various incident and investigative reports on spills that have occurred
Hazardous Materials Information	Listing of hazardous materials, their properties, response data, and related information
Dry Runs	Training exercises for testing the adequacy of the plan, training personnel, and introducing changes

Sources: Brauer, R.L., *Safety and Health for Engineers*, Van Nostrand Reinhold, New York, 1994; FEMA, *Planning Guide and Checklist for Hazardous Materials Contingency Plans*, FEMA-10, Federal Emergency Management Agency, Washington, DC, 1981.

Safety and Health Considerations 225

In the safety official's effort to incorporate and manage a facility emergency response plan, and in the response itself, two elements—security and public relations—must be given special attention. If not handled correctly, a lack of effective security measures and improper public relations can turn an already disastrous incident into a mega-disaster. Provisions should be made to have a well-trained security team limit site access to only the people and equipment that can assist in coping with and resolving the emergency. Public relations can be a tricky enterprise. The person identified to interface with the media must have thorough knowledge of the site, process, and personnel involved. The public relations representative must also have access to the highest levels of site management; otherwise, that person will not be able to deal with the public and media in an effective manner.

Egress Requirements

Design and construction requirements for exit routes

An exit door must be unlocked

1910.36(d)(1)—Employees must be able to open an exit route door from the inside at all times without keys, tools, or special knowledge, even in the dark. A device such as a panic bar that locks only from the outside is permitted on exit discharge doors.

A slide-hinged exit door must be used

1910.36(e)(2)—The door that connects any room to an exit route must swing out in the direction of exit travel if the room is designed to be occupied by more than 50 people or if the room is a high hazard area.

An exit route must meet minimum height and width requirements

1910.36(g)(2)—An exit access must be at least 28 inches (71.1 cm) wide at all points. Where there is only one exit access leading to an exit or exit discharge, the width of the exit and exit discharge must be at least equal to the width of the exit access.

1910.36(g)(4)—Objects that project into the exit route must not reduce the width of the exit route to less than the minimum width requirements for exit routes.

An outdoor exit route is permitted

1910.36(h)(1)—The outdoor exit route must have guardrails to protect unenclosed sides if a fall hazard exists (three or more rise treads).

Maintenance, safeguards, and operational features for exit routes

The dangers to employees must be minimized

1910.37(a)(3)—Exit routes must be free and unobstructed. No materials or equipment may be placed, either permanently or temporarily, within the exit route.

1910.37(a)(4)—Safeguards designed to protect employees during an emergency must be in proper working order at all times (e.g., emergency lighting, alarm systems, sprinkler systems, fire doors, exhaust systems).

Lighting and marking must be adequate and appropriate

1910.37(b)(1)—Each exit route must be adequately lighted so that a person with normal vision can see along the exit route (including exterior lights to a safe location).

1910.37(b)(2)—Each exit must be clearly visible and marked by a sign reading "EXIT" (except a main entrance/exit door that is readily obvious).

1910.37(b)(4)—If the direction of travel to the exit or exit discharge is not immediately apparent, signs must be posted along the exit access indicating the direction of travel to the nearest exit and exit discharge. Additionally, the line-of-sight to an exit sign must clearly be visible at all times. (Sample citation might read, "The direction of travel to the exit or exit discharge was not immediately apparent at the south end.")

1910.37(b)(5)—Each doorway or passage along an exit route access that could be mistaken for an exit must be marked "Not an Exit" or similar designation, or be designated by a sign indicating its actual use (e.g., "Closet").

1910.37(b)(6)—If emergency lighting is available in the building, then exit signs must be illuminated by emergency lighting or internally illuminated.

SUBPART F, POWERED PLATFORMS, MANLIFTS, AND VEHICLE-MOUNTED WORK PLATFORMS

1910.66   Powered Platforms for Building Maintenance
1910.66   Appendix C, Personal Fall Arrest/Protection

What Is Fall Protection?

Fall protection is the series of steps taken to cause reasonable elimination or control of the injurious effects of an unintentional fall while accessing or working (Ellis, 1988, p. xvi):

> Fall hazard distance begins and is measured from the level of a workstation on which a worker must initially step and where a fall hazard exists. It ends with the greatest distance of possible continuous fall, including steps, openings, projections, roofs, and direction of the fall (interior or exterior). Protection is required to keep workers from striking objects and to avoid pendulum swings, crushing and impact with any part of the body to which injury could occur. The object of elevated fall protection is to convert the hazard to a slip or minor fall at the very worst—a fall from which hopefully no injury occurs.

Because injuries received from falls in the workplace are such a common occurrence—in a typical year more than 10,000 workers will lose their lives in falls—safety officials need to be aware of not only fall hazards but also the need to institute a fall protection safety program (Kohr, 1989). Just how frequent and serious are accidents related to falls? Let's look at a few telling facts about falls in the workplace. The National Safety Council's annual report typically predicts 1400 or more deaths and more than 400,000 disabling injuries to occur each year due to falls. Falls are the leading cause of disabling injuries in the United States, accounting for close to 18% of all workers' compensation claims. A Bureau of Labor Statistics 24-state survey reported that 60% of elevated falls were under 10 feet, and 50% of those were under 5 feet (Pater, 1985). The primary causes of falls have been identified as the following (Kohr, 1989):

1. A foreign object on the walking surface
2. A design flaw in the walking surface
3. Slippery surfaces
4. An individual's impaired physical condition

According to OSHA (2017), 4836 worker fatalities were reported in private industry in 2015. Of these, 937 occurred in construction. Falls accounted for 364 of these fatalities, or 38.8%. The construction fall protection standard (29 CFR 1926.501) was among the 10 most frequently cited standards by OSHA in 2016. The other leading causes of death in the construction industry are being struck by an object, electrocution, and being caught in or between objects. These "fatal four" were responsible for more than half (64.2%) of the construction worker deaths in 2015. The majority of falls result in at least lost workdays, if not death. It is interesting to note that worker deaths in the United States are down, on average, from about 38 worker deaths a day in 1970 to 13 a day in 2015. Fatal injuries in the private oil and gas extraction industries were 38% lower in 2015 than 2014 (USDOL, 2016).

When attempting to initiate a fall protection safety program at any organization, safety officials must first define the needs of the organization. The actual needs of any type of fall protection program are going to be driven mainly by the type of work the organization does. Obviously, if the company is involved in construction, the needs are rather straightforward, because much of the

Safety and Health Considerations   227

work conducted will include the necessity of doing elevated work; however, this might also be the case for various trades as well, such as carpentry. Public utility and transportation work might also require elevated work.

To define the problem associated with all types of falls, let's examine what falls are all about. None of us has a problem understanding what a fall from a high-rise construction project involves—it is simply a fall from elevation. In many workplaces, though, worker injuries result from types of falls other than those from elevations. Falls in the workplace also include slips, trips, and stair falls, as well as elevated falls. *Slips* and *trips* are falls on the same level. *Stair falls* are falls on one or more levels. *Elevated falls* are from one level to another. In the following sections, each of these types of falls is discussed in greater detail, but first we discuss the physical factors at work in causing a fall. Remember that safety officials must address and work to reduce or eliminate *all* types of falls.

Physical Factors at Work in a Fall

We've all heard someone say, "The bigger they are, the harder they fall," or "It's not the fall that's so bad; it's the sudden stop at the end." Many would-be practitioners in the safety field are often surprised, however, to find out that science plays a role in falls and that slips, trips, and falls actually involve three well-known laws of science:

- *Friction* is the resistance between things, such as between work shoes and a walking surface. Without friction, workers are likely to slip and fall. Probably the best example of this phenomenon is slipping on ice. On icy surfaces, shoes can't grip the surface normally, causing a loss of traction and a fall.
- *Momentum* is the product of the mass of a body and its linear velocity. Simply put, momentum is affected by the speed and size of the moving object. Momentum is best understood if we translate the saying above to: "The more you weigh and the faster you move, the harder you'll fall if you slip or trip."
- *Gravity* is the force of attraction between any object in the Earth's gravitational field and the Earth itself. Simply put, gravity is the force that pulls you to the ground once a fall is in progress. If someone loses balance and begins to fall, that person is going to hit the ground. The human body is equipped with mechanisms that work to prevent falls, including the eyes, ears, and muscles, all of which work to keep the human body close to its natural center of balance. When this center of balance shifts too far, a fall will occur if balance is not restored to normal. Because gravity obviously has the same effect on all of us here on Earth, it is always surprising to discover how such a well-known basic law of science is so often and conveniently ignored by various industries. It is not unusual to encounter company owners or workplace foremen who ignore the laws of gravity and require their workers to perform daring (and extremely dangerous) feats in the workplace. Workers (who need the job and the security it provides) are led to believe that somehow gravity is something that is not important to them. Obviously, this is a dangerous mindset and practice that company safety officials must not tolerate.

Slips

In its simplest form, a slip is a loss of balance caused by too little friction between the feet and the surface being walked or worked on. The more technical explanation refers to a slip as resulting in a sliding motion when the friction between the feet (shoe sole surface) and the surface is too little. This slip (loss of traction), in turn, often leads to a loss of balance, resulting in a fall. Slips can be caused by a number of design factors and work practices, individually or in combination. Design factors include footwear, floor surfaces, personal characteristics, and the work task. Footwear is an important consideration in the prevention of a slip or fall. Not only is the condition of the footwear important in fall prevention, but also the composition, shape, and style. For industrial applications, the organizational safety professional should ensure that only approved safety shoes are worn. Safety shoes should have toe protection and slip-resistant soles.

Floor surfaces, design, installation, composition, condition, gradient, modifications by protective coatings and cleaning/waxing agents, and illumination are all important elements that must be taken into consideration in providing safe floor surfaces in the workplace. Common ways to make floor surfaces slip resistant include grooving, gritting, matting, and grating.

Personal characteristics such as physical condition, age, health, emotional state, agility, and attentiveness are also factors to consider when making walking and working surfaces slip resistant for workers. Work-task design also plays an important role in causing and preventing slip-falls. Some work practices can cause walking surfaces to be constantly wet (such as from frequent spills), and weather hazards such as snow and ice can also make walking surfaces slippery. Workplace supervisors and workers (and safety officials) must follow safe work practices and exercise vigilance to reduce the occurrence of such conditions and to remediate them as quickly as possible when they do occur. This type of problem is much more common than we might realize.

Unfortunately, it is not unusual for workers to spill oil or some other slippery substance on the workplace floor and then walk away from the spill, leaving behind a slip hazard for another worker. A common workplace safe work practice and housekeeping rule should be to clean up spills right away. Another unsafe work practice that commonly leads to slips and falls is when a worker is in a hurry, rushes to finish a particular task, and overlooks safe work practices.

Trips

Trips normally occur whenever a worker's foot contacts an object that causes him or her to lose balance; however, you do not always have to come into contact with an object to trip. Trips may also be caused by too much friction between footwear and the walking surface. Like slips, trips commonly occur when a worker is in a hurry. The problem with hurrying, of course, is that the potential victim's attention is usually focused on anything but possible trip hazards. Another common factor that leads to a trip is the practice of carrying objects that are so large that the worker cannot see the walking surface. Lighting also plays a critical role in preventing trips. Inadequate lighting fixtures, burned-out bulbs, and lights that are turned off all increase the opportunity for trips to occur. Again, as in the prevention of slips, housekeeping plays an important role in prevention. Good workplace housekeeping practices include keeping passageways clean and uncluttered; arranging equipment so that it does not interfere with walkways or pedestrian traffic; keeping working areas clear of extension or power tool cords; eliminating loose footing on stairs, steps, and floors; and properly storing gangplanks and ramps.

Stair Falls

For information about falls from stairs, probably the best reference is the Bureau of Labor Statistics' *Injuries Resulting from Falls on Stairs* (Bulletin 2214). This particular booklet is excellent because not only does it provide statistical data but it is also an eye-opener for how these injuries occur. It is widely known and accepted, for example, that stairs are a high-risk area. It is also accepted that a loss of balance can occur from a slip or trip while a worker (or any person) is traveling up or down a stairway. Safety officials must consider why stairs are so hazardous. What are the causal factors?

Bulletin 2214 comes in handy when trying to answer questions like these; for example, it points out that the vast majority of falls on stairs occur when people are traveling down the stairs and are not holding onto the handrail. This is an important point about handrails for two reasons: (1) The safety person will know to focus training on this important topic, and (2) the safety person can ensure that handrails not only are in place in all stairways but are also in good repair. Loss of traction is the common cause of the highest number of stairway slipping and falling accidents. Again, this is where good housekeeping practices come into play. Many of the stairway slipping and falling accidents happen because of water or other liquid on steps. Along with improper housekeeping

Safety and Health Considerations

practices, stairs can also become hazardous whenever they are improperly designed or installed or are neglected. Safe work practices should also be considered. A work practice that allows the worker to carry or reach for objects while climbing stairs is not a good one.

Elevated Falls from One Level to Another

When workers are working from elevated scaffolds, ladders, platforms, and other surfaces, the risk of serious injury from an elevated fall is increased exponentially whenever a worker loses his balance as a result of a slip or trip. Unfortunately, it is the practice of too many companies that require workers to perform work from elevated areas to only provide some type of device (handrail or handline) that workers are supposed to grab onto to break their fall. In the judgment of most experienced safety professionals, this is *not* fall protection. These types of jerry-rigged devices are not acceptable substitutes for guardrails, appropriate midrails, and toeboards. OSHA requires guardrails to be 42 inches nominal, midrails 21 inches, and toe boards 4 inches. Ellis (1988) made a good point in observing that, "unlike many workplace hazards, few, if any, 'near-miss' incidents help people learn to appreciate the seriousness of elevated falls" (p. 28). When you consider that losing one's balance from an elevation of 10 to 200 feet or more usually leaves little chance to avoid serious or fatal injury, Ellis' statement makes a lot of sense.

Fall Protection Measures

Under 29 CFR 1926.501 (Duty to Have Fall Protection), employers must assess the workplace to determine if the walking or working surfaces on which employees are to work have the strength and structural integrity to safely support workers. Accordingly, the real goal should be to prevent slips, trips, and falls from elevation from occurring in the first place. To accomplish this, the following steps are recommended: (1) preplan before beginning any elevated work (e.g., on scaffolds); (2) establish a written policy and develop rules; and (3) implement safe work practices to prevent falls. Preplanning is all about thinking through the job at hand; for example, for exterior refurbishing work on a chemical storage tank that is 80 feet in height, scaffolding will almost certainly be required. Preplanning and a great deal of skill are both necessary to properly erect scaffolding. If scaffolding is to be used, the organization responsible for erecting the scaffolding should have a written scaffold safety program.

SUBPART G, OCCUPATIONAL AND ENVIRONMENTAL CONTROL

1910.95   Occupational Noise Exposure

It is important to note that 1910.95(o) states: "Paragraphs (c) through (n) of this section shall not apply to employers engaged in oil and gas well drilling and servicing operations."

OSHA Noise Hazards Requirements

In 1983, OSHA adopted a hearing conservation amendment to 29 CFR 1910.95 that requires employers to implement *hearing conservation programs* in any work setting where employees are exposed to an 8-hour time-weighted average of 85 dBA and above. Employers are required to implement *hearing conservation procedures* in settings where the noise level exceeds a time-weighted average of 90 dBA. They are also required to provide personal protective equipment for employees who show evidence of hearing loss, regardless of the noise level at their worksites. In addition to concerns over noise levels, the OSHA standard also addresses the issue of the duration of exposure (LaBar, 1989):

> Duration is another key factor in determining the safety of workplace noise. The regulation has a 50 percent 5 dBA logarithmic tradeoff. That is, for every 5-decibel increase in the noise level, the length of exposure must be reduced by 50 percent. For example, at 90 decibels (the sound level of a lawnmower or shop tools), the limit on "safe" exposure is 8 hours. At 95 dBA, the limit on exposure

is 4 hours, and so on. For any sound that is 106 dBA and above—this would include such things as a sandblaster, rock concert, or jet engine—exposure without protection should be less than 1 hour, according to OSHA's rule.

Although not all of the standard's requirements are pertinent to oil and gas drilling operations, the basic requirements of OSHA's hearing conservation standard are explained here (LaBar, 1989):

- *Monitoring noise levels.* Noise levels should be monitored on a regular basis. Whenever a new process is added, an existing process is altered, or new equipment is purchased, special monitoring should be undertaken immediately.
- *Medical surveillance.* The medical surveillance component of the regulation specifies that employees who will be exposed to high noise levels must be tested upon being hired and again at least annually.
- *Noise controls.* The regulation requires that steps be taken to control noise at the source. Noise controls are required in situations where the noise level exceeds 90 dBA. Administrative controls are sufficient until noise levels exceed 100 dBA. Beyond 100 dBA, engineering controls must be used.
- *Personal protection.* Personal protective devices are specified as the next level of protection when administrative and engineering controls do not reduce noise hazards to acceptable levels. They are to be used in addition to rather than instead of administrative and engineering controls.
- *Education and training.* The regulation requires the provision of education and training to ensure that employees understand (1) how the ear works, (2) how to interpret the results of audiometric tests, (3) how to select personal protective devices that will protect them against the types of noise hazards to which they will be exposed, and (4) how to properly use personal protective devices (LaBar, 1989).

Occupational Noise Exposure

Noise is commonly defined as any unwanted sound. Noise literally surrounds us every day and is with us just about everywhere we go; however, the noise we are concerned with here is that produced by industrial processes. Excessive amounts of noise in the work environment (and outside of it) cause many problems for workers, including increased stress levels, interference with communication, disrupted concentration, and, most importantly, varying degrees of hearing loss. Exposure to high noise levels also adversely affects job performance and increases accident rates.

One of the major problems with attempting to protect workers' hearing acuity is the tendency of many workers to ignore the dangers of noise. Because hearing loss, like cancer, is insidious, it is easy to ignore. It sort of sneaks up slowly and often is not apparent until after the damage is done. Alarmingly, hearing loss from occupational noise exposure has been well documented since the 18th century, and since the advent of the industrial revolution the number of exposed workers has greatly increased (Mansdorf, 1993). Today, though, the picture of hearing loss is not as bleak as it has been in the past, as a direct result of OSHA's requirements. Now that noise exposure must be controlled in all industrial environments, well-written and well-managed hearing conservation programs must be put in place, and employee awareness regarding the dangers of exposure to excessive levels of noise has been raised, job-related hearing loss is coming under control (see Table 12.5).

Hearing Protection

The *hearing protection* element of a hearing conservation program provides hearing protection devices for employees and training in how to wear them effectively, as long as hazardous noise levels exist in the workplace. Hearing protection comes in various sizes, shapes, and materials, and the cost of this equipment can vary dramatically. Two general types of hearing protection are used widely in industry: the cup muff (commonly referred to as *Mickey Mouse ears*) and the plug insert

Safety and Health Considerations

TABLE 12.5
Permissible Noise Exposures (29 CFR 1910.95)

Duration per Day (hr)	Sound Level (dBA)
8	90
6	92
4	95
3	97
2	100
1-1/2	102
1	105
1/2	110
1/4 or less	115

Note: When the daily noise exposure is composed of two or more periods of noise exposure of different levels, their combined effect should be considered, rather than the individual effect of each. If the sum of the following fractions $C_1/T_1 + C_2/T_2 + C_n/T_n$ exceeds unity, then the mixed exposure should be considered to exceed the limit value. C_n indicates the total time of exposure at a specified noise level, and T_n indicates the total time of exposure permitted at that level. Exposure to impulsive or impact noise should not exceed 140-dB peak sound pressure level.

type. Because feasible engineering noise controls have not been developed for many types of industrial equipment, hearing protection devices are the best option for preventing noise-induced hearing loss in these situations. As with the other elements of a hearing conservation program, the hearing protective device element must be in writing and included in the program.

SUBPART H, HAZARDOUS MATERIALS

1910.106 Flammable and Combustible Liquids
1910.110 Storage and Handling of Liquefied Petroleum Gases
1910.120 Hazardous Waste Operations and Emergency Response

Hazardous Material

A hazardous material is a substance (gas, liquid, or solid) capable of causing harm to people, property, and the environment. The U.S. Department of Transportation (DOT) uses the term *hazardous materials* to cover nine categories identified by the United Nations Hazard Class Number System, which are as follows:

- Explosives
- Gases (compressed, liquefied, dissolved)
- Flammable liquids
- Flammable solids
- Oxidizers
- Poisonous materials
- Radioactive materials
- Corrosive materials
- Miscellaneous materials

Flammable and Combustible Liquids

In addition to basic fire prevention, emergency response training, and fire extinguisher training, employees must be trained on the hazards involved with flammable and combustible liquids; 29 CFR 1910.106 addresses this area. Industrial facilities typically use all types of flammable and combustible liquids. These dangerous materials must be clearly labeled and stored safely when not in use. The safe handling of flammable and combustible liquids is a topic that needs to be fully addressed by the facility safety engineer and workplace supervisor. Worker awareness of the potential hazards that flammable and combustible liquids pose must be stressed. Employees need to know that flammable and combustible liquid fires burn extremely hot and can produce copious amounts of dense, black smoke. Explosion hazards exist under certain conditions in enclosed, poorly ventilated spaces where vapors can accumulate. A flame or spark can cause vapors to ignite, creating a flash fire with the terrible force of an explosion. One of the keys to reducing the potential spread of flammable and combustible fires is to provide adequate containment. All storage tanks should be surrounded by storage dikes or containment systems, for example. Correctly designed and built dikes will contain spilled liquid. Spilled flammable and combustible liquids that are contained are easier to manage than those that have free run of the workplace. Properly installed containment dikes can prevent environmental contamination of soil and groundwater. Flammable liquids have a flash point below 100°F. Both flammable and combustible liquids are divided into the three classifications shown below:

Flammable Liquids
   Class IA—Flash point below 73°F, boiling point below 100°F
   Class IB—Flash point below 73°F, boiling point at or above 100°F
   Class IC—Flash point at or above 73°F, but below 100°F

Combustible Liquids
   Class II—Flash point at or above 100°F, but below 140°F
   Class IIIA—Flash point at or above 140°F, but below 200°F
   Class IIIB—Flash point at or above 200°F

What Is a Hazardous Waste?

A general rule of thumb states that any hazardous substance that is spilled or released into the environment is no longer classified as a hazardous substance but as a hazardous waste. The USEPA uses the same definition for hazardous wastes as it does for hazardous substances. The four characteristics of reactivity, ignitability, corrosivity, and toxicity can be used to identify hazardous substances as well as hazardous wastes. The USEPA lists substances that it considers to be hazardous waste; these lists take precedence over any other method used to identify and classify a substance as hazardous. If a substance is included on one of the USEPA's lists described below, it is a hazardous substance, no matter what.

Hazardous wastes are organized by the USEPA into three categories: nonspecific source wastes, specific source wastes, and commercial chemical products. All listed wastes are presumed to be hazardous, regardless of their concentrations. USEPA developed these lists by examining different types of wastes and chemical products to determine whether they met any of the following criteria:

- Exhibit one or more of the four characterizations of a hazardous waste
- Meet the statutory definition of hazardous waste
- Are acutely toxic or acutely hazardous
- Are otherwise toxic

These listed wastes can be described briefly as follows:

Safety and Health Considerations

- *Nonspecific source wastes* are generic wastes commonly produced by manufacturing and industrial processes. Examples from this list include spent halogenated solvents used in degreasing and wastewater treatment sludge from electroplating processes, as well as dioxin wastes, most of which are "acutely hazardous" wastes because of the danger they present to human health and the environment.
- *Specific source wastes* are from specially identified industries such as wood preserving, petroleum refining, and organic chemical manufacturing. These wastes typically include sludges, still bottoms, wastewaters, spent catalysts, and residues, such as wastewater treatment sludge from pigment production.
- *Commercial chemical products* (also called "P" or "U" list wastes because their code numbers begin with these letters) include specific commercial chemical products or manufacturing chemical intermediates. This list includes chemicals such as chloroform and creosote, acids such as sulfuric and hydrochloric, and pesticides such as DDT and kepone (40 CFR 261.31, 261.32, 261.33).

The USEPA ruled that any waste mixture containing a listed hazardous waste is also considered a hazardous waste and must be managed accordingly. This applies regardless of what percentage of the waste mixture is composed of listed hazardous wastes. Wastes derived from hazardous wastes (residues from the treatment, storage, and disposal of a listed hazardous waste) are considered hazardous waste as well. Hazardous wastes are derived from several waste generators. Most of these waste generators are in the manufacturing and industrial sectors and include chemical manufacturers, the printing industry, vehicle maintenance shops, leather products manufacturers, the construction industry, and metal manufacturing, among others. These industrial waste generators produce a wide variety of wastes, including strong acids and bases, spent solvents, heavy metal solutions, ignitable wastes, cyanide wastes, and many more.

From the responsible safety official's perspective, any hazardous waste release that could alter the environment or impact the health and safety of employees in any way is a major concern. The specifics of the safety engineer's concern lie in the acute and chronic toxicity to organisms, bioconcentration, biomagnification, genetic change potential, etiology, pathways, change in climate or habitat, extinction, persistence, esthetics such as visual impact, and, most importantly, the impact on the health and safety of employees.

Remember, we have stated consistently that when a hazardous substance or hazardous material is spilled or released into the environment, it becomes a hazardous waste. This is important because specific regulatory legislation has been put in place regarding hazardous wastes, responding to hazardous waste leak and spill contingencies, and the proper handling, storage, transportation, and treatment of hazardous wastes. The goal, of course, is protecting the environment and ultimately the health and safety of our employees and the surrounding community. Why are we so concerned about hazardous substances and hazardous wastes? This question is relatively easy to answer based on experience, publicity, and actual hazardous materials incidents, which have resulted in tragic consequences to the environment and to human life.

Humans are strange in many ways. We may know that a disaster is possible, is likely, could happen, and is predictable, but do we act before someone dies? Not often enough. We often ignore the human element—we forget a victim's demise. We simply do not want to think about it, because if we think about it, we must come face to face with our own mortality. The safety engineer, though, must think constantly about potential disasters to prevent them from ever occurring. Because of the Bhopal incident and other similar but less catastrophic chemical spill events, the U.S. Congress (pushed by public concern) developed and passed certain environmental laws and regulations to regulate hazardous substances and wastes in the United States. Two regulatory acts have been most crucial to the current management programs for hazardous wastes. The first, mentioned already in this text, is the Resource Conservation and Recovery Act (RCRA). Specifically, the RCRA provides guidelines for prudent management of new and future hazardous substances and wastes. The second act is the Comprehensive Environmental Response, Compensation, and Liability Act (CERCLA), otherwise known as Superfund, which deals primarily with mistakes of the past (i.e., inactive and abandoned hazardous waste sites).

SUBPART I, PERSONAL PROTECTIVE EQUIPMENT

1910.132 General Requirements
1910.133 Eye and Face Protection
1910.134 Respiratory Protection
1910.135 Head Protection
1910.136 Occupational Foot Protection

Personal Protective Equipment

Noncompliance with the requirements of OSHA's respiratory protection standard is one of the top ten citations listed in Table 12.2. In the following statement, Mansdorf (1993) makes a number of important statements concerning personal protective equipment (PPE) worth taking some time to consider carefully:

> The primary objective of any health and safety program is worker protection. It is the responsibility of management to carry out this objective. Part of this responsibility includes protecting workers from exposure to hazardous materials and hazardous situations that arise in the workplace. It is best for management to try to eliminate these hazardous exposures through changes in workplace design or engineering controls. When hazardous workplace exposures cannot be controlled by these measures, personal protective equipment (PPE) becomes necessary. When looking at hazardous workplace exposures, keep in mind that government regulations consider PPE the last alternative in worker protection because it does not eliminate the hazards. PPE only provides a barrier between the worker and the hazard. If PPE must be used as a control alternative, a positive attitude and strong commitment by management is required.

"It is best for management to try to eliminate these hazardous exposures through changes in workplace design or engineering controls." Sound familiar? We consistently make this same point throughout this text. A hazard, any hazard, if possible, should be engineered out of the system or process. Determining when and how to engineer out a hazard is one of the safety official's primary functions; however, the safety official can much more effectively accomplish this if he or she is included in the earliest stages of design. Remember, it does little good (and is often very expensive) to attempt to engineer out any hazard once the hazard is in place.

"When hazardous workplace exposures cannot be controlled by these measures, personal protective equipment (PPE) becomes necessary." Although the goal of safety officials is certainly to engineer out all workplace hazards, we realize that this goal is virtually impossible to achieve. Even in this day of robotics, computers, and other automated equipment and processes, the man–machine–process interface still exists. When people are included in the work equation, the opportunity for their exposure to hazards is very real—as injury statistics make clear.

"[C]onsider PPE the last alternative in worker protection because it does not eliminate the hazards." This is extremely important for two reasons: First, the safety official's primary goal is (as we have said before) to engineer out the problem. If this is not possible, the second alternative is to implement administrative controls. When neither is possible, PPE becomes the final choice. The key words here are "the final choice." Second, PPE is sometimes incorrectly perceived—by both the supervisor and the worker—as their first line of defense against all hazards. This, of course, is incorrect and dangerous. The worker must be made to understand (by means of enforced company rules, policies, and training) that PPE affords only minimal protection against most hazards—*it does not eliminate the hazard.*

Safety and Health Considerations

"PPE only provides a barrier between the worker and the hazard." Experience shows that when some workers put on their PPE, they also don a "Superperson" mentality. What does this mean? Often, when workers use eye, hand, foot, head, hearing, or respiratory protection, they take on an "I can't be touched" attitude. They feel safe, as if the PPE somehow magically protects them from the hazard, so they act as if they are protected, are invincible, are beyond injury. They feel, however illogically, that they are well out of harm's way. Nothing could be further from the truth.

OSHA's PPE Standard

In the past, many OSHA standards have included PPE requirements, ranging from very general to very specific. It may surprise the reader to know, however, that not until relatively recently (1993–1994) did OSHA incorporate a stand-alone PPE standard into its 29 CFR 1910/1926 guidelines. This personal protective equipment standard is covered under 1910.132–138, but PPE requirements can also be found elsewhere in the General Industry Standards. For example, 29 CFR 1910.156, OSHA's Fire Brigade Standard, has requirements for firefighting gear. In addition, 29 CFR 1926.95–106 cover the construction industry. The PPE standard focuses on head, foot, eye, hand, respiratory, and hearing protection. Common PPE classifications and examples include the following:

1. Head protection (hard hats, welding helmets)
2. Eye protection (safety glasses, goggles)
3. Face protection (face shields)
4. Respiratory protection (respirators)
5. Arm protection (protective sleeves)
6. Hearing protection (ear plugs, muffs)
7. Hand protection (gloves)
8. Finger protection (cots)
9. Torso protection (aprons)
10. Leg protection (chaps)
11. Knee protection (kneeling pads)
12. Ankle protection (boots)
13. Foot protection (boots, metatarsal shields)
14. Toe protection (safety shoes)
15. Body protection (coveralls, chemical suits)

Respiratory protection and hearing protection each has its own standard. Respiratory protection is covered under 29 CFR 1910.134 and hearing protection under 1910.95. Using PPE is often essential, but it is generally the last line of defense after engineering controls, work practices, and administrative controls. Engineering controls involve physically changing a machine or work environment. Administrative controls involve changing how or when employees do their jobs, such as scheduling work and rotating employees to reduce exposures. Work practices involve training workers how to perform tasks in ways that reduce their exposure to workplace hazards.

OSHA's PPE Requirements

Several requirements for both the employer and the employee are mandated under OSHA's personal protective equipment standard. OSHA's requirements include the following:

1. Employers are required to provide employees with PPE that is sanitary and in good working condition.
2. The employer is responsible for examining all PPE used on the job to ensure that it is of a safe (and approved) design and in proper condition.
3. The employer must ensure that employees use PPE.

4. The employer must provide a means for obtaining additional and replacement equipment; defective and damaged PPE is not to be used.
5. The employer must ensure that PPE is inspected on a regular basis.
6. The employee must ensure that he or she dons PPE when required.
7. Where employees provide their own PPE, the employer must ensure that it is adequate and that it is properly maintained and sanitized.

Although the employer must ensure that employees wear PPE when required, both employers and employees should factor in three things: (1) the PPE used must not degrade performance unduly, (2) it must be reliable, and (3) it must be suitable for the hazard involved.

Respiratory Protection

The basic purpose of any respirator is, simply, to protect the respiratory system from inhalation of hazardous atmospheres. Respirators provide protection either by removing contaminants from the air before it is inhaled or by supplying an independent source of respirable air. The principal classifications of respirator types are based on these categories.

NIOSH (1987)

Written procedures shall be prepared covering safe use of respirators in dangerous atmospheres that might be encountered in normal operations or in emergencies. Personnel shall be familiar with these procedures and the available respirators.

—OSHA 29 CFR 1910.134(c)

Respirators allow workers to breathe safely without inhaling particles or toxic gases. Two basic types are (1) *air-purifying*, which filter dangerous substances from the air, and (2) *air-supplying*, which deliver a supply of safe breathing air from a tank (SCBA), from a group of tanks (cascade system), or from an uncontaminated area nearby via a hose or airline to the mask. Respiratory protection might be a requirement in ensuring safe confined space entry. Often the organization's safety official holds the responsibility for making this determination. If the safety official determines that respiratory protection is required, then it is incumbent upon him or her to implement a written respiratory protection program that is in compliance with OSHA's respiratory protection standard (29 CFR 1910.134). Remember, though, that respiratory protection is often necessary to protect workers who may not ever be called upon to enter a confined space with an atmosphere containing airborne contaminants. Workers may need protection from airborne contaminants at any worksite where airborne contaminants are health hazards. This text has continuously stressed the vital need to attempt first to engineer out such hazards; however, when engineering and other methods of control or proper selection and use of respiratory protection cannot eliminate airborne hazards, it becomes part of the safety official's responsibility. Unlike past practices, where respiratory protection entailed nothing more than providing respirators to workers who could be exposed to airborne hazards and expecting workers to use the respirator to protect themselves, supplying respirators today without the proper training, paperwork, and testing is illegal. Employers are sometimes unaware that by supplying respirators to their employees

---

DID YOU KNOW?

For permit-required confined space entry operations, respiratory protection is a key piece of safety equipment, one always required for entry into an immediately dangerous to life or health (IDLH) space and one that must be readily available for emergency use and rescue if conditions change in a non-IDLH space. Remember, however, that only air-supplying respirators should be used in confined spaces where there is not enough oxygen.

Safety and Health Considerations

without having a comprehensive respiratory protection program they are making a serious mistake. By issuing respirators, they have implied that a hazard actually exists. In a lawsuit, they then become fodder for the lawyers.

OSHA mandates that an effective program must be put in place. This respiratory protection program must not only follow OSHA's guidelines but must also be well planned and properly managed. A well-planned, well-written respiratory protection program must include all of the basic elements listed in the standard. Selecting the proper respirator for the job, the hazard, and the worker is very important, as is thorough training in the use and limitations of respirators. Compliance with OSHA's respiratory protection standard begins with developing written procedures covering all applicable aspects of respiratory protection.

SUBPART J, GENERAL ENVIRONMENTAL CONTROLS

1910.141 Sanitation
1910.145 Specifications for Accident Prevention Signs and Tags
1910.146 Permit-Required Confined Spaces
1910.147 The Control of Hazardous Energy (Lockout/Tagout)
1910.151 Medical Services and First Aid

Sanitation Facilities

Note that a lack of personnel sanitation facilities for workers is one of the top ten OSHA violations cited in Table 12.2. The site manager must factor into any workplace design several sanitation and personal hygiene requirements (i.e., provisions for potable water for drinking and washing; sewage, solid waste, and garbage disposal; sanitary food services; and drinking fountains, washrooms, locker rooms, toilets, and showers), in addition to providing a facility or plant site with easy-to-use and correct housekeeping activities. Housekeeping and sanitation are closely related. Control of health hazards requires sanitation, and control is usually enacted through good housekeeping practices. Disease transmission and ingestion of toxic or hazardous materials are controlled through a variety of sanitation practices, but if the workplace is not properly designed with appropriate sanitary and storm sewers, safe drinking water, and sanitary dispensing equipment, then sound sanitary practices are made much more difficult to implement within the workplace.

Recommended Color Codes for Accident Prevention Tags

"DANGER"—Red, or predominately red, with lettering or symbols in a contrasting color
"CAUTION"—Yellow, or predominating yellow, with lettering or symbolism in a contrasting color
"WARNING"—Orange, or predominantly orange, with lettering or symbols in a contrasting color
"BIOLOGICAL HAZARD"—Fluorescent orange or orange-red, or predominantly so, with letters or symbols in contrasting colors

OSHA's Confined Space Entry Program

Note that in Table 12.2 the second most frequent OSHA citation issued for oil and gas drilling operations was for violation of 29 CFR 1910.146, the permit-required confined spaces standard. OSHA has a specific standard that mandates specific compliance with its requirements for making confined space entries; however, no matter how many standards and regulations OSHA and other regulators write, promulgate, and attempt to enforce, if employers and employees do not abide by their responsibilities under the act, the requirements are not worth the paper they are written on. OSHA's Confined Space Entry Program (CSEP) is a vital guideline to protect workers and others. CSEP was issued to protect workers who must enter confined spaces. It is designed and intended to protect workers from toxic, explosive, or asphyxiating atmospheres and from possible engulfment from small particles such as sawdust and grain (e.g., wheat, corn, and soybean normally contained in silos). It focuses on areas with immediate health or safety risks—areas with hazards that could potentially cause death or injury. These areas or spaces are classified as *permit-required* confined spaces. Under the standard, employers are required to identify all permit-required spaces in their workplaces, prevent unauthorized entry into them, and protect authorized workers from hazards through an entry-by-permit-only program. CSEP covers all of general industry, including agricultural services (the keyword here is "services" and not agriculture), manufacturing, chemical plants, refineries, transportation, utilities, wholesale and retail trade, and miscellaneous services. It applies to manholes, vaults, digesters, contact tanks, basins, clarifiers, boilers, storage vessels, furnaces, railroad tank cars, cooking and processing vessels, tanks, pipelines, and silos, among other spaces.

Permit-Required Confined Space Written Program

If the employer decides that its employees will enter permit spaces, the employer shall develop and implement a written permit space program .... The written program shall be available for inspection by employees and their authorized representatives.

—29 CFR 1910.146(c)(4)

The first step the employer must take in implementing a permit-required confined space program is to take the measures necessary to prevent unauthorized entry. Typically, this is accomplished by identifying and labeling all confined spaces. The next step is to list all of those confined spaces and clearly communicate to employees that the listed spaces are not to be entered by organizational personnel under any circumstances. Remember that the *employer* is responsible for identifying, labeling, and listing all site permit-required confined spaces, in addition to identifying and evaluating the hazards of each confined space. Once the hazards have been identified and evaluated, the identity and hazards for each confined space must be listed in the organization's written confined space entry program (obviously, it is important that employees are made well aware of all the hazards). The next step is to develop written procedures and practices for those personnel who are required to enter, for any reason, permit-required confined spaces. The procedures and practices used for permit-required confined space entry must be *in writing* and at the very least must include the following:

- Specifying acceptable entry conditions
- Isolating the permit space
- Purging, inerting, flushing, or ventilating the permit space as necessary to protect entrants from external hazards
- Providing pedestrian, vehicle, or other barriers as necessary to protect entrants from external hazards
- Verifying that conditions in the permit space are acceptable for entry throughout the duration of an authorized entry

In its permit-required confined spaces standard (1910.146), OSHA specifies the equipment required to make a safe and approved confined space entry into permit-required confined spaces. Note that the employer, at no cost, must provide this equipment to the employee. The employer is also required not only to procure this equipment at no cost to the employee but also to maintain the equipment properly. Most importantly, the employer is also required to ensure that employees use the equipment properly. The required equipment includes the following:

- *Testing and monitoring equipment*—Numerous makes and models of confined space air monitors (gas detectors or sniffers) are available on the market, and selection should be based on the facility's specific needs; for example, if the permit-required confined space to be entered is a sewer system, then the specific need is a multiple-gas monitor. This type

Safety and Health Considerations

of instrument is best suited for sewer systems, where toxic and combustible gases and oxygen-deficient atmospheres are prevalent. No matter what type of air monitor is selected for a specific use in a particular confined space, any user must be thoroughly trained on how to effectively use the device. Users must also understand the monitor's limitations and how to calibrate the device according to manufacturer's requirements. Having an approved air monitor is useless if workers are not trained in its operation or proper calibration. When choosing an air monitor for use in confined space entry, you must ensure that the monitor selected is suitable for the type of atmosphere to be entered and that it is equipped with audible and visual alarms that can be set, for example, at 19.5% or lower for oxygen and preset for levels of the combustible or toxic gases it is used to detect.

- *Ventilating equipment*—In many cases, it is possible to eliminate, reduce, or modify atmospheric hazards in confined spaces by ventilating—using a special fan or blower to displace the bad air inside a confined space with good air from outside the enclosure. Whatever blower or ventilator type that is chosen, a certain amount of common sense and consideration of the depth of the confined space, size of the enclosure, and number of openings available is required. Keep in mind that the blower must be equipped with a vaporproof, totally enclosed electrical motor or a non-sparking gas engine. Obviously, the size and configuration of the confined space dictate the size and capacity of the blower to be used. Typically, a blower with a large-diameter flexible hose (elephant trunk) is most effective.

- *Personal protective equipment*—Note that noncompliance with the requirements of the OSHA personal protective equipment standard is listed as one of the top ten violations in Table 9.12. OSHA requires personal protective equipment (PPE) for confined space entries. The entrant must be equipped with the standard PPE required to make a vertical entry into a permit-required confined space (a full-body harness combined with a lanyard or lifeline), and also the PPE required to protect him or her from specific hazards. As an example, an employee who is to enter a manhole is typically equipped with (1) an approved hard hat to protect the head; (2) approved gloves to protect the hands; (3) approved footwear (safety shoes) to protect the feet; (4) approved safety eyewear or face protection to protect the eyes and face; (5) full body clothing (long-sleeved shirt and long trousers) to protect the trunk and extremities; and (6) a tight-fitting NIOSH-approved self-contained breathing apparatus (SCBA) or supplied-air hose mask with emergency escape bottle for IDLH atmospheres.

- *Lighting*—Many confined spaces could be described as nothing more than dark (and sometimes foreboding) holes in the ground—often a fitting description. As you might guess, typically many confined spaces are not equipped with installed lighting. To ensure safe entry into such a space, the entrant must be equipped with intrinsically safe lighting. Intrinsically safe? Absolutely. Think about it. The last thing you want to do is to send anyone into a dark space filled with methane with a torch in his or her hand, but a light source that emits sparks might as well be a torch. Confined spaces present enough dangers on their own without adding to the hazards. Even after the space has been properly ventilated (with copious and continuous amounts of outside fresh air) and, for example, the source of methane has been shut off, we still obviously have a space that has the potential for an extremely explosive atmosphere. Do not underestimate the hazards such a confined space presents! So, what do we do? If lighting is required in a confined space, we need to ensure that it is provided to the entrant—for his or her safety as well as to enable work to be done. For confined space entries, explosion-proof lanterns or flashlights (intrinsically safe devices) are recommended. Such devices, if NIOSH and OSHA approved, are equipped with spring-loaded bulbs that, upon breaking, eject themselves from the electrical circuit, preventing ignition of hazardous atmospheres. Another safe, low-cost, instant light source now readily available for confined space entry is lightsticks. They can be used safely near explosive materials because they contain no source of ignition. Lightsticks are available with illumination times ranging from 1/2 to 12 hours. Another common work light used for confined space entry is the droplight. UL-approved droplights that are vaporproof, explosion-proof, and equipped with ground-fault circuit interrupters (GFCIs) are the recommended type for confined space entry.

Note: If you have a confined space that has the potential for an explosive atmosphere and has light fixtures permanently installed in place, remember that these lights must be certified for use in hazardous locations and maintained in excellent condition.

- *Barriers and shields*—We must be concerned with not only the safety of the confined space entrant but also the safety of those outside the confined space. An open manhole, for example, obviously presents a pedestrian and traffic hazard. To prevent accidents in areas where manhole work is in progress, we can use several safety devices, including manhole guard rail assemblies, guard rail tents, barrier tape, fences, and manhole shields. Remember that we want to prevent someone from falling into a manhole (or other type of confined space opening) and we also want to prevent unauthorized entry. Occasionally, manholes or ordinarily inaccessible areas, when open for work crews, present an attractive nuisance—even ordinary curiosity may lead people (especially children) to put themselves at risk by attempting to enter a confined space. Along with protecting the confined space opening from someone falling into it or entering it illegally, we must also control traffic around or near the opening. To do this we may need to employ the use of cones, signs, or stationed guard personnel. Don't forget the nighttime hours. After dark, it is obviously difficult to see an open confined space opening or guard device; these devices should be lighted with vehicle strobes or beacon lights.
- *Ingress and egress equipment: ladders*—Have you ever peered down a 40-foot deep, 24-inch diameter vertical manhole? Not pleasant? Depends on your point of view. If the manhole has no lighting (as most do not) then you are peering into what appears to be a bottomless pit (and maybe it is). Have you been there? If so, no further explanation is needed. You know that, at best, entering any manhole can be a perilous undertaking. If you have never faced entering a manhole, let's consider an important point. If you are tasked to enter such a confined space, you will obviously be interested in entering it (ingressing) safely (taking all required precautions) and returning (egressing) safely. Experience with assessing safety considerations in confined space areas has shown that many of the installed ladders (in place to allow entry and exit inside confined spaces) are not always in the best condition due to the environment to which they are constantly exposed year after year. Confined spaces may be shrouded in moist, chemical-laden atmospheres—conditions excellent for corroding most metals. Most ladders installed in confined spaces are made of metal. Not only do we require our workers to enter dangerous permit-required confined spaces, but without properly evaluating all of the confined-space's conditions we may also be asking them to enter them in a totally unsafe manner—on equipment that may fail. Don't forget about the devices used to hold the ladders in place—the securing or attachment bolts or screws. Most of these are also made of metal as well—metal that will corrode and weaken with time. How about those spaces that do not have installed ladders? For confined spaces not equipped with ladders, stairways, or some other installed means of ingress and egress, we often employ the use of portable ladders. One way or another, we are required to provide a safe way in and out of a confined space—ladders often fit this need. Occasionally, though, ladders or stairways for safe entry or exit are not available, practical, or practicable. When such a situation arises, winches and hoisting devices are commonly used to raise and lower entrants. Remember that any lowering and lifting devices must be OSHA-approved as safe to use. Using a rope attached to the bumper of a vehicle to lower or raise an entrant, for example, is strictly prohibited. Only hand-operated lifting/hoisting

Safety and Health Considerations

devices should be employed. Motorized devices are unforgiving—especially whenever the entrant gets caught up in an obstruction (machinery, pipe, angle iron, etc.) that prevents his or her body from moving. The motorized device doesn't care—it just continues to pull the entrant out (sometimes by body parts only). On a motorized device, a person stuck in a confined space could literally be pulled apart. OSHA regulations were created to prevent just such gruesome incidents from occurring. But gruesome and fatal events (sometimes involving multiple fatalities) do occur. When an entrant gets into trouble while inside a confined space, what do we do? When this is the case, OSHA is quite specific on what should and should not be done in rescuing a confined space entrant who is in trouble.

- *Rescue equipment*—When confined space rescue is to be effected by any agency other than the facility itself (e.g., emergency rescue service, fire department), the facility is not required to provide the rescue equipment; however, when confined space rescue is to be performed by facility personnel, proper rescue equipment is required. Proper rescue equipment consists of the equipment needed to remove personnel from confined spaces in a safe manner. "In a safe manner" means "to prevent further injury to the entrants and *any* injury to the rescuers." Confined space rescue equipment (commonly called *retrieval equipment*) typically consists of three components: *safety harness*, *rescue and retrieval line*, and a *means of retrieval*. Let's take a closer look at each of these components.
  - A full-body harness combined with a lanyard or lifeline evenly distributes the fall-arresting forces among the worker's shoulders, legs, and buttocks, reducing the chance of further internal injuries. A harness also keeps the worker upright and more comfortable while awaiting rescue. The full-body harness used for confined space rescue should consist of flexible straps that continually flex and give with movement, conforming to the wearer's body—eliminating the need to frequently stop and adjust the harness. Usually constructed of a combination of nylon, polyester, and specially formulated elastomer, the proper harness resists the effects of sun, heat, and moisture to maintain its performance on the job. The full-body harness should include a sliding back D-ring (to attach the retrieval line hook), and a non-slip adjustable chest strap.
  - The heavy-duty rescue and retrieval line is usually a component of a winch system. Both ends of the retrieval lines should be equipped with approved locking mechanisms of at least the same strength as the lines for attaching to the entrant's harness and anchor point. The winch systems used today are either an approved two-way system or three-way system. The two-way system is used for raising and lowering rescue operations whenever a retractable lifeline is not needed. Typical systems feature three independent braking systems, a tough two-speed gear drive, and approximately 60 feet of steel cable. Three-way systems offer additional protection when a self-retracting lifeline is used. The winch is usually a heavy-duty model (usually rated at 500 lb or 225 kg) with disc brakes to stop falls within inches, and it is equipped with a shock-absorption feature to minimize injuries. The proper winch should allow the user to raise and lower loads at an average speed of 10 to 32 feet per minute in an emergency.
  - The means of retrieval usually includes the proper winch with built-in fall protection attached to a 7- or 9-foot tripod. The tripod should be of sufficient height to allow the victim to be brought above the rim of the manhole or other opening and placed on the ground.
- *Other equipment*—If tools are to be used during a confined space entry or rescue, it may be necessary to use non-sparking tools if flammable vapors or combustible residues are present. These non-sparking, non-magnetic, and corrosion resistant tools are usually fashioned from copper or aluminum. A fire extinguisher, additional radios for communication, spare oxygen bottles (for SCBA and cascade systems as needed), a first-aid kit, and any other equipment required for safe entry into and rescue from permit spaces may also be necessary.

Pre-Entry Requirements

Before anyone is allowed to enter a permit-required confined space, certain space conditions must first be evaluated. The first step taken should be to determine whether workers must enter the permit-required space to complete the task at hand. You should ask yourself, "Do we really need to enter the permit-required confined space?" If the answer is yes, then before initiating a confined space entry the space should be tested with a calibrated air monitor to determine if acceptable entry conditions exist before entry is authorized. If air monitoring indicates that entry can be made safely without respiratory protection or if appropriate respiratory protection must be worn, then the supervisor (qualified or competent person) must decide how to effect the entry in the safest manner possible. Whether the atmosphere is safe or unsafe without proper respiratory protection, monitoring must be continuous. Taking only one reading and basing decisions on that reading is not wise; in fact, it is unsafe. Conditions can change within a confined space at any time—it is critical to the wellbeing of the entrant to know when these changes take place and what the changes are. When conducting the air test for atmospheric hazards, a standard testing protocol should be followed:

1. Test for oxygen.
2. Test for combustible gases and vapors.
3. Test for toxic gases and vapors.

You should also test the atmosphere within a confined space at different levels. For example, if you are about to authorize the entry of workers into a manhole that is 30 feet deep, you should test top to bottom for a stratified atmosphere. Remember that some toxic gases (methane, for example) are lighter than air. They tend to accumulate at higher levels within the manhole. If the manhole may contain carbon monoxide (which has a vapor density similar to air) you should test at the middle level. Hydrogen sulfide (a deadly killer) is heavier than air; therefore, you should test close to the bottom of the manhole if hydrogen sulfide may be present. Along with testing at different levels for stratification of toxic gases, you should also check in all directions as much as possible.

The key point to remember is that atmospheric testing should be continuous, especially when entrants are inside the confined space. To ensure that continuous atmospheric testing is conducted while an entrant is inside the confined space, an attendant (at least one) must be stationed outside the space to conduct the testing. In addition to continuously monitoring the atmosphere of the permit-required confined space, the attendant or some other designated person must be familiar with the procedure for summoning rescue and emergency services. For those facilities having fully trained and equipped onsite rescue teams, it is common (and prudent) practice to have the rescue team standing outside the confined space to be immediately available if required.

Another important function of the attendant or other designated person involved in permit-required confined space entry is to ensure that unauthorized entry into the confined space is prevented. Before any permit-required confined space entry can be made, a proper confined space entry permit must be used. When employees from more than one work center (e.g., electricians, machinists, painters, and others from different work centers) or more than one employer are involved in confined space entry, an entry procedure to ensure the safety of all entrants must be developed and implemented.

After the confined space entry is completed, procedures must be in place and used to ensure that the space has been closed off and the permit canceled. The final step that should be taken after any confined space entry has been made and is completed is to critique the procedure. Questions should be asked and answers given. Did anything go wrong during the entry procedure? Did an unauthorized person make an entry into the space? Did any of the equipment used fail? Was anyone injured? Were there any employee complaints about the procedure? If such questions do come up, steps must be taken to make sure they are answered or that corrections are made to ensure that the next entry into a permit-required confined space is a safer one. At least once each year, the permits accumulated during the year (confined space permits must be retained by the employer for one year) should be reviewed. If it is apparent from the review that the procedure should be changed, then it should be changed as needed.

Safety and Health Considerations 243

Permit System

A permit system for permit-required confined space entry is required by the confined space entry standard. An entry supervisor (qualified or competent person) must authorize entry, prepare and sign written permits, order corrective measures if necessary, and cancel permits when work is completed. Permits must be available to all permit space entrants at the time of entry and should extend only for the duration of the task. They must be retained for a year to facilitate review of the confined space program. Specifically, OSHA's requirements for a permit-required confined space entry are intended to ensure that

1. A permit is actually used for entry into permit-required confined spaces.
2. An entry supervisor (the qualified or competent person) authorizes the entry.
3. The entry permit is signed.
4. Any corrective measures are taken if found necessary.
5. The permit is canceled when work is completed.

Confined space entry permits must be available to all permit space entrants at the time of entry and should extend only for the duration of the task. As we stated previously, the permits must be retained for a year to facilitate review of the confined space program. According to OSHA, an entry permit must include the following:

1. Identification of the permit space to be entered
2. The purpose of the entry
3. The date and authorized duration of the entry permit
4. The authorized entrants within the permit space by name, or by such other means as will enable the attendant to determine quickly and accurately for the duration of the permit, which authorized entrants are inside the permit space
5. The personnel, by name, currently serving as attendants
6. The individual, by name, currently serving as the entry supervisor (qualified or competent person), with a space for the signature or initials of the entry supervisor who originally authorized entry
7. The hazards of the permit space to be entered
8. The measures used to isolate the permit space and to eliminate or control permit space hazards before entry (i.e., lockout/tagout must be completed)
9. The acceptable entry conditions
10. The results of initial and periodic tests performed, accompanied by the names or initials of the testers and by an indication of when the tests were performed
11. The rescue and emergency services that can be summoned and the means (such as the equipment to use and the numbers to call) for summoning those services
12. The communication procedures used by authorized entrants and attendants to maintain contact during the entry
13. Equipment, such as personal protective equipment, testing equipment, communications equipment, alarm systems, and rescue equipment
14. Any other information whose inclusion is necessary, given the circumstances of the particular confined space, to ensure employee safety
15. Any additional permits, such as for hot work, that has been issued to authorize work in the permit space

Confined Space Training

According to 29 CFR 1910.146(g) (Training), the employer must provide training so that all employees whose work is regulated by the standard acquire the understanding, knowledge, and skills necessary for the safe performance of the duties assigned. Any work requirement is easier to perform if the person doing the task is fully trained on the proper way to accomplish it. Training offers another advantage as well—increased safety. In accomplishing any work task safely, proper training is critical.

Confined space entry operations are extremely dangerous undertakings. We stated earlier that confined spaces are very unforgiving, and this is the case even for those workers who have been well trained; however, training helps to reduce the severity of any incident. When something goes wrong (as is often the case) it is better to have fully trained personnel standing by than to have people standing by who are not trained and do not know how to properly rescue an entrant, let alone how to rescue themselves. When you get right down to it, having fully trained workers for any job just makes good common sense.

OSHA is very clear on its requirement to train confined space entry personnel. Both initial and refresher training must be provided. This training must provide employees with the necessary understanding, skills, and knowledge to perform confined space entry safely. Refresher training must be provided and conducted whenever an employee's duties change, when hazards in the confined space change, or whenever an evaluation of the confined space entry program identifies inadequacies in the employee's knowledge. The training must establish employee proficiency in the duties required and introduce new or revised procedures as necessary for compliance with the standard.

OSHA also requires the employer to certify *in writing* that the employee has been trained. This certification must include the employee's name, the signature of the trainer, and the dates of training. Typically, employers certify this training by conducting written and practical examinations (including training dry runs or drills). When an employee meets the certification requirements, the employee is normally awarded a certificate stating that he or she has been trained and certified (by whatever means). These written certifications should be filed in the employee's personnel record and training records.

Any time you conduct safety training, you must keep accurate records of the training. OSHA will want to see these records when they audit your facility (for whatever reason). Any supervisor or training official that provides critically important and possibly life-saving training would be foolish not to keep and maintain accurate training records, as they may be needed in a legal action. .

Remember, not only does OSHA require training on its confined space entry standard and other associated standards (i.e., Lockout/Tagout, Respiratory Protection, and Hot Work Permits), but this training is also critically important to the wellbeing of workers. Making sure that they know that their work organization is taking all possible steps to ensure their safety should encourage them to buy into the required safe work practices themselves. You must be able to demonstrate that this training was actually conducted.

Control of Hazardous Energy—Lockout/Tagout

> When maintenance and servicing are required on equipment and machines, the energy sources must be isolated and lockout/tagout procedures implemented. The terms *zero mechanical state* or *zero energy state* have often been used to describe machines with all energy sources neutralized. These terms have been incorporated in many standards. The current term indicating a machine at total rest is *energy isolation*. Machine energy can be electrical, pneumatic, steam, hydraulic, chemical, thermal, and others. Energy is also the potential energy from suspended parts or springs.
>
> **NSC (1992)**

OSHA's 29 CFR 1910.147 states that employers are required to develop, document, and utilize an energy control procedures program to control potentially hazardous energy. The energy control procedures must specifically outline the scope, purpose, authorization, rules, and techniques to be utilized for the control of hazardous energy and the means to enforce compliance including, but not limited to, the following:

Safety and Health Considerations

- Specific statement of the intended use of the procedure
- Specific procedural steps for shutting down, isolating, blocking, and securing machines and equipment to control hazardous energy
- Specific procedural steps for the placement, removal, and transfer of lockout devices or tagout devices and the responsibility for them
- Specific requirements for testing a machine or equipment to determine and verify the effectiveness of lockout devices, tagout devices, and other energy control measures

It has been estimated by OSHA that full compliance with the lockout/tagout standard can prevent 120 accidental deaths, 29,000 serious injuries, and 32,000 minor injuries every year (Carney, 1991). Experience has shown that many workers mistake the results of atmospheric testing that show no hazard exists in a particular confined space as meaning that the space is totally safe for entry. Indeed, this might be the case; however, many other dangers inherent to confined spaces make entry into them hazardous. If the confined space has some type of open liquid stream flowing through it, the chance for engulfment exists. If the space has electrical devices and circuitry inside, an electrocution hazard exists. If hazardous chemicals are stored and taken into the space, the potential for a hazardous atmosphere exists. Many confined spaces contain physical hazards, including piping and other obstructions; for example, rotating machinery is often housed within confined spaces.

To ensure that the confined space is indeed safe, any and all sources of hazardous energy must be isolated before entry is made. The primary method employed to accomplish this is through lockout/tagout procedures; however, the intent of employing lockout/tagout procedures goes far beyond just providing for safe confined space entry. The control of hazardous energies by locking or tagging out also applies to most servicing, adjusting, or maintenance activities involving machines and processes that place personnel at elevated risk. In addition to the sources of machine energy mentioned earlier (electrical, pneumatic, steam, and so forth), of particular concern is inadvertent activation when personnel are in contact with the hazards.

Safety professionals employed in major industrial groups recognize that the need to incorporate a viable, fully compliant lockout/tagout program (one that includes all elements of 29 CFR 1910.147) cannot be overstated. Review the historical data. It has been estimated that 7% of all workplace deaths and nearly 10% of serious accidents in many major industrial groups are associated with the failure to properly restrain or de-energize equipment during maintenance. Maintenance workers account for one third of injuries, even though they are familiar with the machines they are working on. Statistical records show that most injuries involve machines that are still running or that have been accidentally activated. In the sawmill industry, start-ups and unwanted movements have been involved in about a third of accidents that occur, and, surprisingly, it has been found that no emergency shutoffs are available about 50% of the time.

SUBPART K, MEDICAL AND FIRST AID

1910.151 Medical Services and First Aid

Note that the lack of medical and first aid services at oil and gas drilling sites is listed as one of the top ten OSHA citations in Table 12.2. Subpart K of 29 CFR 1910 directly addresses eye-flushing capabilities in the workplace and indirectly the need to have medical personnel readily available. "Readily available" can mean that there is a clinic or hospital nearby. If such a facility is not located nearby, employers must have a person onsite that has had first-aid training. Because of these OSHA requirements, the organization's safety official must, as with all other regulatory requirements, ensure that the organization is in full compliance. First-aid awareness and training in the workplace usually require providing lectures, interactive video presentations, discussions, and hands-on training to teach participants how to

- Recognize emergency situations.
- Check the scene and call for help.
- Avoid bloodborne pathogen exposure.
- Care for wounds, bone and soft-tissue injuries, head and spinal injuries, burns, and heat and cold emergencies.
- Manage sudden illnesses, stroke, seizure, bites, and poisoning.
- Minimize stroke.

First-aid services in the workplace typically include training and certification of selected individuals to perform cardiopulmonary resuscitation (CPR) when necessary. This training usually combines lectures, video demonstrations, and hands-on manikin training. This training teaches participants to

- Call and work with emergency medical services (EMS).
- Recognize breathing and cardiac emergencies that call for CPR.
- Perform CPR and care for breathing and cardiac emergencies.
- Avoid bloodborne pathogen exposure.
- Know the role of automated external defibrillators in the cardiac chain of survival.

The American Red Cross points out that typical first-aid and CPR training for the workplace has been enhanced to include training on the automated external defibrillator (AED): "Although the idea of using a handheld device to deliver a shock directly into a coworker's heart may seem daunting, the American Red Cross hopes this life-saving practice becomes more common over the next year" (Orfinger, 2002). AED training focuses on typical AED equipment with hands-on simulation, lectures, and live and video demonstrations. Participants learn to

- Call and work with EMS.
- Care for conscious and unconscious choking victims.
- Perform rescue breathing and CPR.
- Use an AED safely on a victim of sudden cardiac arrest.

SUBPART L, FIRE PROTECTION

> 1910.157 Portable Fire Extinguishers
> 1910.165 Employee Alarm Systems

Fire Safety

As shown in Table 12.2, noncompliance with OSHA's portable fire extinguisher standard was one of the top ten cited violations in the oil and gas drilling industry. Although technical knowledge about flame, heat, and smoke continues to grow, and although additional information continues to be acquired concerning the ignition, combustibility, and flame propagation of various solids, liquids, and gases, it still is not possible to predict with any degree of accuracy the probability of fire initiation or consequences of such initiation. Thus, while the study of controlled fires in laboratory situations provides much useful information, most unwanted fires happen and develop under widely varying conditions, making it virtually impossible to compile complete bodies of information from actual unwanted fire situations. This fact is further complicated because the progress of any unwanted fire varies from the time of discovery to the time when control measures are applied (Cote and Bugbee, 1991).

Industrial facilities are not immune to fire and its terrible consequences. Each year fire-related losses in the United States are considerable. According to conservative figures reported by Brauer (1994), about 1 million fires involving structures and about 8000 deaths occur each year. The total annual property loss is more than $7 billion. Complicating the fire problem is the point that Cote

Safety and Health Considerations

and Bugbee (1991) made above—the unpredictability of fire. Fortunately, facility safety officials are aided in their efforts in fire prevention and control by the authoritative and professional guidance readily available from the National Fire Protection Association (NFPA), the National Safety Council (NSC), fire code agencies, local fire authorities, and OSHA regulations. In this section, we discuss fire prevention and control and fire protection provided by the use of fire extinguishers.

Along with providing fire prevention guidance, OSHA regulates several aspects of fire prevention and emergency response in the workplace. Emergency response and evacuation and fire prevention plans are required under 29 CFR 1910.38. The requirements for fire extinguishers and worker training are addressed in 1910.157. Along with state and municipal authorities, OSHA has listed several fire safety requirements for general industry.

All of the advisory and regulatory authorities approach fire safety in much the same manner; for example, they all agree that electrical short circuits or malfunctions usually start fires in the workplace. Other leading causes of workplace fires are friction heat, welding and cutting of metals, improperly stored flammable/combustible materials, open flames, and cigarette smoking.

For fire to start, three components must be present: *temperature (heat)*, *fuel*, and *oxygen*. Because oxygen is naturally present in most environments on Earth, fire hazards usually involve the mishandling of fuel or heat. The fire triangle helps us understand fire prevention, because the objective of fire prevention and firefighting is to separate any one of the fire ingredients from the other two. To prevent fires, it is necessary to keep fuel (combustible materials) away from heat (as in airtight containers), thus isolating the fuel from oxygen in the air. To gain a better perspective of the chemical reaction known as *fire*, remember that the combustion reaction normally occurs in the gas phase; generally, the oxidizer is air. If a flammable gas is mixed with air, there is a minimum gas concentration below which ignition will not occur. That concentration is known as the *lower flammable limit* (LFL). When trying to visualize the LFL and its counterpart, the *upper flammable limit* (UFL), it helps to use an example that most people are familiar with—the combustion process that occurs in the automobile engine. When an automobile engine has a gas/air mixture that is below the LFL, the engine will not start because the mixture is too lean. When the same engine has a gas/air mixture that is above the UFL, it will not start because the mixture is too rich (the engine is flooded). When the gas/air mixture is between the LFL and UFL levels, however, the engine should start (Spellman, 1996b).

Fire Prevention and Control

The best way to prevent and control fires in the workplace is to institute a facility Fire Safety Program. Safety experts agree that the best way to reduce the possibility of fire in the workplace is prevention. For the facility safety official this begins with developing a fire prevention plan, which must be in writing and must list fire hazards and fire controls and specify the control jobs and personnel responsible and emergency actions to be taken. More specifically, in accordance with OSHA 29 CFR 1910.38, the elements that make up the plan must include the following:

1. A list of the major workplace fire hazards and their proper handling and storage procedures, potential ignition sources (such as welding, smoking, and others), their control procedures, and the type of fire protection equipment or systems that can control a fire involving them.
2. Names or regular job titles of those personnel responsible for maintenance of equipment and systems installed to prevent or control ignitions or fires.
3. Names or regular job titles of those personnel responsible for control of fuel source hazards.
4. Control of accumulation of flammable and combustible waste materials and residues so that they do not contribute to a fire emergency. These housekeeping procedures must be included in the written fire prevention plan.
5. All workplace employees must be apprised of the fire hazards of the materials and processes to which they are exposed.

6. All new employees must be made aware of those parts of the fire prevention plan that the employee must know to protect the employee in the event of an emergency. The written plan must be kept in the workplace and made available for employee review.

7. The employer is required to regularly and properly maintain, according to established procedures, equipment and systems installed on heat-producing equipment to prevent accidental ignition of combustible materials. The maintenance procedure must be included in the written fire prevention plan.

Fire prevention and control measures are those taken *before* fires start and include the following:

- Elimination of heat and ignition sources
- Separation of incompatible materials
- Adequate means of firefighting (e.g., sprinklers, extinguishers, hoses)
- Proper construction and choices of storage containers
- Proper ventilation systems for venting and reducing vapor buildup
- In the event of fire emergency, maintaining unobstructed means of egress for workers, as well as adequate aisle and fire-lane clearance for firefighters and equipment

In the event of a fire emergency, all employees need to know what to do; they need a plan to follow. The fire emergency plan normally is the protocol to follow for fire emergency response and evacuation. Typically, the facility safety official is charged with developing fire prevention and emergency response plans that spell out everyone's role. In this effort, the safety official's goal should be to make the plan as simple as possible. In addition to a fire emergency response plan, each facility needs to have a well-thought-out fire emergency evacuation plan.

Fire Protection Using Fire Extinguishers

OSHA, under 29 CFR 1910.157, requires employers to provide portable fire extinguishers that are mounted, located, and identified so they are readily accessible to employees without subjecting the employee to possible injury. OSHA also requires that each workplace institute a portable fire extinguisher maintenance plan. Fire extinguisher maintenance service must take place at least once a year, and a written record must be kept to show the maintenance or recharge date. Note that, when the facility provides portable fire extinguishers for employee use in the facility, the employee must be provided with training to learn the general principles of fire extinguisher use and the hazards involved in firefighting. Employees who are expected to use fire extinguishers in the workplace must be trained on the types of fire extinguishers available to them, the different classes of fires, and where the fire extinguishers are located. The ABC type of fire extinguisher is probably best suited for most industrial applications because it can be used on Class A, B, and C fires. Class A is used for common combustibles (such as paper, wood, and most plastics); Class B is for flammable liquids (such as solvents, gasoline, and oils); and Class C is for fires in or near live electrical circuits. In areas such as electrical substations and switchgear rooms, only Class C (carbon dioxide, CO_2) should be used. Though combination Class A, B, and C extinguishers will extinguish most electrical fires, the chemical residue left behind can damage delicate electrical/electronic components; thus, the CO_2 type of extinguisher is more suitable for extinguishing electrical fires. Each employee must know how to use the fire extinguisher. Most importantly, employees must know when it is not safe to use fire extinguishers—that is, when the fire is beyond being extinguishable with a portable fire extinguisher. Emergency telephone numbers should be strategically placed throughout the workplace. Employees need to know where they are posted. Workers should be trained on the information they need to provide to the 911 operator (or other emergency service number) in case of fire.

Safety and Health Considerations

Miscellaneous Fire Prevention Measures

In addition to basic fire prevention, emergency response, and fire extinguisher training, employees must be trained on the hazards involved with flammable and combustible liquids. 29 CFR 1910.106 addresses this area. Industrial facilities typically use all types of flammable and combustible liquids. These dangerous materials must be clearly labeled and stored safely when not in use. The safe handling of flammable and combustible liquids is a topic that needs to be fully addressed by the facility safety official and workplace supervisor. Worker awareness of the potential hazards that flammable and combustible liquids pose must be stressed. Employees need to know that flammable and combustible liquid fires burn extremely hot and can produce copious amounts of dense, black smoke. Explosion hazards exist under certain conditions in enclosed, poorly ventilated spaces where vapors can accumulate. A flame or spark can cause vapors to ignite, creating a flash fire with the terrible force of an explosion. One of the keys to reducing the potential spread of flammable and combustible fires is to provide adequate containment. All storage tanks should be surrounded by storage dikes or containment systems, for example. Correctly designed and built dikes will contain spilled liquid. Spilled flammable and combustible liquids that are contained are easier to manage than those that have free run of the workplace. Properly installed containment dikes can prevent environmental contamination of soil and groundwater.

SUBPART N, MATERIALS HANDLING AND STORAGE

1910.176 Materials Handling and Storage
1910.184 Slings

Rigging Safety Program

In lifting the various materials and supplies, a number of standard chokers, slings, bridle hitches, and basket hitches can be used. Because loads vary in physical dimension, shape, and weight, the rigger needs to know what method of attachment can be safely used. It is estimated that 15% to 35% of crane accidents may involve improper rigging. The employer needs to train those employees who are responsible for rigging loads. They need to be able to (1) know the load, (2) judge distances, (3) properly select tackle and lifting gear, and (4) direct the operation. The single most important rigging precaution is to determine the weight of the load before attempting to lift it. The weight of the load will in turn determine the lifting device, such as a crane, and the rigging gear to be used. It is also important to rig a load so that it will be stable, that is, it does not move as it is lifted.

NSC (1992)

The facility safety professional needs to realize that special safety precautions apply to rigging operations and to using and storing fiber ropes, rope slings, wire ropes, chains, and chain slings. The safety official should know the properties of the various types used, the precautions for use, and the maintenance required. In addition, the safety official must be familiar with the requirements of OSHA's rigging equipment for material handling standard (29 CFR 1926.251). Rigging operations are inherently dangerous. Any time any type of load is lifted, the operation is dangerous in itself. When heavy loads are lifted several feet and suspended in air while they are moved from one place to another, the dangers are increased exponentially. Although rigging and lifting operations include the use of several different types of mechanical devices such as cranes, winches, chain falls, and come-alongs, in this section we focus on those components that form the interface between the load and the lifting or hoisting equipment—the ropes, chains, and slings. We place our focus on these devices not only because they are the most commonly used rigging devices found in industrial applications but also because the safety professional is directly responsible for ensuring that they are safe to use—and are used safely.

250 Hydraulic Fracturing Wastewater: Treatment, Reuse, and Disposal

Safety: Ropes, Slings, and Chains

Because of the dangers inherent in any rigging and lifting operation, the safety official must check out and ensure the safety of every element involved. This may seem like common sense to some, but others might be surprised to find out how often rigging mistakes are made, by assuming that the only factor that need be considered is the safe operation of hoisting equipment to lift a given load. Experience has shown that the attachments used to secure the hook to the load are often overlooked and thus the cause of failure and injuries. In this section, we discuss OSHA's general requirements and the main rigging attachments: ropes, slings, and chains.

Rigging Equipment and Attachments: General

In 29 CFR 1926.251, the point is made that rigging equipment for handling material must not be loaded in excess of its recommended safe working load (see Tables H-1 through H-20 in the standard). All such equipment must be inspected prior to use on each shift and as necessary during use to ensure safety. Any rigging equipment found to be defective must be immediately removed from service. Rigging equipment not in use that presents a hazard must be removed from the immediate working area to ensure the safety of employees. The safety official must ensure that all custom-designed grabs, hooks, clamps, or other lifting accessories are marked to indicate their safety working loads. Each device must be proof tested to 125% of its rated load before use. Whenever a sling is used, the following practices must be observed:

- Slings must not be shortened with knots, bolts, or other makeshift devices.
- Sling legs must not be kinked.
- Slings used in a basket hitch must have the loads balanced to prevent slippage.
- Slings must be padded or protected from the sharp edges of their loads.
- Shock loading is prohibited
- A sling must not be pulled from under a load when the load is resting on the sling.
- Hands or fingers must not be placed between the sling and its load while the sling is being tightened around the load.

Rope Slings

Ropes used in rigging (for slings) are usually divided into two main classes: fiber rope slings and wire rope slings. Fiber ropes are further divided into natural and synthetic fibers depending on their construction. There are many types of slings. Slings normally have a fixed length. They may be made from various materials and have the form of rope, belts, mesh, or fabric. Natural fiber ropes and slings are usually made from manila, sisal, or henequen fibers. Most natural fiber ropes and slings used in industry today are made from manila fibers because of its superior breaking strength, consistency between grades, excellent wear properties in both freshwater and saltwater atmospheres, and elasticity. The main advantages of natural fiber ropes are their price and their ability to form or bend around angles of the object being lifted. The disadvantages of using natural fiber ropes include increased susceptibility to cuts and abrasions, their reduced capability or inability to be used to lift materials at elevated temperatures, and that hot or humid conditions may reduce their service life. Fiber ropes should never be used in atmospheres where they may come in contact with acids and caustics, as these substances will degrade the fibers. Safe working loads of various sizes and classifications of natural fiber ropes can be determined from tables in 29 CFR 1926.251.

Synthetic fiber rope slings are made from synthetic fibers (such as nylon, polyester, polypropylene, polyethylene, or a combination of these) to obtain the desired properties. Synthetic fiber ropes have many of the same qualities as natural fiber rope slings, but are in much wider use throughout the industry because they can be engineered to fit a particular operation. Synthetic fiber ropes have many advantages, including increased strength and elasticity, over natural fiber rope. Synthetic fiber rope also stands up better to shock loading and has better resistance to abrasion than natural fiber rope. One of the key advantages of synthetic fiber rope is that it does not swell when wet. It is also

Safety and Health Considerations

more resistant to acids, caustics, alcohol-based solvents, bleaching solutions, and their atmospheres. As with the use of natural fiber rope, synthetic fiber rope also has some disadvantages, including damage from excessive heat (they can melt) or from alkalis and susceptibility to abrasion damage. They also cost more than natural ones.

Wire Rope

The most widely used type of rope sling in industry is the cable laid 6 × 19 and 6 × 37 wire rope. By definition, wire rope is a twisted bundle of cold-drawn steel wires, usually composed of wires, strands, and a core. When used in rope slings, wire ropes must have a minimum clear length of wire rope 10 times the component rope diameter between splices, sleeves, or end fittings. The main reasons for the wider usage of wire compared to fiber rope are its greater strength, durability, predictability of stretch characteristics when placed under heavy stresses, and stable physical characteristics over a wide variety of environmental conditions. The main advantages of wire rope that is preformed are its lessened tendency to unwind, set, kink, or generate sharp protruding wires.

Chains and Chain Slings

Steel and alloys (stainless steel, monel, bronze, and other metals) are commonly used for lifting slings made of chain. The safety official needs to know a number of facts related to chain slings and the type of chain that is authorized for use in slings; for example, the rated capacity (working load limit) for welded alloy steel chain slings must conform to the values in the appropriate tables in 29 CFR 1926.251. Whenever wear at any point of any chain link exceeds that specified, the assembly must be removed from service. All such slings have permanently affixed durable identification that states size, grade, rated capacity, and the sling manufacturer. Finally, regular hardware chain or other chain not specifically designed for use in slings should not be used for load lifting.

Proof Testing Rigging Equipment

One of the safety official's primary duties involving rigging operations is to ensure that the equipment used is safe to use. Ropes, slings, chains, and other lifting devices must be certified via proof testing to verify their soundness and safety for use. *Proof testing* is a nondestructive tension test performed by the sling manufacturer or an equivalent entity to verify construction and workmanship of a sling or other lifting device. During proof testing, a *proof load* is applied to test the lifting device. The safety official is responsible for ensuring that, before each use, each new, repaired, or reconditioned lifting device (rope, chain, or sling)—including all welded components in the sling assembly—is proof tested by the sling manufacturer or equivalent entity, in accordance with American Society for Testing and Materials Specification A391-65 (ANSI G61.1-1968). The safety official should ensure that a written certification of the proof test is provided and that such records are available for review by regulatory auditors. Typically, sling proof test or load test results are stamped, marked, or labeled right on the sling itself. In addition to verifying the satisfactory condition of each sling or other rigging component, the safety official should ensure that certification labels and identification tags are attached and visible and that test data (e.g., load rating) are current.

Rigging Inspections

Each day before being used, the sling and all rigging fastenings and attachments must be inspected for damage or defects by a competent person designated by the employer. A few of the kinds of items that should be inspected to ensure that slings are safe to use include the following:

1. Alloy steel chain slings must have permanently affixed, durable identification stating size, grade, rated capacity, and reach.
2. A thorough periodic inspection of alloy steel chain slings in use must be made on a regular basis (at least once every 12 months).

3. A record must be maintained of the most recent month in which each alloy steel chain sling was thoroughly inspected.
4. Alloy steel chains slings must be permanently removed from service if they are heated above 1000°F.
5. Worn or damaged alloy steel chain slings and attachments must be taken out of service until repaired.
6. Wire rope slings must be used only with loads that do not exceed the rated capacities.
7. Fiber-core wire rope slings of all grades must be permanently removed from service if they are exposed to temperatures in excess of 200°F.
8. Welding of end attachments, except covers to thimbles, must be performed prior to the assembly of the sling.
9. Welded end attachments must be proof tested by the manufacturer or equivalent entity at twice their rated capacity prior to initial use.
10. All synthetic web slings must be marked or coded to show the rated capacities for each type of hitch and type of synthetic web material.

Additional inspection must also be performed during sling use where service conditions warrant. Damaged or defective slings must be immediately removed from service. Make them unusable by burning or cutting them before they are discarded; otherwise, they may mysteriously reappear and be used again.

SUBPART O, MACHINERY AND MACHINE GUARDING

1910.212   General requirements for all machines
1910.215   Abrasive wheel machinery
1910.219   Mechanical power-transmission apparatus

Machine Guarding

[S]afety and health on the job begin with sound engineering and design. The engineer and designer will be familiar with most of the common hazards to be dealt with in the design phase. For the senior manager, however, highlighting the most common hazards found in equipment and the ones requiring particular alertness [is called for here]. The most common sources of mechanical hazards are unguarded shafting, shaft ends, belt drives, gear trains, and projections on rotating parts. Where a moving part passes a stationary part or another moving part, there can be a scissor-like effect on anything caught between the parts. A machine component which moves rapidly with power or a point of operation where the machine performs its work are also typical hazard sources.

There are probably over 2 million metalworking machines and half that many woodworking machines in use that are at least 10 years old. Most are poorly guarded, if at all. Even the newer ones may have substandard guards, in spite of OSHA requirements. ...The basic objective of machine guarding is to prevent personnel from coming in contact with revolving or moving parts such as belts, chains, pulleys, gears, flywheels, shafts, spindles, and any working part that creates a shearing or crushing action or that may entangle the worker.

Machine guarding is visible evidence of management's interest in the worker and its commitment to a safe work environment. It is also to management's benefit, as unguarded machinery is a principal source of costly accidents, waste, compensation claims, and lost time.

Ferry (1990)

The basic purpose of machine guarding is to prevent contact of the human body with dangerous parts of machines. Moving machine parts have the potential for causing severe workplace injuries, such as crushed fingers or hands, amputations, burns, and blindness, just to name a few. Machine guards are essential for protecting workers from these needless and preventable injuries. Any machine part, function, or process that may cause injury must be safeguarded. When the operation of a machine

Safety and Health Considerations 253

or accidental contact with it can injure the operator or others in the vicinity, the hazards must be either eliminated or controlled (OSHA, 2003). Our experience has clearly (and much too frequently) demonstrated that when an arm, finger, hair, or any body part enters into or makes contact with moving machinery, the results can be not only gory, bloody, and disastrous but also sometimes fatal.

Depending on the machine and the types of hazards it presents, methods of machine guarding vary greatly. The intent of this section is to familiarize safety professionals with the hazards of unguarded machines, common safeguarding methods, and the safeguarding of machines—all of which, if followed, combine to ensure that Ferry's main point—"Machine guarding is visible evidence of management's interest in the worker and its commitment to a safe work environment"— becomes a reality. It logically follows that if the employer provides a safe workplace then all sides benefit from the results.

SUBPART P, HAND AND PORTABLE POWERED TOOLS

1910.242 Hand and Portable Power Tools and Equipment

Safe Work Practice for Hand Tools, Power Tools, and Portable Power Equipment

1. Use care and caution when using hand tools, power tools, and portable power equipment.
2. Do not use tools and equipment unless trained and experienced in the proper use and operation of the tools and equipment.
3. Use the proper tools and equipment for the required task. *Never* use tools or equipment in a misapplication.
4. Inspect tools carefully before using them and discard any tool that appears unsafe.
5. Use care and caution when using tools with sharp points or edges such as saws, knives, chisels, punches, and screwdrivers. Hand tools of this type are not to be set down on surfaces where they can be tripped over, stepped on, or bumped.
6. Use equipment guards and other safety devices at all times when operating tools and equipment. *Never* bypass a safety guard or switch.
7. Use safety glasses, goggles, and face shields as appropriate.
8. Inspect tools on a regular basis and before each use to ensure that tools and equipment are in good working order.
9. Keep tools and power equipment clean and in good operating condition. *Never* use broken hand tools and power tools.
10. Replace worn-out tools and equipment.
11. Use only grounded or double-insulated electrical tools.
12. Never use electrical tools in or near water without a ground fault interrupter circuit. Never stand in water when using an electrical tool or equipment.
13. Have frayed or broken electrical cords repaired or replaced immediately.
14. Shut off gasoline or diesel engines before refueling whenever possible.
15. Direct exhaust fumes from gasoline or diesel engines away from work areas.
16. Apply working force away from the body to minimize the chance for injury if the hand tool slips.
17. Be sure tool handles are fitted to tools and free of grease and other slippery substances.
18. Dress cold chisels, punches, hammers, drift pins, and other similar tools that have a tendency to mushroom from repeated poundings. As soon as they begin to crack and curl, grind a slight bevel (approximately 3/16 inch or 4.7 mm) around the head to prevent it from mushrooming.
19. Do not carry sharp edges or pointed tools in clothing pockets.
20. Do not use defective wrenches, such as open-end and adjustable wrenches with spur jaws or pipe wrenches with dull teeth.

254 Hydraulic Fracturing Wastewater: Treatment, Reuse, and Disposal

21. Do not apply hand tools to moving machinery except tools designed for the purpose and necessary in the operation.
22. Do not throw tools and material from one employee to another, or from one location to another. Use a suitable container to raise or lower small equipment or tools between elevations.

SUBPART Q, WELDING, CUTTING, AND BRAZING

1910.252 General Requirements (Hot Work Permit)
1910.253 Oxygen–Fuel Gas Welding and Cutting

Hot Work Permit Procedure

Many organizations use a permit procedure for all hot work, except that involving normal operations or processes. Hot work is any kind of welding, cutting, burning, or activity that involves or generates sparks or open flame. It includes heated equipment that may provide an ignition source for a fire. Hot work often involves people from a maintenance department going to other departments to perform activities. The main idea in a hot work permit procedure is to ensure that supervisors of all departments involved and workers who may be involved in any way in the work participate in the decision to start work and to conduct it safely.

Brauer (1996)

Exactly what is accomplished by employing the use of a hot work permitting system? A hot work permitting procedure works primarily to ensure that work areas and all adjacent areas to which sparks and heat might be spread (including floors above and below and on opposite sides of walls) are inspected during the work and again 30 minutes after the work is completed, to ensure they are firesafe. During the inspection, work areas and surrounding areas should be inspected to ensure that

- Sprinklers are in service.
- Cutting and welding equipment is in good repair.
- Floors are swept clean of combustibles.
- Combustible floors are wetted down and covered with damp sand, metal, or other shields.
- No combustible material or flammable liquids are within 35 feet of the work.
- Combustibles and flammable liquids within 35 feet of work are protected with covers, guards, or metal shields.
- All wall and floor openings within 35 feet of work are covered.
- Covers are suspended beneath the work to collect sparks.
- For work on walls or ceilings, construction is of noncombustible materials.
- Combustibles are moved away from the opposite side of the wall.
- For work on or in enclosed tanks, containers, ducts, etc., equipment is cleaned of all combustibles and purged of flammable vapors.
- Fire watch is provided during and 30 minutes after operation.
- The assigned fire watch is properly trained and equipped.

Fire Watch Requirements

A fire watch must be assigned whenever hot work operations are being performed around hazardous materials, in confined spaces, and other times when there is the danger of fire or explosion from such work. OSHA has specific requirements regarding fire watch duties. Fire watchers are required whenever welding or cutting is performed in locations where other than a minor fire might develop or where any of the following conditions exist:

1. Appreciable combustible material, in building construction or contents, are closer than 35 feet (10.7 m) to the point of operation.
2. Appreciable combustibles are more than 35 feet (10.7 m) away but are easily ignited by sparks.

Safety and Health Considerations

3. Wall or floor openings within a 35-foot (10.7-m) radius expose combustible materials in adjacent areas, including concealed spaces in walls or floors.
4. Combustible materials are adjacent to the opposite side of metal partitions, walls, ceilings, or roofs and are likely to be ignited by conduction or radiation.

Fire watchers must have fire-extinguishing equipment readily available and be trained in its use. They should be familiar with facilities for sounding an alarm in the event of a fire. They should watch for fires in all exposed areas and try to extinguish them only when obviously within the capacity of the equipment available; otherwise, they should sound the alarm. A fire watch should be maintained for at least a half hour after completion of welding or cutting operations to detect and extinguish possible smoldering fires.

Welding and Cutting Safety

Welding is typically thought of as the electric arc and gas (fuel gas/oxygen) welding process; however, welding can involve many types of processes. Some of these other processes include inductive welding, thermite welding, flash welding, percussive welding, and plasma welding, among others. The most common type of electric arc welding also has many variants, including gas shielded welding, metal arc welding, gas–metal arc welding, gas–tungsten arc welding, and flux-cored arc welding (McElroy, 1980). Welding, cutting, and brazing are widely used processes. 29 CFR 1910 Subpart Q contains the standards relating to these processes in all of their various forms. The primary health and safety concerns are fire protection, employee personal protection, and ventilation. The standards contained in this subpart are as follows:

1910.251 Definitions
1910.252 General Requirements
1910.253 Oxygen–Fuel Gas Welding and Cutting
1910.254 Arc Welding and Cutting
1910.255 Resistance Welding
1910.256 Sources of Standards
1910.257 Standards Organization

A study on deaths related to welding/cutting incidents (OSHA, 1989) revealed that of 200 deaths over an 11-year period, 80% were caused by failure to practice safe work procedures. Surprisingly, only 11% of deaths involved malfunctioning or failed equipment, and only 4% were related to environmental factors. The implications of this study should be obvious: Equipment malfunctions or failures are not the primary causal factor of hazards presented to workers. Instead, the safety official's emphasis should be on establishing and ensuring safe work practices for welding tasks.

General Welding Safety

A viable Welding Safety Program should consist of the elements provided in 29 CFR 1910.252, Welding, Cutting, and Brazing. The fire prevention and protection element of any welding safety program begins with basic precautions, including the following:

1. *Fire hazards*—If the material or object cannot be readily moved, all movable fire hazards in the area must be moved to a safe location.
2. *Guards*—If the object to be welded or cut cannot be moved, and if all the fire hazards cannot be removed, then guards should be used to confine the heat, sparks, and slag and to protect the immovable fire hazards.
3. *Restrictions*—If the welding or cutting cannot be performed without removing or guarding against fire hazards, then the welding and cutting should not be performed.

4. *Combustible material*—Wherever floor openings or cracks in the flooring cannot be closed, precautions must be taken so that no readily combustible materials on the floor below will be exposed to sparks that might drop through the floor. The same precautions should be taken with cracks or holes in walls, open doorways, and open or broken windows.

5. *Fire extinguishers*—Suitable fire extinguishing equipment must be maintained in a state of readiness for instant use. Such equipment may consist of pails of water, buckets of sand, hoses, or portable extinguishers, depending on the nature and quantity of the combustible material exposed.

6. *Fire watch*—Fire watchers are required whenever welding or cutting is performed in locations where other than a minor fire might develop. Fire watchers are required to have fire-extinguishing equipment readily available and must be trained in its use. They must be familiar with facilities for sounding an alarm in the event of fire. They must watch for fires in all exposed areas, try to extinguish them only when obviously within the capacity of the equipment available, or otherwise sound the alarm. A fire watch must be maintained for at least a half-hour after completion of welding or cutting operations to detect and extinguish possible smoldering fires.

7. *Authorization*—Before cutting or welding is permitted, the individual responsible for authorizing cutting and welding operations must inspect the area. The responsible individual must designate precautions to be followed in granting authorization to proceed, preferably in the form of a written permit (hot work permit).

8. *Floors*—Where combustible materials such as paper clippings, wood shavings, or textile fibers are on the floor, the floor must be swept clean for a radius of at least 35 feet (OSHA requirement). Combustible floors must be kept wet, covered with damp sand, or protected by fire-resistant shields. Where floors have been wet down, personnel operating arc welding or cutting equipment must be protected from possible shock.

9. *Prohibited areas*—Welding or cutting must not be permitted in areas that are not authorized by management. Such areas include in sprinklered buildings while such protection is impaired; in the presence of explosive atmospheres, or explosive atmospheres that may develop inside uncleaned or improperly prepared tanks or equipment that have previously contained such materials, or that may develop in areas with an accumulation of combustible dusts; and in areas near the storage of large quantities of exposed, readily ignitable materials such as bulk sulfur, baled paper, or cotton.

10. *Relocation of combustibles*—Where practicable, all combustibles must be relocated at least 35 feet from the work site. Where relocation is impracticable, combustibles must be protected with fireproofed covers, or otherwise shielded with metal of fire-resistant guards or curtains.

11. *Ducts*—Ducts and conveyor systems that might carry sparks to distant combustibles must be suitably protected or shut down.

12. *Combustible walls*—Where cutting or welding is done near walls, partitions, ceilings, or roofs of combustible construction, fire-resistant shields or guards must be provided to prevent ignition.

13. *Noncombustible walls*—If welding is to be done on a metal wall, partition, ceiling, or roof, precautions must be taken to prevent ignition of combustibles on the other side from conduction or radiation, preferably by relocating the combustibles. Where combustibles are not relocated, a fire watch on the opposite side from the work must be provided.

14. *Combustible cover*—Welding must not be attempted on a metal partition wall, ceiling, or roof that has combustible coverings, nor on any walls or partitions, ceilings, or roofs that have combustible coverings or on walls or partitions of combustible sandwich-type panel construction.

Safety and Health Considerations

15. *Pipes*—Cutting or welding on pipes or other metal in contact with combustible walls, partitions, ceilings, or roofs must not be undertaken if the work is close enough to cause ignition by conduction.

16. *Management*—Management must recognize its responsibility for the safe usage of cutting and welding equipment on its property, must establish specific areas for cutting and welding, and must establish procedures for cutting and welding in other areas. Management must also designate an individual responsible for authorizing cutting and welding operations in areas that are not specifically designed for such processes. Management must also insist that cutters or welders and their supervisors are suitably trained in the safe operation of their equipment and the safe use of the process. Management has a duty to inform contractors about flammable materials or hazardous conditions of which they may not be aware.

17. *Supervisor*—The supervisor has many responsibilities in welding and cutting operations, including the following:
    * Is responsible for the safe handling of the cutting or welding equipment and the safe use of the cutting or welding process.
    * Must determine the combustible materials and hazardous area present or likely to be present in the work location.
    * Must protect combustibles from ignition by whatever means necessary.
    * Must secure authorization for the cutting or welding operations from the designated management representative.
    * Must ensure that the welder or cutter secures his or her approval that conditions are safe before going ahead.
    * Must determine that fire protection and extinguishing equipment is properly located at the site.
    * Where fire watches are required, must ensure that they are available at the site.

18. *Fire prevention precautions*—Cutting and welding must be restricted to areas that are or have been made fire safe. When work cannot be moved practically, as in most construction work, the area must be made safe by removing combustibles or protecting combustibles from ignition sources.

19. *Welding and cutting used containers*—No welding, cutting, or other hot work is to be performed on used drums, barrels, tanks, or other containers until they have been cleaned so thoroughly as to make absolutely certain that no flammable materials are present, or any substances such as greases, tars, acids, or other materials that when subjected to heat might produce flammable or toxic vapors. Any pipelines or connections to the drum or vessel must be disconnected or blanked.

20. *Venting and purging*—All hollow spaces, cavities, or containers must be vented to permit the escape of air or gases before preheating, cutting, or welding. Purging with inert gas (e.g., nitrogen) is recommended.

21. *Confined spaces*—To prevent accidental contact in confined space operations involving hot work, when arc welding is to be suspended for any substantial period of time (such as during breaks or overnight), all electrodes are to be removed from the holders and the holders carefully located so that accidental contact cannot occur. The machine must be disconnected from the power source. To eliminate the possibility of gas escaping through leaks or improperly closed valves, when gas welding or cutting, the torch valves must be closed and the gas supply to the torch positively shut off at some point outside the confined area whenever the torch is not to be used for a substantial period of time (such as during breaks or overnight). Where practicable, the torch and hose must also be removed from the confined space.

Note: The safety official should use the proceeding information as guidance in preparing the organizational Welding Safety Program.

Personal Protective Equipment and Other Protection

Personnel involved in welding or cutting operations not only must learn and abide by safe work practices but must also be aware of possible bodily dangers during such operations. They must learn about the personal protective equipment (PPE) and other protective devices and measures designed to protect them.

Arc Welding Safety

In 29 CFR 1910.254 (Arc Welding and Cutting), OSHA specifically lists various safety requirements that must be followed when arc welding; for example, in equipment selection, OSHA stipulates that welding equipment must be chosen for safe application to the work to be done. Welding equipment must also be installed safely as per the manufacturer's guidelines and recommendations. Finally, OSHA specifies that workpersons designated to operate arc-welding equipment must have been properly trained and qualified to operate such equipment. Training and qualification procedures are important elements that must be included in any welding safety program. Along with OSHA's requirements above, the safety official must ensure that the facility's welding safety program includes written safe work practices detailing and explaining safety requirements that must be followed whenever arc welding is performed.

Gas Welding and Cutting

Specific safety requirements for oxygen–fuel gas welding and cutting are covered under 29 CFR 1910.253 and are listed in the units involving oxyacetylene welding. These safety requirements (precautions) cover proper handling of cylinders, operation of regulators, use of oxygen and acetylene, welding hose, testing for leaks, and lighting a torch. All of these safety requirements are extremely important and should be followed with the utmost care and regularity. Along with the normal precautions to be observed in gas welding operations, a very important safety procedure involves the piping of gas. All piping and fittings used to convey gases from a central supply system to work stations must withstand a minimum pressure of 150 psi. Oxygen piping can be of black steel, wrought iron, copper, or brass. Only oil-free compounds should be used on oxygen threaded connections. Piping for acetylene must be of wrought iron. (*Note:* Acetylene gas must never come into contact with unalloyed copper, except in a torch; any such contact could result in a violent explosion.) After assembly, all piping must be blown out with air or nitrogen to remove foreign materials. Five basic rules contribute to the safe handling of oxyacetylene equipment (Giachino and Weeks, 1985):

1. Keep oxyacetylene equipment clean, free of oil, and in good condition.
2. Avoid oxygen and acetylene leaks.
3. Open cylinder valves slowly.
4. Purge oxygen and acetylene lines before lighting a torch.
5. Keep heat, flame, and sparks away from combustibles.

Torch Cutting Safety

Whenever torch cutting operations are conducted, the possibility of fire is very real, because proper precautions are often not taken. Torch cutting is particularly dangerous because sparks and slag can travel several feet and can pass through cracks out of sight of the operator. The safety official must ensure that the persons responsible for supervising or performing cutting of any kind follow accepted safe work practices. Accepted safe work practices for torch cutting operations typically include the following:

Safety and Health Considerations

1. Use of a cutting torch where sparks will be a hazard is prohibited.
2. If cutting is to be over a wooden floor, the floor must be swept clean and wet down before starting the cutting.
3. A fire extinguisher must be kept in reach any time torch-cutting operations are conducted.
4. Cutting operations should be performed in wide-open areas so sparks and slag will not become lodged in crevices or cracks.
5. In areas where flammable materials are stored and cannot be removed, suitable fire-resistant guards, partitions, or screens must be used.
6. Sparks and flame must be kept away from oxygen cylinders and hoses.
7. Never perform cutting near ventilators.
8. Fire watchers with fire extinguishers should be used.
9. Never use oxygen to dust off clothing or work.
10. Never substitute oxygen for compressed air.

SUBPART Z, TOXIC AND HAZARDOUS SUBSTANCES

1900.1200   Hazard Communication

Fracking fluids can contain a toxic mix of hundreds of chemicals. Surprisingly, the exact chemical makeup of the chemicals used in fracking is not public knowledge, because disclosure of these fluids is protected as proprietary trade secrets. However, when fracking fluids contain chemicals that can harm workers, the employer has an obligation under OSHA's Hazard Communication Standard (29 CFR 1910.1200) to provide information on how workers can protect themselves with proper handling techniques and personal protective equipment, if necessary. Hazard Communication (HazCom) is a dynamic standard that is constantly upgraded as needed. For example, in an effort to provide better worker protection from hazardous chemicals and to help American businesses compete in a global economy, OSHA revised its HazCom standard to align with the United Nations' Globally Harmonized System of Classification and Labeling of Chemicals (referred to as GHS). These modifications improved the quality, consistency, and clarity of hazard information that workers receive by providing harmonized criteria for classifying and labeling hazardous chemicals and for preparing safety data sheets for these chemicals. The GHS system was developed through international negotiations and embodies the knowledge gained in the field of chemical hazard communication since the HazCom standard was first introduced in 1983. Simply, HazCom with GHS means better communication of chemical hazards for workers on the job.

Benefits of HazCom with GHS[*]

Practicing occupational safety and health professionals are familiar with OSHA's original 1983 Hazard Communication Standard. Many are now becoming familiar with the phase-in of the new combined HazCom and GHS standard. The Globally Harmonized System (GHS) is an international approach to hazard communication that provides agreed criteria for the classification of chemical hazards and a standardized approach to label elements and safety data sheets. The GHS was negotiated in a multi-year process by hazard communication experts from many different countries, international organizations, and stakeholder groups. It is based on major existing systems around the world, including OSHA's Hazard Communication Standard and the chemical classification and labeling systems of other U.S. agencies.

---

[*] Based on OSHA, *Modification of the Hazardous Communication Standard (HCS) to Conform with the United Nations' (UN) Globally Harmonized System of Classification and Labeling of Chemicals (GHS)*, Occupational Safety and Health Administration, Washington, DC, 2014 (https://www.osha.gov/dsg/hazcom/hazcom-faq.html).

The result of this negotiation process is the United Nations' document entitled *Globally Harmonized System of Classification and Labeling of Chemicals*, commonly referred to as *The Purple Book*. This document provides harmonized classification criteria for health, physical, and environmental hazards of chemicals. It also includes standardized label elements that are assigned to these hazard classes and categories and provide the appropriate signal words, pictograms, and hazard and precautionary statements to convey the hazards to users. A standardized order of information for safety data sheets is also provided. These recommendations can be used by regulatory authorities such as OSHA to establish mandatory requirements of hazard communication but do not constitute a model regulation.

OSHA's motive for modifying the Hazard Communication Standard was to improve the safety and health of workers through more effective communications on chemical hazards. Since it was first promulgated in 1983, the Hazard Communication Standard has provided employers and employees extensive information about the chemicals in their workplaces. The original standard is performance oriented, allowing chemical manufacturers and importers to convey information on labels and material data sheets in whatever format they choose. Although the available information has been helpful in improving employee safety and health, a more standardized approach to classifying the hazards and conveying the information will be more effective and provide further improvements in American workplaces. The GHS provides such a standardized approach, including detailed criteria for determining what hazardous effects a chemical poses, as well as standardized label elements assigned by hazard class and category. This will enhance both employer and worker comprehension of the hazards, which will help to ensure appropriate handling and safe use of workplace chemicals. In addition, the safety data sheet requirements establish an order of information that is standardized. The harmonized format of the safety data sheets will enable employers, workers, health professionals, and emergency responders to access the information more efficiently and effectively, thus increasing their utility.

Adoption of the GHS in the United States and around the world will also help to improve information received from other countries. Because the United States is both a major importer and exporter of chemicals, American workers often see labels and safety data sheets from other countries. The diverse and sometimes conflicting national and international requirements can create confusion among those who seek to use hazard information effectively. For example, labels and safety data sheets may include symbols and hazard statements that are unfamiliar to readers or not well understood. Containers may be labeled with such a large volume of information that important statements are not easily recognized. Given the differences in hazard classification criteria, labels may also be incorrect when used in other countries. If countries around the world adopt the GHS, these problems will be minimized, and chemicals crossing borders will have consistent information, thus improving communication globally.

Major Changes to the Hazard Communication Standard

The three major areas of change in the modified Hazard Communication Standard involve hazard classification, labels, and safety data sheets:

- *Hazard classification*—The definitions of hazards have been changed to provide specific criteria for the classification of health and physical hazards, as well as the classification of mixtures. These specific criteria will help to ensure that evaluations of hazardous effects are consistent across manufacturers and that labels and safety data sheets are more accurate as a result.
- *Labels*—Chemical manufacturers and importers will be required to provide a label that includes a harmonized signal word, pictogram, and hazard statement for each hazard class and category. Precautionary statements must be provided.
- *Safety data sheets*—Safety data sheets will have a 16-section format.

Safety and Health Considerations

Note: The GHS does not include harmonized training provisions but recognizes that training is essential to an effective hazard communication approach. The revised Hazard Communication Standard requires that workers be retrained within 2 years of the publication of the final result to facilitate recognition and understanding of the new labels and safety data sheets.

Hazard Classification

Not all HCS provisions are changed in the revised Hazard Communication Standard, which is simply a modification to the existing standard, designed to make it universal and worker friendly. The parts of the standard that did not relate to the GHS (such as the basic framework, scope, and exemptions) remain largely unchanged. There have been some modifications in terminology in order to align the revised Hazard Communication Standard with language used in the GHS; for example, the term "hazard determination" has been changed to "hazard classification," and "material safety data sheet" was changed to "safety data sheet." Under both the current Hazard Communication Standard and the revised version, an evaluation of chemical hazards must be performed considering the available scientific evidence concerning such hazards. Under the current Hazard Communication Standard, the hazard determination provisions have definitions of hazards and the evaluator determines whether or not the data on a chemical meet those definitions. It is a performance-oriented approach that provides parameters for the evaluation but not specific, detailed criteria. The hazard classification approach in the revised Hazard Communication Standard is quite different. The revised standard has specific criteria for each health and physical hazard, along with detailed instructions for hazard evaluation and determinations as to whether mixtures or substances are covered. It also establishes both hazard classes and hazard categories for most of the effects; the classes are divided into categories that reflect the relative severity of the effect. The current Hazard Communication Standard does not include categories for most of the health hazards covered, so this new approach provides additional information that can be related to the appropriate response to address the hazard. OSHA has included the general provisions for hazard classification in paragraph (d) of the revised rule and added extensive appendices that address the criteria for each health or physical effect.

Label Changes Under the Revised Hazard Communication Standard

Under the current Hazard Communication Standard, the label preparer must provide the identity of the chemical and the appropriate hazard warnings. This may be done in a variety of ways, and the method to convey the information is left to the preparer. Under the revised Hazard Communication Standard, once the hazard classification is completed, the standard specifies what information is to be provided for each hazard class and category. Labels will require the following elements:

- *Pictogram*—A symbol plus other graphic elements, such as a border, background pattern, or color that is intended to convey specific information about the hazards of a chemical. Each pictogram consists of a different symbol on a white background within a red square frame set on a point (i.e., a red diamond). There are nine pictograms under the GHS; however, only eight pictograms were required under the Hazard Communication Standard.
- *Signal words*—A single word used to indicate the relative level of severity of hazard and to alert the reader to a potential hazard on the label. The signal words used are "danger" and "warning." "Danger" is used for the more severe hazards, and "warning" is used for less severe hazards.
- *Hazard statement*—A statement assigned to a hazard class and category that describes the nature of the hazards of a chemical, including, where appropriate, the degree of hazard.
- *Precautionary statement*—A phrase that describes recommended measures to be taken to minimize or prevent adverse effects resulting from exposure to a hazardous chemical or improper storage or handling of a hazardous chemical.

In the revised Hazard Communication Standard, OSHA is lifting the stay on enforcement regarding the provision to update labels when new information on hazards becomes available. Chemical manufacturers, importers, distributors, or employers who become newly aware of any significant information regarding the hazards of a chemical must revise the labels for the chemical within six months of becoming aware of the new information. If the chemical is not currently produced or imported, the chemical manufacturer, importer, distributor, or employer must add the information to the label before the chemical is shipped or introduced into the workplace again.

The current standard provides employers with flexibility regarding the type of system to be used in their workplaces, and OSHA has retained that flexibility in the revised Hazard Communication Standard. Employers may choose to label workplace containers either with the same label that would be on shipped containers for the chemical under the revised rule or with label alternatives that meet the requirements for the standard. Alternative labeling systems such as the National Fire Protection Association (NFPA) 704 Hazard Rating and the Hazardous Material Identification System (HMIS) are permitted for workplace containers. However, the information supplied on these labels must be consistent with the revised HCS (e.g., no conflicting hazard warnings or pictograms).

Safety Data Sheet Changes Under the Revised Hazard Communication Standard

The information required on a safety data sheet (SDS) remains essentially the same as in the current Hazard Communication Standard, which indicates what information has to be included on an SDS but does not specify a format for presentation or order of information. The revised Hazard Communication Standard requires that the information on the SDS be presented in a specified sequence. The revised SDS should contain 16 headings (see Table 12.6).

Worker Training

The employer must provide employee training on the hazard communication program. Training on the hazardous chemicals in their work areas must be provided to employees upon their initial assignment. Whenever a new physical or health hazard is introduced into the workplace (one for which training has not previously been accomplished), the employer must provide the training. Specifically, employee training topics must include the following:

1. Methods and observations that may be used to detect the presence or release of a hazardous chemical in the work area
2. Physical and health hazards of the chemicals in the work area
3. Measures employees can take to protect themselves from these hazards, including specific procedures the employer has implemented to protect employees from exposure to hazardous chemicals, such as appropriate work practices, emergency procedures, and personal protective equipment to be used
4. Details of the hazard communication program developed by the employer, including an explanation of the labeling system and the safety data sheet and how employees can obtain and use the appropriate hazard information

Note: As with all OSHA-required training, it is necessary not only to ensure that the training is conducted but also to ensure that it has been properly documented.

Labeling Requirements

The employer's responsibilities include signs, placards, process sheets, batch tickets, operating procedures, or other such written materials in lieu of affixing labels to individual stationary process containers—as long as the alternative method identifies the containers to which it is applicable and conveys the information required on the label. The written materials must be readily accessible to employees in their work areas throughout each shift.

TABLE 12.6
Minimum Information for a Safety Data Sheet

Heading	Information Provided
1. Identification of the substance or mixture and of the supplier	GHS product identifier or other means of identification Recommended use of the chemical and restrictions on use Supplier's details (e.g., name, address, phone number) Emergency phone number
2. Hazards identification	GHS classification of substance or mixture and any national or regional information GHS label elements, including precautionary statements (hazard symbols may be provided as a graphical reproduction of the symbols in black and white or the name of the symbol, such as flame or skull and crossbones) Other hazards that do not result in classification (e.g., dust explosion hazard) or are not covered by the GHS
3. Composition/information on ingredients	*Substance:* Chemical identity; common name, synonyms, etc.; CAS number, EC number, etc.; impurities and stabilizing additives that are themselves classified and which contribute to the classification of the substance *Mixture:* Chemical identity and concentration or concentration ranges of all ingredients that are hazardous within the meaning of the GHS and are present above their cutoff levels
4. First aid measures	Description of necessary measures, subdivided according to the different routes of exposure (i.e., inhalation, skin and eye contact, and ingestion) Most important symptoms/effects, acute and delayed Indication of immediate medical attention and special treatment needed, if necessary
5. Firefighting measures	Suitable (and unsuitable) extinguishing media Specific hazards arising from the chemical (e.g., nature of any hazardous combustion products) Special protective equipment and precautions for firefighters
6. Accidental release measures	Personal precautions, protective equipment, and emergency procedures Environmental precautions Methods and materials for containment and cleanup
7. Handling and storage	Precautions for safe handling Conditions for safe storage, including any incompatibilities
8. Exposure controls/personal protection	Control parameters (e.g., occupational exposure limit values, biological limit values) Appropriate engineering controls Individual protection measures, such as personal protective equipment
9. Physical and chemical properties	Appearance (e.g., physical state, color) Odor and odor threshold pH Melting point/freezing point Initial boiling point and boiling range Flash point Evaporation rate Flammability (solid, gas) Upper and lower flammability or explosive limits Vapor pressure Vapor density Relative density Solubility or solubilities Partition coefficient (*n*-octanol/water) Autoignition temperature Decomposition temperature Viscosity

(continued)

TABLE 12.6 (continued)
Minimum Information for a Safety Data Sheet

	Heading	Information Provided
10.	Stability and reactivity	Reactivity Chemical stability Possibility of hazardous reactions Conditions to avoid (e.g., static discharge, shock or vibration) Incompatible materials Hazardous composition products
11.	Toxicological information	Concise but complete and comprehensible description of the various toxicological (health) effects and available data used to identify those effects, including information on the likely routes of exposure (inhalation, ingestion, skin and eye contact; symptoms related to the physical, chemical, and toxicological characteristics; delayed and immediate effects and also chronic effects from short- and long-term exposure
12.	Ecological information	Ecotoxicity (aquatic and terrestrial, where available) Persistence and degradability Bioaccumulative potential Mobility in soil Other adverse effects
13.	Disposal considerations	Description of waste residues and information on their safe handling and methods of disposal, including the disposal of any contaminated packaging
14.	Transport information	UN number Transport hazard class(es) Packaging group, if applicable Marine pollutant (yes/no) Special precautions that a user needs to be aware of or needs to comply with in connection with transport or conveyance either within or outside their premises
15.	Regulatory information	Safety, health, and environmental regulations specific for the product in question
16.	Other information including information on preparation and revision of SDS	

The employer must not remove or deface existing labels on incoming containers of hazardous chemicals, unless the container is immediately marked with the required information. Safety and health practitioners must ensure that labels or warnings in the workplace are legible, are in English, and are prominently displayed on the container or are readily available in the work area throughout each work shift. Employers with employees who speak other languages may need to add the information in those languages to the material presented, as long as the information is also presented in English.

If existing labels already convey the required information, the employee need not affix new labels. If the employer becomes newly aware of any significant information regarding the hazards of a chemical, the employer must revise the labels for the chemical within 3 months of becoming aware of the new information. Labels on containers of hazardous chemicals shipped after that time must contain the new information.

Note: Hazard warnings or labels represent an area where facility safety engineers, supervisors, and employees must maintain constant vigilance to ensure that they are in place and legible.

Employers are required to develop a written hazard communication program. This particular requirement is often cited as the most common noncompliance violation found in industry today. The written hazard communication program must be present, maintained, and readily available to

Safety and Health Considerations 265

all workers and visitors in each workplace. The written program must contain a section for labels and other warning devices and a section for safety data sheets, and employee information must be provided and training conducted. The written program must include a list of hazardous chemicals known to be present using an identity that is referenced on the appropriate safety data sheet, the methods the employer uses to inform employees of the hazards of non-routine tasks, and the hazards associated with chemicals contained in unlabeled pipes in their work areas.

Hazard Communication Program Audit Items

A facility that has a written hazard communication program as described above is well on its way to compliance. If the program is audited by OSHA, the goal, of course, is for any auditor who might visit the facility to be able to readily observe that the employer is in compliance. Often, auditors will not even review a written hazard communication program if they can plainly see that the company is in compliance. Let's take a look at some of the items that OSHA will be checking on. Employers must be able to answer "yes" to each of the following items, where applicable:

- Do all of the chemical containers clearly indicate the contents and associated hazards?
- Are storage cabinets used to hold flammable liquids labeled "Flammable—Keep Fire Away?"
- For a fixed extinguishing system, is a sign posted warning of the hazards presented by the extinguishing medium?
- Are all aboveground storage tanks properly labeled?
- If hazardous materials (including gasoline) are stored in aboveground storage tanks, are the tanks or other containers holding hazardous materials appropriately labeled with the chemical name and hazard warnings?
- Are all chemicals used in spray painting operations correctly labeled?
- If chemicals are stored on the premises, are all containers properly labeled with the chemical name and hazard warnings?

OSHA auditors will note any chemicals found in the workplace. During their walk-arounds, auditors are likely to seek out any flammable material storage lockers that might be in the workplace and will list items stored in the lockers. Later, when the walk-around is completed, the auditor will ask the employer to provide a copy of the SDS for each chemical.

To avoid a citation, the employer must not fail this major test. If the auditor, for example, noticed during the walk-around that employees were using some type of solvent or cleaning agent in the performance of their work, the auditor will want to see a copy of the SDS for that particular chemical. If a copy of that SDS cannot be produced, the company will be cited. Not being able to produce a relevant SDS is one of the most commonly cited offenses. Obviously, the only solution to this problem is to ensure that the facility has an SDS for each chemical used, stored, or produced and that the chemical inventory list is current and accurate. Be sure that safety data sheets are available to employees for every chemical used onsite.

Keep in mind that OSHA auditors will look at each work center within a company, and that each different work center will present its own specialized requirements. If a company has an environmental laboratory, for example, the auditor will spend considerable time in the lab to ensure that the company is in compliance with OSHA's Laboratory Standard and that it has a written chemical hygiene plan.

THOUGHT-PROVOKING DISCUSSION QUESTIONS

1. Safety is all about fear. Explain.
2. Safety compliance is all about fear. Explain.
3. Exposure to fracking produced wastewater can be hazardous. How do you protect yourself from being contaminated?

REFERENCES AND RECOMMENDED READING

AAA. (2003). AAA encourages motorists to prepare now for safe winter driving. *PR Newswire*, October 1.

AAAdvantage. (2003). Nine vital driving tips for mature operators. *Home & Away Magazine*, November/December.

ALL Consulting and Montana Board of Oil and Gas Conservation (MBOGC). (2002). *Handbook on Best Management Practices and Mitigation Strategies for Coal Bed Methane in the Montana Portion of the Powder River Basin*, prepared for National Energy Technology Laboratory, National Petroleum Technology Office, U.S. Department of Energy, Tulsa, OK, 44 pp.

ALL Consulting and Montana Board of Oil and Gas Conservation (MBOGC). (2004). *Coal Bed Natural Gas Handbook: Resources for the Preparation and Review of Project Planning Elements and Environmental Documents*, prepared for National Petroleum Technology Office, U.S. Department of Energy, Washington, DC, 182 pp.

Allaby, A. and Allaby, M. (1991). *The Concise Dictionary of Earth Sciences*. Oxford University Press, Oxford, UK.

Archea, J., Collins, B.L., and Stahl, F.I. (1976). *Guidelines for Stair Safety*, Building Sciences Series 120. National Bureau of Standards, Washington, DC.

Arms, K. (1994). *Environmental Science*, 2nd ed. Saunders College Publishing, Fort Worth, TX.

Associated Press. (1998). Town evacuated after acid spill. *Lancaster New Era*, September 6.

AWWA. (1995). *Water Quality*, 2nd ed. American Water Works Association, Denver, CO.

Baden, J. and Stroup, R.C., Eds. (1981). *Bureaucracy vs. Environment*. University of Michigan Press, Ann Arbor.

Bangs, R. and Kallen, C. (1985). *Rivergods: Exploring the World's Great Wild Rivers*. Sierra Club Books, San Francisco, CA.

Barker, J.R. and Tingey, D.T. (1991). *Air Pollution Effects on Biodiversity*. Van Nostrand Reinhold, New York.

BLS. (1984). *Injuries Resulting from Falls from Elevation*, Bulletin 2195. Bureau of Labor Statistics, Washington, DC.

Botkin, D.B. (1995). *Environmental Science: Earth as a Living Planet*. John Wiley & Sons, New York.

Boyce, A. (1997). *Introduction to Environmental Technology*. Van Nostrand Reinhold, New York.

Brahic, C. (2009). Fish "an ally" against climate change. *New Scientist*, January 16 (www.newscientist.com/article/dn16432-fish-an-ally-against-clmate-cahnge.html).

Brauer, R.L. (1994). *Safety and Health for Engineers*. Van Nostrand Reinhold, New York.

Caldeira, K. and Wickett, M.E. (2003). Anthropogenic carbon and ocean pH. *Nature*, 425(6956):365.

Carney, A. (1991). Lock out the chance for injury. *Safety & Health*, 143(5):46.

Carson, J.E. and Moses, H. (1969). The validity of several plume rise formulas. *Journal of the Air Pollution and Control Association*, 19(11):862–866.

CDPHE. (2007). *Pit Monitoring Data for Air Quality*. Colorado Department of Public Health and Environment, Denver.

CH2M HILL. (2007). *Review of Oil and Gas Operation Emissions and Control Options, Final Report*, prepared for Air Pollution Control Division, Colorado Department of Public Health and Environment, Denver.

CICA. (2008). *The Construction Industry Compliance Assistance Center*, www.cicacenter.org/index.cfm.

Coastal Video. (1993). *Confined Space Rescue Booklet*. Coastal Video Communication Corporation, Virginia Beach, VA.

Cobb, R.W. and Elder, C.D. (1983). *Participation in American Politics*, 2nd ed. The Johns Hopkins University Press, Baltimore, MD.

Cohen, H.H. and Compton, D.M.J. (1982). Fall accident patterns. *Professional Safety*, 27(6):16–22.

Cote, A. and Bugbee, P. (1991). *Principles of Fire Protection*. National Fire Protection Association, Batterymarch Park, MA.

CoVan, J. (1995). *Safety Engineering*. John Wiley & Sons, New York.

Davis, M.L. and Cornwell, D.A. (1991). *Introduction to Environmental Engineering*. McGraw-Hill, New York.

Diamond, J. (2006). *Guns, Germs, and Steel: The Fates of Human Societies*. Norton, New York.

Downing, P.B. (1984). *Environmental Economics and Policy*. Little, Brown, Boston.

DRBC. (2010). *Administrative Manual—Part III: Water Quality Regulations: 18 CFR Part 410*. Delaware River Basin Commission, West Trenton, NJ.

Easterbrook, G. (1995). *A Moment on the Earth: The Coming Age of Environmental Optimism*. Viking Penguin, Bergenfield, NJ.

Eisma, T.L. (1990). Rules change: worker training helps simplify fall protection. *Occupational Health & Safety*. 59(3):52–55.

Safety and Health Considerations

Ellis, J.E. (1988). *Introduction to Fall Protection*. American Society of Safety Engineers, Des Plaines, IL.

FEMA. (1981). *Planning Guide and Checklist for Hazardous Materials Contingency Plans*, FEMA-10. Federal Emergency Management Agency, Washington, DC.

Ferry, T. (1990). *Safety and Health Management Planning*. Van Nostrand Reinhold, New York.

Field, B.C. (1996). *Environmental Economics: An Introduction*, 2nd ed. McGraw-Hill, New York.

Franck, I. and Brownstone, D. (1992). *The Green Encyclopedia*. Prentice Hall, New York.

Freedman, B. (1989). *Environmental Ecology*. Academic Press, New York.

Gasaway, D.C. (1985). *Hearing Conservation: A Practical Manual and Guide*. Prentice Hall, Englewood Cliffs, NJ.

Giachino, J. and Weeks, W. (1985). *Welding Skills*. American Technical Publications, Homewood, IL.

Godish, T. (1997). *Air Quality*, 3rd ed. Lewis Publishers, Boca Raton, FL.

Hammer, W. (1989). *Occupational Safety Management and Engineering*. Prentice Hall, Englewood Cliffs, NJ.

Hardin, G. (1968). The tragedy of the commons. *Science*, 162:1243–1248.

Healy, R.J. (1969). *Emergency and Disaster Planning*. John Wiley & Sons, New York.

Henry, J.G. and Heinke, G.W. (1995). *Environmental Science and Engineering*, 2nd ed. Prentice Hall, New York.

Holmes, G., Singh, B.R., and Theodore, L. (1993). *Handbook of Environmental Management and Technology*. John Wiley & Sons, New York.

Hoover, R.L., Hancock, R.L., Hylton, K.L., Dickerson, O.B., and Harris, G.E. (1989). *Health, Safety and Environmental Control*. Van Nostrand Reinhold, New York.

Ingram, C. (1991). *The Drinking Water Book*. Ten Speed Press, Berkeley, CA.

Institute for Southern Studies. (2012). Fracking's danger for workers. *Facing South*, May 5 (https://www.facingsouth.org/2012/05/institute-index-frackings-dangers-for-workers.html).

IOGCC. (1996). *Review of Existing Reporting Requirements for Oil and Gas Exploration and Production Operators in Five Key States*. Interstate Oil and Gas Compact Commission, Washington, DC.

IOGCC. (2008). *Issues: States' Rights*. Interstate Oil and Gas Compact Commission, Washington, DC.

IPAA. (2000). *IPAA Opposes EPA's Possible Expansion of TRI*, Environment and Safety Fact Sheets. Independent Petroleum Association of America, Washington, DC (www.ipaa.org/issues/factsheets/environment_safety/tri.php).

IPAA. (2004). *Reasonable and Prudent Practices for Stabilization (RAPPS) of Oil and Gas Construction Sites*, Guidance Document. Independent Petroleum Association of America, Washington, DC.

Jackson, A.R. and Jackson, J.M. (1996). *Environmental Science: The Natural Environment and Human Impact*. Longman, New York.

Jacobson, J. (1998). *The Supervisor's Tough Job: Dealing with Drug and Alcohol Abusers*, Supervisors' Safety Update 97. Eagle Insurance Group, Largo, FL.

Jacus, J.R. (2011). Air quality constraints on shale development activities, in *Proceedings of The Institute for Energy Law Second Conference on the Law of Shale Plays*, Fort Worth, TX, September 7–8.

Karliner, J. (1998). Earth predators. *Dollars and Sense*, 218:7.

Keller, E.A. (1988). *Environmental Geology*. Merrill, Columbus, OH.

Kimberlin, J. (2009). That reeking paper mill keeps Franklin running. *The Virginian-Pilot*, February 1.

Kohr, R.L. (1989). Slip slidin' away. *Safety & Health*, 140(5):52.

LaBar, G. (1989). Sound policies for protecting workers' hearing. *Occupational Hazards*, July, p. 46.

Lave, L.B. (1981). *The Strategy of Social Regulations: Decision Frameworks for Policy*. Brookings Institution, Washington, DC.

Lee, R.H. (1969). Electrical grounding: safe or hazardous? *Chemical Engineering*, 76:158.

Leopold, A. (1970). *A Sand County Almanac*. Ballentine Books, New York.

Lewis, S.A. (1996). *Safe Drinking Water*. Sierra Club Books, San Francisco, CA.

MacKenzie, J.J. and El-Ashry, T. (1988). *Ill Winds: Airborne Pollutant's Toll on Trees and Crops*. World Resource Institute, Washington, DC.

Mansdorf, S.Z. (1993). *Complete Manual of Industrial Safety*. Prentice Hall, Englewood Cliffs, NJ.

Masters, G.M. (1991). *Introduction to Environmental Engineering and Science*. Prentice-Hall, Englewood Cliffs, NJ.

McElroy, F.E., Ed. (1980). *NSC Accident Prevention Manual for Industrial Operations: Engineering and Technology*, 8th ed. International Fire Chiefs Association, Merrifield, VA.

McHibben, B. (1995). *Hope, Human and Wild: True Stories of Living Lightly on the Earth*. Little, Brown & Company, Boston, MA.

Miller, G.T. (1997). *Environmental Science: Working with the Earth*, 5th ed. Wadsworth, Belmont, CA.

Miller, G.T. (2004). *Environmental Science*, 10 ed. Brooks/Cole, Belmont, CA.

Morrison, A. (1983). In Third World villages, a simple handpump saves lives. *Civil Engineering*, 52:68–72.

Nathanson, J.A. (1997). *Basic Environmental Technology: Water Supply, Waste Management, and Pollution Control*. Prentice Hall, Upper Saddle River, NJ.

National Institute on Aging. (2002). *Practice Common Sense Safety Rules*. National Institutes of Health, U.S. Department of Health and Human Services, Washington, DC.

NFPA. (1989). *Cutting and Welding Processes*, NFPA 51B-1989. National Fire Protection Association, Quincy, MA.

NFPA. (1991). *Fire Protection Handbook*, 16th ed. National Fire Protection Association, Quincy, MA.

NHTSA. (1992). *Sudden Impact*. National Highway Traffic Safety Administration, Washington DC.

NIOSH. (1987). *Guide to Industrial Respiratory Protection*, NIOSH Publication No. 87-116. National Institute for Occupational Safety and Health, Cincinnati, OH.

NIOSH. (2005). *NIOSH Alert: Preventing Worker Injuries and Deaths from Traffic-Related Crashes*, NIOSH Publication No. 98-142. National Institute for Occupational Safety and Health, Cincinnati, OH.

Norse, E.A. (1985). The value of animal and plant species for agriculture, medicine, and industry, in Hoage, R.J., Ed., *Animal Extinctions*. Smithsonian Institution Press, Washington, DC, pp. 59–70.

NSC. (1985). *Accident Facts 1984*. National Safety Council, Chicago, IL.

NSC. (1986). *Accident Facts 1985*. National Safety Council, Chicago, IL.

NSC. (1987). *Guards: Safeguarding Concepts Illustrated*, 5th ed. National Safety Council, Chicago, IL.

NSC. (1992). *Accident Prevention Manual for Business and Industry: Engineering and Technology*. National Safety Council, Chicago, IL.

NSC. (2017). *Injury Facts*. National Safety Council, Chicago, IL.

Ophuls, W. (1977). *Ecology and the Politics of Scarcity*. W.H. Freeman, New York.

Orfinger, B. (2002). *Saving More Lives: Red Cross Adds AED Training to CPR Course*. American Red Cross, Washington, DC.

OSHA. (1989). OSHA studies workplace deaths involving welding. *OSHA News*, February 8.

OSHA. (1992). *Concepts and Techniques of Machine Safeguarding*, OSHA 3067. Occupational Safety and Health Administration, Washington, DC.

OSHA. (2003). *Network of Employers for Traffic Safety Align with OSHA to Reduce Job-Related Traffic Injuries and Fatalities*, OSHA Trade Release. Occupational Safety and Health Administration, Washington, DC.

OSHA. (2012). *Weekly Fatality/Catastrophe Report*. Occupational Safety and Health Administration, Washington, DC (https://www.osha.gov/dep/fatcat/fatcat_weekly_rpt_02252012.html).

OSHA. (2017). *Commonly Used Statistics*. Washington, DC: Occupational Safety and Health Administration (https://www.osha.gov/oshstats/commonstats.html).

OTA. (1989). *Catching Our Breath: Next Steps for Reducing Urban Ozone*, OTA-O-412. Office of Technology Assessment, Washington, DC.

Pater, R. (1985). Fallsafe: reducing injuries from slips and falls. *Professional Safety*, 30(10):15–18.

Peavy, H.S., Rowe, D.R., and Tchobanoglous, G. (1985). *Environmental Engineering*. McGraw-Hill, New York.

Penton. (1986). *Right Off the Docket: Case Summaries and Court Rulings Involving Disputed Workers' Compensation and Occupational Disease Claims*. Penton Media, Cleveland, OH.

Pepper, I.L., Gerba, C.P., and Brusseau, M.L. (1996). *Pollution Science*. Academic Press, San Diego, CA.

Postel, S. (1987). Stabilizing chemical cycles, in Brown, L.R., Ed., *State of the World: A Worldwatch Institute Report on Progress Toward a Sustainable Society*. Norton, New York.

Rosen, S.I. (1983). *The Slip and Fall Handbook: Case Evaluation, Preparation, and Settlement*. Hanrow Press, Columbia, MD.

Royster, J.D. and Royster, L.H. (1990). *Hearing Conservation Programs: Practical Guidelines for Success*. Lewis Publishers, Chelsea, MI.

Ryan, M. (1998). How a little headwork saves a lot of children. *Parade Magazine*, May 24, pp. 4–5.

Santoro, R.L., Howarth, R.W., and Ingraffea, A.R. (2011). *Indirect Emissions of Carbon Dioxide from Marcellus Shale Gas Development*, Technical Report. Agriculture, Energy, and Environment Program, Cornell University, New York.

Sayers, D.L. and Walsh, J.L. (1998). *Thrones, Dominations*. St. Martin's Press, New York.

Smith, A.J. (1980). *Managing Hazardous Substances Accidents*. McGraw-Hill, New York.

Sonnenstuhl, W. and Trice, H. (1986). The social construction of alcohol problems in a union's peer counseling program. *Journal of Drug Issues*, 17(3):223–254.

Spellman, F.R. (1996a). *Stream Ecology and Self-Purification: An Introduction for Wastewater and Water Specialists*. Technomic, Lancaster, PA.

Spellman, F.R. (1996b). *Safe Work Practices for Wastewater Treatment Plants*. Technomic, Lancaster, PA.

Safety and Health Considerations

Spellman, F.R. (1997). *A Guide to Compliance for Process Safety Management Planning (PSM/RMP)*. Technomic, Lancaster, PA.

Spellman, F.R. (1998). *Surviving an OSHA Audit*. Technomic, Lancaster, PA.

Spellman, F.R. (1999). *Confined Space Entry*. Technomic, Lancaster, PA.

Spellman, F.R. and Whiting, N. (2006). *Environmental Science and Technology: Concepts and Applications*. CRC Press, Boca Raton, FL.

STRONGER. (2008a). *History of STRONGER—Helping to Make an Experiment Work*. State Review of Oil and Natural Gas Regulations, Oklahoma City, OK (www.strongerinc.org/about/history.asp).

STRONGER. (2008b). *List of State Reviews*. State Review of Oil and Natural Gas Regulations, Oklahoma City, OK (www.strongerinc.org/reviews/reviews.asp).

Tower, E. (1995). *Environmental and Natural Resource Economics*. Eno River Press, New York.

Urone, P. (1976). The primary air pollutants—gaseous: their occurrence, sources, and effects, in Stern, A.C., Ed., *Air Pollution*, Vol. 1. Academic Press, New York, pp. 23–75.

URS. (2008). *Sampling and Analysis Plan for Exploration and Production: Pit Solids and Fluids in Colorado Energy Basins*, Revision B. URS Corporation, Denver, CO.

USACE. (1987). *Safety and Health Requirements Manual*, rev. ed., EM 385-1-1. U.S. Army Corps of Engineers, Washington, DC.

USCC. (2009). *Analysis of Workers' Compensation Laws*. U.S. Chamber of Commerce, Washington, DC.

USDHHS. (2000). *Worker Deaths by Falls: A Summary of Surveillance Findings and Investigative Case Reports*. U.S. Department of Health and Human Services, Washington, DC.

USDOL. (2012). *Oil and Gas Well Drilling, Servicing and Storage Standards*. U.S. Department of Labor, Washington, DC (http://www.osha.gov/SLTC/oilgaswelldrilling/index.html).

USDOL. (2016). *Census of Fatal Occupational Injuries Summary, 2015*. Washington, DC: U.S. Department of Labor (https://www.bls.gov/news.release/cfoi.nr0.htm).

USEPA. (1988). *Regulatory Determination for Oil and Gas and Geothermal Exploration, Development and Production Wastes*. U.S. Environmental Protection Agency, Washington, DC.

USEPA. (1998). *RCRA: Superfund and EPCRA Hotline Training Module*, EPA 540-R-98-022. U.S. Environmental Protection Agency, Washington, DC.

USEPA. (1999). *USEPA's Program to Regulate the Placement of Waste Water and Other Fluids Underground*, EPA 810-F-99-019. U.S. Environmental Protection Agency, Washington, DC.

USEPA. (2002). *Exemption of Oil and Gas Exploration and Production Wastes from Federal Hazardous Waste Regulations*. U.S. Environmental Protection Agency, Washington, DC (epa.gov/osw/nonhaz/industrial/special/oil/oil-gas.pdf).

USEPA. (2003). *Underground Injection Control Program*. U.S. Environmental Protection Agency, Washington, DC (http://water.epa.gov/type/groundwater/uic/index.cfm).

USEPA. (2004). *Understanding the Safe Drinking Water Act*, EPA 816-F-04-030. U.S. Environmental Protection Agency, Washington, DC (http://water.epa.gov/lawsregs/guidance/sdwa/basicinformation.cfm).

USEPA. (2005). *Basic Air Pollution Meteorology*. U.S. Environmental Protection Agency, Washington, DC (www.epa.gov/apti).

USEPA. (2007a). *National Ambient Air Quality Standards (NAAQS)*. U.S. Environmental Protection Agency, Washington, DC (www.epa.gov/air/criteria.html).

USEPA. (2007b). *Final Emission Standards of Performance for Stationary Spark Ignition Internal Combustion Engines; and Final Air Toxics Standards for Reciprocating Internal Combustion Engines*. U.S. Environmental Protection Agency, Washington, DC (www.epa.gov/ttn/atw/nsps/sinsps/sinspspg.html).

USEPA. (2008a). *Summary of the Clean Water Act*. U.S. Environmental Protection Agency, Washington, DC (www.epa.gov/lawsregs/laws/cwa.html).

USEPA. (2008b). *Water Quality and Technology-Based Permitting*. U.S. Environmental Protection Agency, Washington, DC (http://cfpub.epa.gov/npdes/generalissues/watertechnology.cfm).

USEPA. (2008c). *Effluent Limitations Guidelines and Standards*. U.S. Environmental Protection Agency, Washington, DC (http://cfpub.epa.gov/npdes/techbasedpermitting/effguide.cfm).

USEPA. (2008d). *Final Rule: Amendments to the Storm Water Regulations for Discharges Associated with Oil and Gas Construction Activities*. U.S. Environmental Protection Agency, Washington, DC (www.epa.gov/npdes/regulations/final_oil_gas_factsheet.pdf).

USEPA. (2008e). *The Green Book Nonattainment Areas for Criteria Pollutants*. U.S. Environmental Protection Agency, Washington, DC (www.epa.gov/oar/oaqps/greenbk/).

USEPA. (2008f). *Air Trends: Basic Information*. U.S. Environmental Protection Agency, Washington, DC (www.epa.gov/airtrends/sixpoll.html).

USEPA. (2008g). *Region 2. Solid Waste; RCRA Subtitle D*. U.S. Environmental Protection Agency, Washington, DC (www.epa.Gov/region2/waste/dsummary.htm).

USEPA. (2008h). *Region 9. Superfund*. U.S. Environmental Protection Agency, Washington, DC (http://www.epa.gov/region9/superfund/).

USEPA. (2008i). *Water. State, Tribal & Territorial Standards*. U.S. Environmental Protection Agency, Washington, DC (www.epa.gov/waterscience/standards/wqslibrary/).

USEPA. (2008j). *Summary of the Oil Pollution Act*. U.S. Environmental Protection Agency, Washington, DC (www.epa.gov/lawsregs/laws/opa.html).

USEPA. (2010). *Ozone Science: The Facts Behind the Phase-Out*. U.S. Environmental Protection Agency, Washington, DC (www.epa.gov/ozone/science/sc_fact.html).

USFWS. (2011). *Endangered Species Program: Laws and Policies*. U.S. Fish and Wildlife Service, Arlington, VA (http://www.fws.gov/endangered/laws-policies/regulations-and-policies.html).

Walker, M. (1963). *The Nature of Scientific Thought*. Prentice Hall, Englewood Cliffs, NJ.

WEF. (1994). *Confined Space Entry*. Water Environment Federation, Alexandria, VA.

WRI. (1988). *World Resources 1988–1989: An Assessment of the Resource Base that Supports the Global Economy*. World Resources Institute, Washington, DC.

Zurer, P.S. (1988). Studies on ozone destruction expand beyond Antarctic. *C & E News*, 66:16–25.

Glossary[*]

A

Abandonment pressure: The lowest gas pressure before a gas well must be abandoned.

Accelerator: An additive that increases the rate of a process such as cement setting.

Acid gas: A corrosive gas such as hydrogen sulfide or carbon dioxide that forms an acid with water.

Acid job: Process where acid is poured or pumped down a well to dissolve limestone and increase fluid flow.

Adsorption: Adhesion of gas molecules, ions, or molecules in solution to the surface of solid bodies with which they are in contact.

Air drilling (pneumatic drilling): Rotary drilling with air pumped down the drill string instead of circulating drilling mud.

Alluvial aquifer: A water-bearing deposit of unconsolidated material (e.g., sand, gravel) left behind by a river or other flowing water.

Amphoteric: Having both basic and acidic properties.

Anaerobic bacteria: Bacteria that thrive in oxygen-poor environments.

Anisotropic: Having some physical property that varies with direction from a given location.

Annulus: The space between the casing (the material, typically steel, that is used to keep the well stable) in a well and the wall of the hole, or between two concentric strings of casing, or between the casing and tubing.

Anticline: A fold in the Earth's crust, convex upward, whose core contains stratigraphically older rocks.

Antifoam: An additive used to reduce foam.

Aquifer: Formation, group of formations, or part of a formation that contains sufficient saturated permeable material to yield significant quantities of water to wells and springs. The definition is based on a geological concept in which water bodies are classified in accordance with stratigraphy or rock types; the definition clearly intends that an aquifer include the unsaturated part of the permeable unit.

Argillaceous: Shaly.

Artesian: Synonymous with confined. The terms *confined groundwater* and *confined water body* are equivalent, respectively, to *artesian water* and *artesian water body*.

Artesian well: A well deriving its water from an artesian or confined water body. The water level in an artesian well stands above the top of the artesian water body it taps. If the water level in an artesian well stands above the land surface, the well is a flowing artesian well. If the water level in the well stands above the water table, that indicates that the artesian water can and probably does discharge to the unconfined water body. It should be noted that, in groundwater discharge areas, wells having heads higher than the water table, or even flowing wells, may exist without confinement of the water body, due to vertical components of gradient in the flow field.

Associated gas: Natural gas that is in contact with crude oil in the reservoir.

Attenuate: To reduce the amplitude of sound pressure (noise).

Audible range: The frequency range over which normal ears hear—approximately 20 to 20,000 Hz.

---

[*] Adapted from Spellman, F.R., *The Science of Water*, 3rd ed., CRC Press, Boca Raton, FL, 2015; Spellman, F.R., *Environmental Impacts of Hydraulic Fracturing*, CRC Press, Boca Raton, FL, 2012; USGS, *Glossary of Selected Terms Useful in Studies of the Mechanics of Aquifer Systems and Land Subsidence Due to Fluid Withdrawal*, U.S. Geological Survey, Washington, DC, 1972.

272   Glossary

Audiogram: A chart, graph, or table resulting from an audiometric test showing an individual's hearing threshold levels as a function of frequency.

Audiologist: A professional, specializing in the study and rehabilitation of hearing, who is certified by the American Speech–Language–Hearing Association or licensed by a state board of examiners.

Aulacogen: A long, narrow rift in a continent, often filled with thick sediments.

Aureole: A ring surrounding a volcanic intrusion where the surrounding rock has been altered.

Authigenic: Type of mineral formed by a chemical reaction in the subsurface.

B

Backflush: Flushing an injected fluid back out of a well.

Background noise: Noise coming from sources other than the particular noise sources being monitored.

Backoff operation: Method used to remove stuck pipe from a well.

Barrel of oil equivalent (BOE): The amount of energy equivalent to the amount of energy found in a barrel of crude oil. For natural gas, 1 BOE = 6000 ft^3 of gas.

Baseline audiogram: An audiogram against which future audiograms are compared.

Bedrock aquifer: An aquifer located in the solid rock underlying unconsolidated surface materials (i.e., sediment). Solid rock can bear water when it is fractured.

Bentonite: A clay mineral used to make common drilling mud.

Billion cubic feet (Bcf): A unit typically used to define gas production volumes in the coalbed methane industry; 1 Bcf is roughly equivalent to the volume of gas required to heat approximately 12,000 households for one year.

Biogenic: Refers to a direct product of the physiological activities of organisms.

Biotite: Black mica.

Bitumen: Solid hydrocarbons such as tar in sedimentary rocks.

Bituminous: From the base word *bitumen*, a general term describing various solid and semisolid hydrocarbons that are able to join together and are soluble in carbon bisulfide (e.g., asphalts).

Blowout: An uncontrolled flow of fluid from a well.

Bottom water: A mixture of freshwater and brine.

Breaker: A fracturing fluid additive that is added to break down the viscosity of the fluid.

Breccia: A coarse-grained clastic rock composed of angular broken rock fragments held together by a mineral cement or a fine-grained matrix.

Brecciated: Consisting of angular fragments cemented together.

British thermal unit (Btu): A unit of measure used to define energy.

Butt cleat: A short, poorly defined vertical cleavage plane in a coal seam, usually at right angles to the long face cleat; the coal cleat set that abuts into face cleats.

Buttress sand: Sand deposited on top of an unconformity.

C

Capillary fringe: The zone above the water table in which water is held by surface tension. Water in the capillary fringe is under a pressure less than atmospheric. The water is held above the water table by interfacial forces (e.g., surface tension). The capillary fringe is typically saturated to some distance above its base at the water table; upward from the saturated part only, progressively smaller pores are filled and the upper limit is indistinct. The upper limit can be defined more or less arbitrarily; for example, this limit may be defined as the level at which 50% of the pore space is filled with water. Some lateral flow generally occurs throughout the capillary fringe, but because the effective hydraulic conductivity decreases

Glossary

rapidly with moisture content, the lateral flow in the capillary fringe generally is negligible compared with that in the saturated zone, except where the capillary fringe and the saturated zone are of comparable thickness.

Caprock: (1) Impermeable rock layer that forms the seal on top of an oil or gas reservoir (seal); (2) insoluble rock on the top of a salt plug.

Capture: Water withdrawn artificially from an aquifer is derived from a decrease in storage in the aquifer, a reduction in the previous discharge from the aquifer, an increase in the recharge, or a combination of these changes. The decrease in discharge plus the increase in recharge is termed *capture*. Capture may occur in the form of decreases in the groundwater discharge into streams, lakes, and the ocean, or from decreases in the component of evapotranspiration derived from the saturated zone. After a new artificial withdrawal from the aquifer has begun, the head in the aquifer will continue to decline until the new withdrawal is balanced by capture.

Capture zone: The portion of an aquifer that contributes water to a particular pumping well.

Casing: Relatively slim-walled, large-diameter (5.5 to 13.37 inches) steel pipe. Joints of casing are screwed together to form a casing string, which is run into a well and cemented to the sides of the well.

Casinghead gas: Natural gas that bubbles out of oil on the surface of the well.

Catheads: A hub on a shaft (catshaft) on the drawworks of a drilling rig that is used to pull a line (catline) to lift or pull equipment.

Cavitation cycling: Also known as *cavity completion*, an alternative completion technique for hydraulic fracturing in which a cavity is generated by alternately pumping in nitrogen and blowing down pressure.

Cement: (1) Minerals that naturally grow between clastic grains and solidify a sedimentary rock; (2) Portland cement used to bind the casing strings to the well walls.

Cement job: The cementing of casing into a well.

Centralizer: An attachment to the outside of a casing string that uses steel bands to keep the string central in the well.

Charcoal test: A test used to measure the amount of condensate in natural gas. Activated charcoal is used to absorb the condensate from a volume of natural gas.

Christmas tree: The fittings, valves, and gauges that are bolted to the wellhead of a flowing well to control the flow from the well.

Clean sands: Well-sorted sands.

Cleats: Natural fractures in coal that often occur in systematic sets through which gas and water can flow.

CMHPG: Carboxymethyl hydroxypropylguar, a form of guar gel.

Compounder: A system of pulleys, belts, shafts, chains, and gears that transmit power from the prime movers to the drilling rig.

Conductivity, effective hydraulic (K_e): The rate of flow of water through a porous medium that contains more than one fluid, such as water and air in the unsaturated zone, and which should be specified in terms of both the fluid type and content and the existing pressure. Effective hydraulic conductivity has been called *capillary conductivity* by many soil physicists and *effective permeability* by many petroleum engineers.

Conductivity, hydraulic (K): Replaces the term *field coefficient of permeability*, which embodies the inconsistent units of gallon, foot, and mile. If a porous medium is isotropic and the fluid is homogeneous, the hydraulic conductivity of the medium is the volume of water at the existing kinematic viscosity that will move in unit time under a unit hydraulic gradient through a unit area measured at right angles to the direction of flow. Darcy's law can be expressed as

$$q = Q/A = -K(dh/dl) \tag{1}$$

where K, the constant of proportionality, is the hydraulic conductivity and may be expressed as

$$K = -[q(dh/dl)] \tag{2}$$

Hydraulic conductivity can have any units of L/T suitable to the problem involved. Hydraulic conductivity can be expressed in feet per day or in meters per day. Thus,

$$K = -[\text{ft}^3/(\text{day ft}^2 \ (-\text{ft ft}^{-1}))] = \text{ft day}^{-1} \tag{3}$$

$$K = -[\text{m}^3/(\text{day m}^2 \ (-\text{m m}^{-1}))] = \text{m day}^{-1} \tag{4}$$

Hydraulic conductivity is dependent primarily on the nature of the pore space, the type of liquid occupying it, and the strength of the gravitational field. For comparing the hydraulic conductivities of aquifers at different locations that contain water of appreciably different kinematic viscosity, it is only necessary to relate them by the dimensionless ratio of the kinematic viscosities and values of the acceleration due to gravity; thus, for the same intrinsic permeability,

$$K_1 = (v_2 g_1/v_1 g_2)K_2 \tag{5}$$

Ordinarily, differences in the acceleration due to gravity are negligible; hence, Equation 5 is closely approximated by

$$K_1 = (v_2/v_1)K_2 \tag{6}$$

In anisotrophic media (not isotropic) the direction of the specific discharge q is not generally parallel to that of the gradient (dh/dl) of the head. In such media, the Cartesian components of the specific discharge are related to those of the gradient by

$$-q_x = K_{xx}(\partial_h/\partial_x) + K_{xy}(\partial_h/\partial_y) + K_{xz}(\partial_h/\partial_z) \tag{7}$$

$$-q_y = K_{yx}(\partial_h/\partial_x) + K_{yy}(\partial_h/\partial_y) + K_{yz}(\partial_h/\partial_z) \tag{8}$$

$$-q_z = K_{zx}(\partial_h/\partial_x) + K_{zy}(\partial_h/\partial_y) + K_{zz}(\partial_h/\partial_z) \tag{9}$$

The quantities in the form K_{xx}, K_{xy}, K_{xz}, and so forth, called *conductivity coefficients*, are the second-order tensor, generally symmetric.

Confining bed: A term that supplants the terms *aquiclude*, *aquitard*, and *aquifuge* and is defined as a body of impermeable material stratigraphically adjacent to one or more aquifers. In nature, its hydraulic conductivity may range from nearly zero to some value distinctly lower than that of the aquifer. Its conductivity relative to that of the aquifer it confines should be specified or indicated by a suitable modifier such as slightly permeable or moderately permeable.

Connate: Saline, subsurface water.

Coquina: A sedimentary rock composed of broken shells.

Craton: A part of the Earth's crust that has attained stability and has been relatively underformed for a long time; the term is restricted to continents and includes both shields and platforms.

Criterion sound level: A sound level of 90 decibels.

Crosslinked gel: A gel to which a crosslinker has been added.

Crosslinker: An additive that, when added to a linear gel, will create a complex, high-viscosity, pseudoplastic fracturing fluid.

Glossary 275

Crude stream: Crude oil from a single field or a mixture from fields that is offered for sale by an exporting country.

Cyclotherm: Alternating marine and nonmarine sedimentary rocks.

D

Darcy (D): A measure of the permeability of rock or sediment that was originally defined as the volume (in cubic centimeters) of water of 1-centipoise viscosity flowing in 1 second through an area of 1 square centimeter under a pressure gradient of 1 atmosphere per square centimeter per centimeter. The direction of flow must be horizontal to negate gravitation effects. The volume was later changed to 1 milliliter; thus, the darcy now involves the inconsistent units of centimeter, milliliter, and atmosphere. The darcy has a value of 0.987 $(\mu m)^2$.

Decibel (dB): Unit of measurement of sound level.

Demulsifier: A chemical used to break an emulsion.

Desorption: Liberation of tightly held methane gas molecules previously bound to the solid surface of the coal.

Detrital: A sediment grain that has been transported and deposited to a whole particle such as a sand grain.

Dip-slip fault: A fault with predominately vertical displacement. It can be either a normal or reverse dip-slip fault.

Doghouse: The room or vehicle that houses seismic recording equipment.

Dolomite: A mineral composed of calcium magnesium carbonate, or $CaMg(CO_3)_2$. It is formed by the natural alteration of calcite. A rock composed of dolomite is called *dolostone* and can be a reservoir rock.

Double hearing protection: A combination of both earplug and earmuff types of hearing protection devices that is required for employees who have demonstrated temporary threshold shift during audiometric examination and for those who have been advised by a medical doctor to wear double protection in work areas that exceed 104 dBA.

Drawworks: A drum in a steel frame used to raise and lower equipment in a bore hole. Power is provided by the prime movers. Hoisting line is wound around the reel.

Drilling mud: A viscous mixture of clay and additives with freshwater, diesel oil, synthetic oil, or an emulsion of water with droplets of oil.

Duster: A well that did not encounter commercial amounts of petroleum.

E

Edge water: Water located in the reservoir to the side of the oil.

Emulsion: Droplets of one liquid suspended in a different liquid, such as water in oil.

Epiclastic: Formed from the fragments or particles broken away (by weathering and erosion) from pre-existing rocks to form an altogether new rock in a new place.

Evapotranspiration: The process by which water is discharged to the atmosphere as a result of evaporation from the soil and surface-water bodies and transpiration by plants.

Explosive fracturing: Involves exploding nitroglycerin in a torpedo at reservoir depth in a well to fracture the reservoir and stimulate production.

F

Face cleat: A coal cleat set that is through-going and continuous.

Fairway (trend): The area along which the play has been proven and more fields could be found.

Flow, steady: Occurs when at any point the magnitude and direction of the specific discharger are constant in time.

Flow, uniform: Occurs when at every point the specific discharge has the same magnitude and direction.

Flow, unsteady or nonsteady: Occurs when at any point the magnitude or direction of the specific discharge changes with time. In practice, the term *transient* is used in reference to the temporary features of unsteady flow; thus, in unsteady flow, the specific discharge, the head, and perhaps other factors consist of a steady component plus a transient component.

Flowback: The process of causing fluid to flow back to the well of a fracture after a hydraulic fracturing event is complete.

Fluid potential: The mechanical energy per unit mass of a fluid at any given point in space and time with respect to an arbitrary state and datum. Loss of fluid potential incurred as the fluid moves from a region of high potential to one of low potential represents a loss of mechanical energy which is converted to heat by friction. In groundwater movement, the kinetic energy term $v^2/2$ ordinarily is negligible. If the expansion and contraction of the fluid due to changes in pressure are unimportant to the problem being considered, the fluid can be assumed to be incompressible. At a given point in a body of liquid, the fluid potential is proportional to the head (h), with the constant of proportionality being the acceleration due to gravity (g); that is,

$$\text{Fluid potential } (\phi) = gh$$

Formation: A mappable rock layer with a sharp top and bottom.

Fracture conductivity: The capability of the fracture to conduct fluids under a given hydraulic head difference.

Frequency: Rate at which pressure oscillations are produced; measured in hertz (Hz).

G

Gas cap: The uppermost portion of a saturated oil reservoir.

Gathering system: A system of flowlines that conducts produced fluids from wells to a central processing unit.

Geophone: A seismic detector, placed on or in the ground, that responds to ground motion at its point of location.

Geothermal gradient: Rate of temperature increase with increasing depth in the Earth's interior.

Graben: An elongated, down-dropped block of land that is bounded by nearly parallel faults on both sides.

Gradient of head: A mathematical term referring to the vector denoted by Δh or grad h whose magnitude is equal to the maximum rate of change in head and whose direction is that in which the maximum rate of increase occurs. The hydraulic gradient and the gradient of the head are equal but of opposite sign.

Greywacke: A poorly sorted, dark-colored sandstone.

Groundwater, confined: Groundwater that is under pressure significantly greater than atmospheric; its upper limit is the bottom of a bed of distinctly lower hydraulic conductivity than that of the material in which the confined water occurs.

Groundwater, perched: Unconfined groundwater separated from an underlying body of groundwater by an unsaturated zone. Its water table is a perched water table. It is not able to bring water in the underlying unsaturated zone above atmospheric pressure. Perched groundwater may be either *permanent*, where recharge is frequent enough to maintain a saturated zone above the perching bed, or *temporary*, where intermittent recharge is not great or frequent enough to prevent the perched water from disappearing from time to time as a result of drainage over the edge of or through the perching bed.

Groundwater, unconfined: Water in an aquifer that has a water table.

Guar: Organic powder thickener typically used to make viscous fracturing fluids; it is completely soluble in hot and cold water but insoluble in oils, grease, and hydrocarbons.

Glossary

H

Head, static: The height above a standard datum of the surface of a column of water (or other liquid) that can be supported by the static pressure at a given point. The static head (h) is the sum of the elevation head (h_e) and the pressure head (h_p); that is, $h = h_e + h_p$. Under conditions in which Darcy's law may be applied, the velocity of groundwater is so small that the velocity head, $h_v = v^2/2g$, is negligible. *Head*, when used alone, is understood to mean static head. Head is proportional to the fluid potential; therefore, head is a measure of the potential.

Head, total: The total head of a liquid at a given point is the sum of three components: (1) elevation head (h_e), which is equal to the elevation of the point above a datum; (2) pressure head (h_p), which is the height of a column of static water that can be supported by the static pressure at the point; and (3) velocity head (h_v), which is the height the kinetic energy of the liquid is capable of lifting the liquid.

Hearing conservation record: An employee's audiometric record that includes name, age, job classification, time-weighted average (TWA) exposure, date of audiogram, and name of audiometric technician. OSHA requires it to be retained for the duration of employment, and it should be kept indefinitely for workers' compensation.

Heavy oil: Viscous high-density oil with an API gravity less than 25.

Hertz (Hz): Unit of measurement of frequency, numerically equal to cycles per second.

Homogeneity: Synonymous with *uniformity*. A material is homogeneous if the hydrologic properties are identical everywhere. Although no known aquifer is homogeneous in detail, models based on the assumption of homogeneity have been shown to be valuable tools for predicting the approximate relationship between discharge and potential in many aquifers.

Hydraulic conductivity: See *conductivity* and *permeability*.

Hydraulic diffusivity: The parameter T/S or K/S_s, which is the conductivity of the saturated medium when the unit volume of water moving is that involved in changing the head a unit amount in a unit volume of medium. By analogy with Maxwell's nomenclature in heat conduction theory (thermometric conductivity), it may be considered potentiometric conductivity. Similar diffusivities characterize the flow of heat and of electricity by conduction and the movement of a dissolved substance in a liquid by diffusion. The parameter arises from the fundamental differential equation for liquid flow in a porous medium. In any isotropic homogeneous system, the time involved for a given head change to occur at a particular point in response to a greater change in head at another point is inversely proportional to the diffusivity. As a common example, the cone of depression affects moderately distant wells by measurable amounts in a short time in confined groundwater bodies for which the diffusivities are commonly large and only after a longer time in unconfined water bodies for which the diffusivities are commonly much smaller.

Hydraulic fracturing: A well stimulation method in which liquid under high pressure is pumped down a well to fracture the reservoir rock adjacent to the wellbore. Propping agents are used to keep the fractures open.

Hydraulic gradient: The change in static head per unit of distance in a given direction. If not specified, the direction generally is understood to be that of the maximum rate of decrease in head.

Hydrochloric acid (HCl): Can be used in diluted form in the hydraulic fracturing process to fracture limestone formations and to clean up perforations in coalbed methane fracturing treatments.

Hydroxyethylcellulose (HEC): A form of guar gel.

I

Injectate: Relative to coalbed methane, the fracturing fluid injected into a coalbed methane well.

Interfinger: A boundary between two rock types in which both form distinctive wedges protruding into each other.

Isopach: A line drawn on a map through points of equal true thickness of a designated stratigraphic unit or group of stratigraphic units.

Isotopic: Rocks formed in the same environment (i.e., in the same sedimentary basin or geologic province).

Isotrophy: Condition in which all significant properties are independent of direction. Although no aquifers are isotropic in detail, models based on the assumption of isotropy have been shown to be valuable tools for predicting the approximate relationships between discharge and potential in many aquifers.

Isotropic: A medium, such as unconsolidated sediments or a rock formation, whose properties are the same in all directions.

J

Junk: A tool or broken pipe that has fallen to the bottom of a well.

K

KCl: Molecular formula for potassium chloride.

Kelly: Strong, four- or six-sided steel pipe that is located at the top of the drill-string. It runs through the kelly bushing.

Kerogen: Insoluble organic matter in sedimentary rocks.

L

Lacustrine: Pertaining to, produced by, or formed in a lake or lakes.

Laminar flow: Water flow in which the stream lines remain distinct and the flow direction at every point remains unchanged with time; non-turbulent flow.

Leakoff: The magnitude of pressure exerted on a formation that causes fluid to be forced into the formation. In common usage, leakoff is often considered the movement of fluid out of primary fractures and into a geologic formation, either through small existing permeable paths (connected pores and natural fracture networks) or through small pathways created or enlarged in the rock through the fracturing process.

Lenticular: Pertaining to a discontinuous lens-shaped (saucer-shaped) stratigraphic body.

Lift gas: Inert gas, usually natural gas, that is used for gas lift (which involves lifting fluids by lowering the combined fluid density and allowing the fluids to flow).

Linear gel: A simple guar-based fracturing fluid usually formulated using guar and water with additives or guar with diesel fuel.

Lithology: The study of rocks based on their mineralogic composition and texture.

M

Make up: To screw together pipe.

Medical pathology: A disorder or disease, such as a condition or disease affecting the ear, which a physician specialist should treat.

Millidarcy: The customary unit of measurement of fluid permeability; equivalent to 0.001 darcy.

Milligrams per liter (mg/L): Typically used to define the concentration of a dissolved compound in a fluid.

Million cubic feet (Mcf): A unit typically used to define gas production volumes in the coalbed methane industry; 1 Mcf is roughly equivalent to the volume of gas required to heat approximately 12 households for one year. Mcf can also represent 1000 cubic feet.

Glossary

Mined-through studies: Mined-through studies are projects in which coalbeds have been actually mined through (i.e., the coal has been removed) so that remaining coal and surrounding rock can be inspected, after the coalbeds have been hydraulically fractured. These studies provide unique subsurface access to investigate coalbeds and surrounding rock after hydraulic fracturing.

Moduli: Plural of modulus, often referred to as *bulk modulus*, which is the ratio of stress to strain (K). The bulk modulus is an elastic constant equal to the applied stress divided by the ratio of the change in volume to the original volume of a body.

N

Natural gas: A gas composed of a mixture of hydrocarbon molecules that have one, two, three, and four carbon atoms.

NIOSH: National Institute for Occupational Safety and Health.

Noise dose: The ratio, expressed as a percentage, of (1) the time integral, over a stated time or event, of the 0.6 power of the measured slow exponential time-averaged, squared A-weighted sound pressure, and (2) the product of the criterion duration (8 hours) and the 0.6 power of the squared sound pressure corresponding to the criterion sound level (90 dB).

Noise dosimeter: An instrument that integrates a function of sound pressure over a period of time to directly indicates a noise dose.

Noise hazard area: Any area where noise levels are equal to or exceed 85 dBA. OSHA requires employers to designate work areas as "noise hazard areas," post warning signs, and warn employees when work practices exceed 90 dBA. Hearing protection must be worn whenever 90 dBA is reached or exceeded.

Noise hazard work practice: Performing or observing work where 90 dBA is equaled or exceeded. Some work practices, however, will be specified as a "rule of thumb." Whenever attempting to hold normal conversation with someone who is 1 foot away and shouting must be employed to be heard, one can assume that a 90-dBA noise level or greater exists and hearing protection is required. Typical examples of work practices where hearing protection is required are jackhammering, heavy grinding, heavy equipment operations, and similar activities.

Noise-level measurement: Total sound level within an area; includes workplace measurements indicating the combined sound levels of tool noise (from ventilation systems, cooling compressors, circulation pumps, etc.).

Noise reduction ratio: The number of decibels of sound reduction actually achieved by a particular hearing protection device.

O

Oilfield brine: A very saline water that is produced with oil.

Otolaryngologist: A physician specializing in diagnosis and treatment of disorders of the ear, nose, and throat.

Otoscopic examination: Inspection of external ear canal and tympanic membrane.

Overthrust: A low-angle thrust fault of large scale, with total displacement (lateral or vertical) generally measured in kilometers.

P

Pad: An initial volume of fluid that is used to initiate and propagate a fracture before a proppant is placed.

Paleochannels: Old or ancient river channels preserved in the subsurface as lenticular sandstones.

Paraffin: A member of the hydrocarbon series of molecules. They are straight chains with single bonds. All hydrocarbon molecules in natural gas and some in crude oil are paraffins.

Parts per million (ppm): Number of weight or volume units of a constituent present with each 1 million units of a solution or mixture. Formerly used to express the results of most water and wastewater analyses, parts per million is being replaced by milligrams per liter (mg/L). For drinking water analyses, concentration in parts per million and milligrams per liter are equivalent. By way of comparison, a single ppm can be compared to a shot glass full of water inside a swimming pool.

Permanent threshold shift (PTS): Hearing loss with less than normal recovery.

Permeability: The capacity of a porous rock, sediment, or soil to transmit a fluid; it is a measure of the relative ease of fluid flow under equal pressure and from equal elevations.

Permeability, intrinsic: A measure of the relative ease with which a porous medium can transmit a liquid under a potential gradient. It is a property of the medium alone and is independent of the nature of the liquid and of the force field causing movement. It is a property of the medium that is dependent on the shape and size of the pores.

Personal protective device: Items such as earplugs or earmuffs used as protection against hazardous noise.

Physiographic: Refers to a region where all parts are similar in geologic structure and climate and which has had a unified geomorphic history; the relief features of the region differ significantly from those of adjacent regions.

Play: A productive coalbed methane formation or a productive oil or gas deposit; a combination of trap, reservoir rock, and seal that has been shown by previously discovered fields to contain natural gas and oil.

Porosity: The property of a rock or soil of containing interstices or voids; it may be expressed quantitatively as the ratio of the volume of its interstices to its total volume. It may be expressed as a decimal fraction or as a percentage. With respect to the movement of water, only the system of interconnected interstices is significant.

Porosity, effective: The amount of interconnected pore space available for fluid transmission; it is expressed as a percentage of the total volume occupied by the interconnecting interstices. Although effective porosity has been used to mean about the same thing as specific yield, such use is discouraged.

Potentiometric surface: A surface that represents the total head of groundwater. The potentiometric surface represents the static head. As related to an aquifer, it is defined by the levels to which water will rise in tightly cased wells. As related to an aquifer, it is defined by the levels to which water will rise in tightly cased wells. Where the head varies appreciably with depth in the aquifer, a potentiometric surface is meaningful only if it describes the static head along a particular specified surface or stratum in the aquifer. More than one potentiometric surface is then required to describe the distribution of head. The water table is a particular potentiometric surface.

Pounds per square inch (psi): Pounds per square inch; a unit of pressure.

Presbycusis: Hearing loss due to age.

Pressure, static: Pressure exerted by the fluid. It is the mean normal compressive stress on the surface of a small sphere around a given point. The static pressure does not include the dynamic pressure ($pv^2/2$) and therefore is distinguished from *total pressure*. The velocity of groundwater ordinarily is so small that the dynamic pressure is negligible. Pressure, when used alone, is understood to mean static pressure.

Primacy: The right to self-establish, self-enforce, and self-regulate environmental standards; this enforcement responsibility is granted by the USEPA to states and Indian tribes.

Primary porosity: The porosity preserved from some time between sediment deposition and the final rock-forming process (e.g., spaces between grains of sediment).

Glossary   **281**

Proppant: Granules of sand, ceramic, or other minerals that are wedged within the fracture and act to "prop" it open after the fluid pressure from fracture injection has dissipated.

Prospect: Location where the geological and economical conditions are favorable for drilling an exploratory well.

Q

Quartz: A common mineral composed of SiO_2. Sandstones are usually composed of quartz sand grains.

R

Rank: The degree of metamorphism in coal; the basis of coal classification into a natural series from lignite to anthracite.

Recovery factor: The percentage of oil or gas in place that will be produced from a reservoir.

Representative exposure: Measurements of an employee's noise dose or 8-hour time-weighted average (TWA) sound level that the employer deems to be representative of the exposures of other employees in the workplace.

S

Sample log: A record of the physical properties of rocks in a well. It includes composition, texture, color, presence of pore spaces, and oil staining.

Scale: Salts that have precipitated out of water. Calcium carbonate, barium sulfate, and calcium sulfate are common in oil fields.

Screen-out: Refers to a fracturing job where proppant placement has failed.

Secondary porosity: The porosity created through alteration of rock, commonly by processes such as dissolution and fracturing.

Semianthracite: Term used to identify coal rank; specifically refers to coal that possesses a fixed-carbon content of 86 to 92%.

Sensorineural: Type of hearing loss characterized as having been induced by industrial noise exposure; this type of hearing loss is permanent.

Shale: A very common sedimentary rock composed of clay-sized particles. Black shales are source rocks for petroleum.

Siltstone: A sedimentary rock composed primarily of silt-sized particles.

Soil: The layer of bonded particles of sand, silt, and clay that covers the land surface of the Earth. Most soils develop multiple layers. The topmost layer (*topsoil*) is the layer in which plants grow. This layer is actually an ecosystem composed of both biotic and abiotic components—inorganic chemicals, air, water, decaying organic material that provides vital nutrients for plant photosynthesis, and living organisms. Below the topmost layer (usually no more than a meter in thickness), is the *subsoil*, which is much less productive, partly because it contains much less organic matter. Below that is the *parent material*, the bedrock or other geologic material from which the soil is ultimately formed. The general rule of thumb is that it takes about 30 years to form one inch of topsoil from subsoil; it takes much longer than that for subsoil to be formed from parent material, with the length of time depending on the nature of the underlying matter.

Soil formation: Soil is formed as a result of physical, chemical, and biological interactions in specific locations. Just as vegetation varies among biomes, so do the soil types that support that vegetation. The vegetation of the tundra and that of the rain forest differ vastly from each other and from vegetation of the prairie and coniferous forest; soils differ in similar ways. In the soil-forming process, two related, but fundamentally different, processes are occurring simultaneously. The first is the formation of soil parent materials by weathering of rocks, rock fragments, and sediments. This set of processes is carried out in the zone of weathering. The end point is producing parent material for the soil to develop in and is referred to as C horizon material. It applies in the same way for glacial deposits as for rocks. The second set of processes is the formation of the soil profile by soil-forming processes, which gradually change the C horizon material into A, E, and B horizons. Soil development takes time and is the result of two major processes: weathering and morphogenesis. *Weathering*, the breaking down of bedrock and other sediments that have been deposited on the bedrock by wind, water, volcanic eruptions, or melting glaciers, happens physically, chemically, or by a combination of both. *Physical weathering* involves the breaking down of rock primarily by temperature changes and the physical action of water, ice, and wind. When a geographical location is characterized as having an arid desert biome, the repeated exposure to very high temperatures during the day, followed by low temperatures at night, causes rocks to expand and contract and eventually to crack and shatter. At the other extreme, in cold climates, rock can crack and break as a result of repeated cycles of expansion of water in rock cracks and pores during freezing and contraction during thawing. Another example of physical weathering occurs when various vegetation types spread their roots and grow, and the roots exert enough pressure to enlarge cracks in solid rock, eventually splitting the rock. Plants such as mosses and lichens also penetrate rock and loosen particles. In addition to physical weathering, bare rocks are subjected to *chemical weathering*, which involves chemical attack and dissolution of rock. Accomplished primarily through oxidation via exposure to oxygen gas in the atmosphere, acidic precipitation (after having dissolved small amounts of carbon dioxide gas from the atmosphere), and acidic secretions of microorganisms (bacteria, fungi, and lichens), chemical weathering speeds up in warm climates and slows down in cold ones. Physical and chemical weathering do not always (if ever) occur independently of each other. Instead, they normally work in combination, and the results can be striking. A classic example of the effect and the power of their simultaneous actions can be seen in the ecological process known as *bare rock succession*. The final stages of soil formation consist of the processes of *morphogenesis*, or the production of a distinctive *soil profile* with its constituent layers or *horizons*. The soil profile (the vertical section of the soil from the surface through all its horizons, including C horizons) gives the environmental scientist critical information. When properly interpreted, soil horizons can warn about potential problems with using the land and tell much about the environment and history of a region. The soil profile allows us to describe, sample, and map soils. *Soil horizons* are distinct layers, roughly parallel to the surface, which differ in color, texture, structure, and content of organic matter. The clarity with which horizons can be recognized depends on the relative balance of the migration, stratification, aggregation, and mixing processes that take place in the soil during morphogenesis. In *podzol-type soils*, striking horizonation is quite apparent; in *vertisol-type soils*, the horizons are less distinct. When horizons are studied, they are each given a letter symbol to reflect the genesis of the horizon. Certain processes work to create and destroy clear soil horizons. Various formations of soil horizons that tend to create clear horizons by vertical redistribution of soil materials include the leaching of ions in the soil solutions, movement of clay-sized particles, upward movement of water by capillary action, and surface deposition of dust and aerosols. Clear soil horizons are destroyed by mixing processes that occur because of organisms, cultivation practices, creep processes on slopes, frost heave, and swelling and shrinkage of clays—all part of the natural soil formation process.

Glossary 283

Soil properties: From the soil engineer's view (regarding land conservation and remediation methodologies for contaminated soil remediation through reuse and recycling), four major properties of soil are of interest: texture, slope, structure, and organic matter.

Soil structure (tilth): Soil structure, or tilth, should not be confused with soil texture—they are different. In fact, in the field, the properties determined by soil texture may be considerably modified by soil structure. Soil structure refers to the way various soil particles clump together. Clusters of soil particles, called *aggregates*, can vary in size, shape, and arrangement; they combine naturally to form larger clumps called *peds*. Sand particles do not clump because sandy soils lack structure. Clay soils tend to stick together in large clumps. Good soil develops small *friable* (crumble easily) clumps. Soil develops a unique, fairly stable structure in undisturbed landscapes, but agricultural practices break down the aggregates and peds, lessening erosion resistance. The presence of decomposed or decomposing remains of plants and animals (*organic matter*) in soil helps not only fertility but also soil structure—especially the ability of the soil to store water. Live organisms such as protozoa, nematodes, earthworms, insects, fungi, and bacteria are typical inhabitants of soil. These organisms work to either control the population of organisms in the soil or aid in the recycling of dead organic matter. All soil organisms, in one way or another, work to release nutrients from the organic matter, changing complex organic materials into products that can be used by plants.

Soil texture: Soil texture is a given and cannot be easily or practically changed in any significant way. It is determined by the size of the rock particles (sand, silt, and clay particles) within the soil. The largest soil particles are gravel, which consists of fragments larger than 2.0 mm in diameter. Particles between 0.05 and 2.0 mm are classified as sand. Silt particles range from 0.002 to 0.05 mm in diameter, and the smallest particles (clay particles) are less than 0.002 mm in diameter. Though clays are composed of the smallest particles, those particles have stronger bonds than silt or sand, though once broken apart, they erode more readily. Particle size has a direct impact on erodibility. Rarely does a soil consist of only one single size of particle—most are a mixture of various sizes. The *slope* (or steepness of the soil layer) is another given, important because the erosive power of runoff increases with the steepness of the slope. Slope also allows runoff to exert increased force on soil particles, which breaks them apart more readily and carries them farther away.

Solution gas: The dissolved natural gas that bubbles out of crude oil on the surface when the pressure drops during production.

Sound level: Ten times the common logarithm of the ratio of the square of the measured A-weighted sound pressure to the square of the standard reference pressure of 20 micropascals; measured in decibels (dB).

Sound level meter: An instrument for the measurement of sound level.

Specific capacity of a well: The rate of discharge of water from the well divided by the drawdown of water level within the well. It varies slowly with duration of discharge which should be stated when known. If the specific capacity is constant except for the time variation, it is roughly proportional to the transmissivity of the aquifer. The relation between discharge and drawdown is affected by the construction of the well, its development, the character of the screen or casing perforation, and the velocity and length of flow up the casing. If the well losses are significant, the ratio between discharge and drawdown decreases with increasing discharge; it is generally possible to roughly separate the effects of the aquifer from those of the well by step drawdown tests. In aquifers with large tubular openings, the ratio between discharge and drawdown may also decrease with increasing discharge because of a departure from laminar flow near the well or, in other words, a departure from Darcy's law.

Specific discharge or specific flux: The specific discharge, or specific flux, for groundwater is the rate of discharge of groundwater per unit area of the porous medium measured at right angles to the direction of flow. Specific discharge (q) has the dimensions of velocity, as follows:

$$q = Q/A$$

where Q equals total discharge through area A. Specific discharge has sometimes been called the bulk velocity or the Darcian velocity. Specific discharge is a precise term and is preferred to terms involving "velocity" because of possible confusion with actual velocity through the pores if a qualifying term is not constantly repeated.

Specific retention: The ratio of the volume of water that the rock or soil, after being saturated, will retain against the pull of gravity to the volume of the rock or soil. Ideally, the definition implies that gravity drainage is complete. However, the amount of water held in pores above the water table during gravity drainage is dependent on particle size, distance above the water table, time of drainage, and other variables. Lowering of the water table and infiltration occur over such short periods of time that gravity drainage is rarely or never complete. Thus, the concepts embodied in specific retention do not adequately recognize the highly complex set of interacting conditions that regulate moisture retention. Nevertheless, specific retention is a useful though approximate measure of the moisture-holding capacity of the unsaturated zone in the region above the capillary fringe.

Specific storage: In problems of three-dimensional transient flow in a compressible groundwater body, it is necessary to consider the amount of water released from or taken into storage per unit volume of the porous medium. The specific storage is the volume of water released from or taken into storage per unit volume of the porous medium per unit change in head.

Specific yield: The ratio of the volume of water which the rock or soil, after being saturated, will yield by gravity to the volume of the rock or soil. The definition implies that gravity drainage is complete. In the natural environment, specific yield is generally observed as the change that occurs in the amount of water in storage per unit area of unconfined aquifer as the result of a unit change in head. Such a change in storage is produced by the draining or filling of pore space and is therefore dependent on particle size, rate of change of the water table, time, and other variables. Hence, specific yield is only an approximate measure of the relation between storage and head in unconfined aquifers. It is equal to porosity minus specific retention.

Storage, bank: The change in storage in an aquifer resulting from a change in stage of an adjacent surface-water body.

Storage coefficient: The volume of water an aquifer releases from or takes into storage per unit surface area of the aquifer per unit change in head. In a confined water body, the water derived from storage with decline in head comes from expansion of the water and compression of the aquifer; similarly, water added to storage with a rise in head is accommodated partly by compression of the water and partly by expansion of the aquifer. In an unconfined water body, the amount of water derived from or added to the aquifer by these processes generally is negligible compared to that involved in gravity drainage or filling of pores; hence, in an unconfined water body, the storage coefficient is virtually equal to the *specific yield*.

Stratigraphy: The study of rock strata: concerning all characteristics and attributes of rocks and their interpretation in terms of mode of origin and geologic history.

Stream flow: Most elementary students learn early in their education process that water on Earth flows downhill—from land to the sea; however, they may or may not be told that water flows downhill toward the sea by various routes. The route of primary concern here is the surface water route taken by surface water runoff. Surface runoff is dependent on various factors; for example, climate, vegetation, topography, geology, soil characteristics, and land use determine how much surface runoff occurs compared with other pathways.

Glossary

285

The primary source of water for total surface runoff is, of course, precipitation. This is the case even though a substantial portion of all precipitation input returns directly to the atmosphere by evapotranspiration, which, as the name suggests, is a combination process whereby water in plant tissues and in the soil evaporates and transpires to water vapor in the atmosphere. Let's take a closer look at the input of precipitation to surface water runoff. A substantial portion of precipitation input returns directly to the atmosphere by evapotranspiration. When precipitation occurs, some rainwater is intercepted by vegetation where it evaporates, never reaching the ground or being absorbed by plants. A large portion of the rainwater that reaches the surface on ground, in lakes and streams also evaporates directly back to the atmosphere. Although plants display a special adaptation to minimize transpiration, plants still lose water to the atmosphere during the exchange of gases necessary for photosynthesis. Notwithstanding the large percentage of precipitation that evaporates, rain- or meltwater that reaches the ground surface follows several pathways in reaching a stream channel or groundwater. Soil can absorb rainfall to its infiltration capacity (i.e., to its maximum rate). During a rain event, this capacity decreases. Any rainfall in excess of infiltration capacity accumulates on the surface. When this surface water exceeds the depression storage capacity of the surface, it moves as an irregular sheet of overland flow. In arid areas, overland flow is likely because of the low permeability of the soil. Overland flow is also likely when the surface is frozen or when human activities have rendered the land surface less permeable. In humid areas, where infiltration capacities are high, overland flow is rare. In rain events where the infiltration capacity of the soil is not exceeded, rain penetrates the soil and eventually reaches the groundwater—from which it discharges to the stream slowly and over a long period. This phenomenon helps to explain why stream flow through a dry weather region remains constant; the flow is continuously augmented by groundwater. This type of stream is known as a *perennial stream*, as opposed to an *intermittent* one, because the flow continues during periods of no rainfall. When a stream courses through a humid region, it is fed water via the water table, which slopes toward the stream channel. Discharge from the water table into the stream accounts for flow during periods without precipitation and explains why this flow increases, even without tributary input, as one proceeds downstream. Such streams are called *gaining* or *effluent*, as opposed to *losing* or *influent streams* that lose water into the ground. The same stream can shift between gaining and losing conditions along its course because of changes in underlying strata and local climate.

Subbituminous: A black coal, intermediate in rank between lignite and bituminous.

Subgreywacke: A sedimentary rock (sandstone) that contains less feldspar and more and better rounded quartz grains than greywacke; it is intermediate in composition between greywacke and orthoquartzite, is lighter colored and better sorted, and has less matrix than greywacke.

Surficial: Pertaining to or lying in or on a surface; specific to the surface of the Earth.

Syncline: A fold of layered, sedimentary rocks whose core contains stratigraphically younger rocks; shape of fold is generally concave upward.

T

Tank battery: Two or more stock tanks connected in line.

Temporary threshold shift (TTS): Temporary loss of normal hearing level brought on by brief exposure to high-level sound. TTS is greatest immediately after exposure to excessive noise and progressively diminishes with increasing rest time.

Thermogenic: A direct product of high temperatures (e.g., thermogenic methane).

Time-weighted average (TWA) sound level: That sound level, which if constant over an 8-hour exposure, would result in the same noise dose as is measured.

Toughness: The point at which enough stress intensity has been applied to a rock formation so that a fracture initiates and propagates.

Transmissivity: A measure of the amount of water that can be transmitted horizontally through a unit width by the full saturated thickness of the aquifer under a hydraulic gradient of one.

Trillion cubic feet (Tcf): A unit typically used to define gas production volumes in the coalbed methane industry; 1 Tcf is roughly equivalent to the volume of gas required to heat approximately 12 million households for one year.

U

Unsaturated pool: An oil reservoir without a free gas cap.

Upwarp: The uplift of a region; usually a result of the release of isostatic pressure (e.g., the melting of an ice sheet).

V

Velocity, average interstitial: Although specific discharge (q) has the dimensions of a velocity, it expresses the average volume rate of flow rather than the particle velocity. To determine the average interstitial velocity (V_i) it is necessary to also know the effective porosity (n_e). The average interstitial velocity can be expressed as

$$V_i = q/n_e = -(Kdh/dl)/n_e$$

Because the hydraulic gradient and effective porosity are dimensionless ratios, V_i has the same dimensions as q and K (hydraulic conductivity). Even within a homogeneous medium there is a wide range of velocities within the pores and from one pore to another, and in a nonhomogeneous medium the velocities may range over several orders of magnitude.

Viscosity: The property of a substance to offer internal resistance to flow; internal friction.

Volcaniclastic: Composed of fragments or particles and related to volcanic processes either by forming as the result of explosive processes or due to the weathering and erosion of volcanic rocks.

W

Water table: The subsurface level below which the pores in the soil or rock are filled with water. The water table is that surface in a groundwater body at which the water pressure is atmospheric. It is defined by the levels at which water stands in wells that penetrate the water body just far enough to hold standing water.

Z

Zone: A rock layer identified by a characteristic microfossil species.

Zone, saturated: That part of the Earth's crust beneath the deepest water table in which all voids, large and small, are ideally filled with water under pressure greater than atmospheric. The saturated zone may depart from the ideal in some respects. A rising water table may cause entrapment of air in the upper part of the zone of saturation, and the lower part may include accumulations of other natural fluids. The saturated zone has been called the *phreatic zone* by some.

Zone, unsaturated: The zone between the land surface and the deepest water table. It includes the capillary fringe. Generally, water in this zone is under less than atmospheric pressure, and some of the voids may contain air or other gases at atmospheric pressure. Beneath flooded areas or in perched water bodies, the water pressure locally may be greater than atmospheric.

Appendix. Chemicals Used in Hydraulic Fracturing

Chemical Component	Chemical Abstract Service Number	No. of Products Containing Chemical
1-(1-Naphthylmethyl)quinolinium chloride	65322-65-8	1
1,2,3-Propanetricarboxylic acid, 2-hydroxy-, trisodium salt, dihydrate	6132-04-3	1
1,2,3-Trimethylbenzene	526-73-8	1
1,2,4-Trimethylbenzene	95-63-6	21
1,2-Benzisothiazol-3	2634-33-5	1
1,2-Dibromo-2,4-dicyanobutane	35691-65-7	1
1,2-Ethanediaminium, *N*,*N'*-*bis*[2-[*bis*(2-hydroxyethyl)methylammonio] ethyl]-*N*,*N'*-*bis*(2-hydroxyethyl)- *N*,*N'*-dimethyl-, tetrachloride	138879-94-4	2
1,3,5-Trimethylbenzene	108-67-8	3
1,6-Hexanediamine dihydrochloride	6055-52-3	1
1,8-Diamino-3,6-dioxaoctane	929-59-9	1
1-Hexanol	111-27-3	1
1-Methoxy-2-propanol	107-98-2	3
2,2'-Azobis (2-amidopropane) dihydrochloride	2997-92-4	1
2,2-Dibromo-3-nitrilopropionamide	10222-01-2	27
2-Acrylamido-2-methylpropanesulphonic acid sodium salt polymer	—[a]	1
2-Bromo-2-nitropropane-1,3-diol	52-51-7	4
2-Butanone oxime	96-29-7	1
2-Hydroxypropionic acid	79-33-4	2
2-Mercaptoethanol (thioglycol)	60-24-2	13
2-Methyl-4-isothiazolin-3-one	2682-20-4	4
2-Monobromo-3-nitrilopropionamide	1113-55-9	1
2-Phosphonobutane-1,2,4-tricarboxylic acid	37971-36-1	2
2-Phosphonobutane-1,2,4-tricarboxylic acid, potassium salt	93858-78-7	1
2-Substituted aromatic amine salt	—[a]	1
4,4'-Diaminodiphenyl sulfone	80-08-0	3
5-Chloro-2-methyl-4-isothiazolin-3-one	26172-55-4	5
Acetaldehyde	75-07-0	1
Acetic acid	64-19-7	56
Acetic anhydride	108-24-7	7
Acetone	67-64-1	3
Acetophenone	98-86-2	1
Acetylenic alcohol	—[a]	1
Acetyltriethyl citrate	77-89-4	1
Acrylamide	79-06-1	2
Acrylamide copolymer	—[a]	1
Acrylamide copolymer	38193-60-1	1
Acrylate copolymer	—[a]	1
Acrylic acid, 2-hydroxyethyl ester	818-61-1	1
Acrylic acid/2-acrylamido-methylpropylsulfonic acid copolymer	37350-42-8	1
Acrylic copolymer	403730-32-5	1
Acrylic polymers	—[a]	1

Appendix. Chemicals Used in Hydraulic Fracturing

Chemical Component	Chemical Abstract Service Number	No. of Products Containing Chemical
Acrylic polymers	26006-22-4	2
Acyclic hydrocarbon blend	—ᵃ	1
Adipic acid	124-04-9	6
Alcohol alkoxylate	—ᵃ	5
Alcohol ethoxylates	—ᵃ	2
Alcohols	—ᵃ	9
Alcohols, C11–C15-secondary, ethoxylated	68131-40-8	1
Alcohols, C12–C14-secondary	126950-60-5	4
Alcohols, C12–C14-secondary, ethoxylated	84133-50-6	19
Alcohols, C12–C15, ethoxylated	68131-39-5	2
Alcohols, C12–C16, ethoxylated	103331-86-8	1
Alcohols, C12–C16, ethoxylated	68551-12-2	3
Alcohols, C14–C15, ethoxylated	68951-67-7	5
Alcohols, C9–C11-iso-, C10-rich, ethoxylated	78330-20-8	4
Alcohols, C9–C22	—ᵃ	1
Aldehyde	—ᵃ	4
Aldol	107-89-1	1
α-Alumina	—ᵃ	5
Aliphatic acid	—ᵃ	1
Aliphatic alcohol polyglycol ether	68015-67-8	1
Aliphatic amine derivative	120086-58-0	2
Alkaline bromide salts	—ᵃ	2
Alkanes, C10–C14	93924-07-3	2
Alkanes, C13–C16-iso	68551-20-2	2
Alkanolamine	150-25-4	3
Alkanolamine chelate of zirconium alkoxide (zirconium complex)	197980-53-3	4
Alkanolamine/aldehyde condensate	—ᵃ	1
Alkenes	—ᵃ	1
Alkenes, C > 10	64743-02-8	3
Alkenes, C > 8	68411-00-7	2
Alkoxylated alcohols	—ᵃ	1
Alkoxylated amines	—ᵃ	6
Alkoxylated phenol formaldehyde resin	63428-92-2	1
Alkyaryl sulfonate	—ᵃ	1
Alkyl (C12–C16) dimethyl benzyl ammonium chloride	68424-85-1	7
Alkyl (C6–C12) alcohol, ethoxylated	68439-45-2	2
Alkyl (C9–C11) alcohol, ethoxylated	68439-46-3	1
Alkyl alkoxylate	—ᵃ	9
Alkyl amine	—ᵃ	2
Alkyl amine blend in a metal salt solution	—ᵃ	1
Alkyl aryl amine sulfonate	255043-08-04	1
Alkyl benzenesulfonic acid	68584-22-5	2
Alkyl esters	—ᵃ	2
Alkyl hexanol	—ᵃ	1
Alkyl orthophosphate ester	—ᵃ	1
Alkyl phosphate ester	—ᵃ	3
Alkyl quaternary ammonium chlorides	—ᵃ	4
Alkylaryl sulfonate	—ᵃ	1
Alkylaryl sulfonic acid	27176-93-9	1

Appendix. Chemicals Used in Hydraulic Fracturing

Chemical Component	Chemical Abstract Service Number	No. of Products Containing Chemical
Alkylated quaternary chloride	—[a]	5
Alkylbenzenesulfonic acid	—[a]	1
Alkylethoammonium sulfates	—[a]	1
Alkylphenol ethoxylates	—[a]	1
Almandite and pyrope garnet	1302-62-1	1
Aluminium isopropoxide	555-31-7	1
Aluminum	7429-90-5	2
Aluminum chloride	—[a]	3
Aluminum chloride	1327-41-9	2
Aluminum oxide (α-alumina)	1344-28-1	24
Aluminum oxide silicate	12068-56-3	1
Aluminum silicate (mullite)	1302-76-7	38
Aluminum sulfate hydrate	10043-01-3	1
Amides, tallow, *n*-[3-(dimethylamino)propyl], *n*-oxides	68647-77-8	4
Amidoamine	—[a]	1
Amine	—[a]	7
Amine bisulfite	13427-63-9	1
Amine oxides	—[a]	1
Amine phosphonate	—[a]	3
Amine salt	—[a]	2
Amines, C14–C18; C16–C18-unsaturated, alkyl, ethoxylated	68155-39-5	1
Amines, coco alkyl, acetate	61790-57-6	3
Amines, polyethylenepoly-, ethoxylated, phosphonomethylated	68966-36-9	1
Amines, tallow alkyl, ethoxylated	61791-26-2	2
Amino compounds	—[a]	1
Amino methylene phosphonic acid salt	—[a]	1
Amino trimethylene phosphonic acid	6419-19-8	2
Ammonia	7664-41-7	7
Ammonium acetate	631-61-8	4
Ammonium alcohol ether sulfate	68037-05-8	1
Ammonium bicarbonate	1066-33-7	1
Ammonium bifluoride (ammonium hydrogen difluoride)	1341-49-7	10
Ammonium bisulfate	7783-20-2	3
Ammonium bisulfite	10192-30-0	15
Ammonium C6–C10 alcohol ethoxysulfate	68187-17-7	4
Ammonium C8–C10 alkyl ether sulfate	68891-29-2	4
Ammonium chloride	12125-02-9	29
Ammonium fluoride	12125-01-8	9
Ammonium hydroxide	1336-21-6	4
Ammonium nitrate	6484-52-2	2
Ammonium persulfate (diammonium peroxidisulfate)	7727-54-0	37
Ammonium salt	—[a]	1
Ammonium salt of ethoxylated alcohol sulfate	—[a]	1
Amorphous silica	99439-28-8	1
Amphoteric alkyl amine	61789-39-7	1
Anionic copolymer	—[a]	3
Anionic polyacrylamide	—[a]	1
Anionic polyacrylamide	25085-02-3	6
Anionic polyacrylamide copolymer	—[a]	3

Appendix. Chemicals Used in Hydraulic Fracturing

Chemical Component	Chemical Abstract Service Number	No. of Products Containing Chemical
Anionic polymer	—[a]	2
Anionic polymer in solution	—[a]	1
Anionic polymer, sodium salt	9003-04-7	1
Anionic water-soluble polymer	—[a]	2
Antifoulant	—[a]	1
Antimonate salt	—[a]	1
Antimony pentoxide	1314-60-9	2
Antimony potassium oxide	29638-69-5	4
Antimony trichloride	10025-91-9	2
a-Organic surfactants	61790-29-8	1
Aromatic alcohol glycol ether	—[a]	2
Aromatic aldehyde	—[a]	2
Aromatic ketones	224635-63-6	2
Aromatic polyglycol ether	—[a]	1
Barium sulfate	7727-43-7	3
Bauxite	1318-16-7	16
Bentonite	1302-78-9	2
Benzene	71-43-2	3
Benzene, C10–C16, alkyl derivatives	68648-87-3	1
Benzenecarboperoxoic acid, 1,1-dimethylethyl ester	614-45-9	1
Benzenemethanaminium	3844-45-9	1
Benzenesulfonic acid, C10–C16-alkyl derivatives, potassium salts	68584-27-0	1
Benzoic acid	65-85-0	11
Benzyl chloride	100-44-7	8
Biocide component	—[a]	3
bis(1-Methylethyl)naphthalenesulfonic acid, cyclohexylamine salt	68425-61-6	1
bis(Hexamethylenetriamine) penta(methylene phosphonic acid)	35657-77-3	1
Bisphenol A/epichlorohydrin resin	25068-38-6	5
Bisphenol A/novolac epoxy resin	28906-96-9	1
Borate	12280-03-4	2
Borate salts	—[a]	5
Boric acid	10043-35-3	18
Boric acid, potassium salt	20786-60-1	1
Boric acid, sodium salt	1333-73-9	2
Boric oxide	1303-86-2	1
β-Tricalcium phosphate	7758-87-4	1
Butanedioic acid	2373-38-8	4
Butanol	71-36-3	3
Butyl glycidyl ether	2426-08-6	5
Butyl lactate	138-22-7	4
C10–C16 ethoxylated alcohol	68002-97-1	4
C11–C14 *n*-alkanes, mixed	—[a]	1
C12–C14 alcohol, ethoxylated	68439-50-9	3
Calcium carbonate	471-34-1	1
Calcium carbonate (limestone)	1317-65-3	9
Calcium chloride	10043-52-4	17
Calcium chloride, dihydrate	10035-04-8	1
Calcium fluoride	7789-75-5	2
Calcium hydroxide	1305-62-0	9

Appendix. Chemicals Used in Hydraulic Fracturing

Chemical Component	Chemical Abstract Service Number	No. of Products Containing Chemical
Calcium hypochlorite	7778-54-3	1
Calcium oxide	1305-78-8	6
Calcium peroxide	1305-79-9	5
Carbohydrates	—[a]	3
Carbon dioxide	124-38-9	4
Carboxymethyl guar gum, sodium salt	39346-76-4	7
Carboxymethyl hydroxypropyl guar	68130-15-4	11
Cellophane	9005-81-6	2
Cellulase	9012-54-8	7
Cellulase enzyme	—[a]	1
Cellulose	9004-34-6	1
Cellulose derivative	—[a]	2
Chloromethylnaphthalene quinoline quaternary amine	15619-48-4	3
Chlorous ion solution	—[a]	2
Choline chloride	67-48-1	3
Chromates	—[a]	1
Chromium (iii) acetate	1066-30-4	1
Cinnamaldehyde (3-phenyl-2-propenal)	104-55-2	5
Citric acid (2-hydroxy-1,2,3 propanetricarboxylic acid)	77-92-9	29
Citrus terpenes	94266-47-4	11
Coal, granular	50815-10-6	1
Cobalt acetate	71-48-7	1
Cocaidopropyl betaine	61789-40-0	2
Cocamidopropylamine oxide	68155-09-9	1
Coco *bis*(2-hydroxyethyl) amine oxide	61791-47-7	1
Cocoamidopropyl betaine	70851-07-9	1
Cocomidopropyl dimethylamine	68140-01-2	1
Coconut fatty acid diethanolamide	68603-42-9	1
Collagen (gelatin)	9000-70-8	6
Complex alkylaryl polyo-ester	—[a]	1
Complex aluminum salt	—[a]	2
Complex organometallic salt	—[a]	2
Complex substituted keto-amine	143106-84-7	1
Complex substituted keto-amine hydrochloride	—[a]	1
Copolymer of acrylamide and sodium acrylate	25987-30-8	1
Copper	7440-50-8	1
Copper iodide	7681-65-4	1
Copper sulfate	7758-98-7	3
Corundum (aluminum oxide)	1302-74-5	48
Crotonaldehyde	123-73-9	1
Crystalline silica—cristobalite	14464-46-1	44
Crystalline silica—quartz (SiO_2)	14808-60-7	207
Crystalline silica, tridymite	15468-32-3	2
Cumene	98-82-8	6
Cupric chloride	7447-39-4	10
Cupric chloride dihydrate	10125-13-0	7
Cuprous chloride	7758-89-6	1
Cured acrylic resin	—[a]	7
Cured resin	—[a]	4

Appendix. Chemicals Used in Hydraulic Fracturing

Chemical Component	Chemical Abstract Service Number	No. of Products Containing Chemical
Cured silicone rubber-polydimethylsiloxane	63148-62-9	1
Cured urethane resin	—ᵃ	3
Cyclic alkanes	—ᵃ	1
Cyclohexane	110-82-7	1
Cyclohexanone	108-94-1	1
Decanol	112-30-1	2
Decyl-dimethyl amine oxide	2605-79-0	4
Dextrose monohydrate	50-99-7	1
D-Glucitol	50-70-4	1
Di(2-ethylhexyl) phthalate	117-81-7	3
Di(ethylene glycol) ethyl ether acetate	112-15-2	4
Diatomaceous earth	61790-53-2	3
Diatomaceous earth, calcined	91053-39-3	7
Dibromoacetonitrile	3252-43-5	1
Dibutylaminoethanol (2-dibutylaminoethanol)	102-81-8	4
Di-calcium silicate	10034-77-2	1
Dicarboxylic acid	—ᵃ	1
Didecyl dimethyl ammonium chloride	7173-51-5	1
Diesel	—ᵃ	1
Diesel	68334-30-5	3
Diesel	68476-30-2	4
Diesel	68476-34-6	43
Diethanolamine (2,2-iminodiethanol)	111-42-2	14
Diethylbenzene	25340-17-4	1
Diethylene glycol	111-46-6	8
Diethylene glycol monomethyl ether	111-77-3	4
Diethylene triaminepenta (methylene phosphonic acid)	15827-60-8	1
Diethylenetriamine	111-40-0	2
Diethylenetriamine, tall oil fatty acids reaction product	61790-69-0	1
Diisopropylnaphthalenesulfonic acid	28757-00-8	2
Dimethyl formamide	68-12-2	5
Dimethyl glutarate	1119-40-0	1
Dimethyl silicone	—ᵃ	2
Dioctyl sodium sulfosuccinate	577-11-7	1
Dipropylene glycol	25265-71-8	1
Dipropylene glycol monomethyl ether (2-methoxymethylethoxy propanol)	34590-94-8	12
Di-secondary-butylphenol	53964-94-6	3
Disodium EDTA	139-33-3	1
Disodium ethylenediaminediacetate	38011-25-5	1
Disodium ethylenediaminetetraacetate dihydrate	6381-92-6	1
Disodium octaborate tetrahydrate	12008-41-2	1
Dispersing agent	—ᵃ	1
D-Limonene	5989-27-5	11
Dodecyl alcohol ammonium sulfate	32612-48-9	2
Dodecylbenzene sulfonic acid	27176-87-0	14
Dodecylbenzene sulfonic acid salts	42615-29-2	2
Dodecylbenzene sulfonic acid salts	68648-81-7	7
Dodecylbenzene sulfonic acid salts	90218-35-2	1
Dodecylbenzenesulfonate isopropanolamine	42504-46-1	1

Appendix. Chemicals Used in Hydraulic Fracturing

Chemical Component	Chemical Abstract Service Number	No. of Products Containing Chemical
Dodecylbenzenesulfonic acid, monoethanolamine salt	26836-07-7	1
Dodecylbenzenesulphonic acid, morpholine salt	12068-08-5	1
EDTA/copper chelate	—[a]	2
EO-C7–C9-iso-, C8-rich alcohols	78330-19-5	5
Epichlorohydrin	25085-99-8	5
Epoxy resin	—[a]	5
Erucic amidopropyl dimethyl betaine	149879-98-1	3
Erythorbic acid	89-65-6	2
Essential oils	—[a]	6
Ethanaminium, *N,N,N*-trimethyl-2-[(1-oxo-2-propenyl)oxy]-, chloride, polymer with 2-propenamide	69418-26-4	4
Ethanol (ethyl alcohol)	64-17-5	36
Ethanol, 2-(hydroxymethylamino)-	34375-28-5	1
Ethanol, 2,2′-(octadecylamino)*bis*-	10213-78-2	1
Ethanoldiglycine disodium salt	135-37-5	1
Ether salt	25446-78-0	2
Ethoxylated 4-nonylphenol (nonylphenol ethoxylate)	26027-38-3	9
Ethoxylated alcohol	104780-82-7	1
Ethoxylated alcohol	78330-21-9	2
Ethoxylated alcohols	—[a]	3
Ethoxylated alkyl amines	—[a]	1
Ethoxylated amine	—[a]	1
Ethoxylated amines	61791-44-4	1
Ethoxylated fatty acid ester	—[a]	1
Ethoxylated nonionic surfactant	—[a]	1
Ethoxylated nonylphenol	—[a]	8
Ethoxylated nonylphenol	68412-54-4	10
Ethoxylated nonylphenol	9016-45-9	38
Ethoxylated octylphenol	68987-90-6	1
Ethoxylated octylphenol	9002-93-1	1
Ethoxylated octylphenol	9036-19-5	3
Ethoxylated oleyl amine	13127-82-7	2
Ethoxylated oleyl amine	26635-93-8	1
Ethoxylated sorbitol esters	—[a]	1
Ethoxylated tridecyl alcohol phosphate	9046-01-9	2
Ethoxylated undecyl alcohol	127036-24-2	2
Ethyl acetate	141-78-6	4
Ethyl acetoacetate	141-97-9	1
Ethyl octynol (1-octyn-3-ol,4-ethyl-)	5877-42-9	5
Ethylbenzene	100-41-4	28
Ethylene glycol (1,2-ethanediol)	107-21-1	119
Ethylene glycol monobutyl ether (2-butoxyethanol)	111-76-2	126
Ethylene oxide	75-21-8	1
Ethylene oxide-nonylphenol polymer	—[a]	1
Ethylenediaminetetraacetic acid	60-00-4	1
Ethylene-vinyl acetate copolymer	24937-78-8	1
Ethylhexanol (2-ethylhexanol)	104-76-7	18
Fatty acid ester	—[a]	1
Fatty acid, tall oil, hexa esters with sorbitol, ethoxylated	61790-90-7	1

Chemical Component	Chemical Abstract Service Number	No. of Products Containing Chemical
Fatty acids	—[a]	1
Fatty alcohol alkoxylate	—[a]	1
Fatty alkyl amine salt	—[a]	1
Fatty amine carboxylates	—[a]	1
Fatty quaternary ammonium chloride	61789-68-2	1
Ferric chloride	7705-08-0	3
Ferric sulfate	10028-22-5	7
Ferrous sulfate, heptahydrate	7782-63-0	4
Fluoroaliphatic polymeric esters	—[a]	1
Formaldehyde	50-00-0	12
Formaldehyde polymer	—[a]	2
Formaldehyde, polymer with 4-(1,1-dimethyl)phenol, methyloxirane, and oxirane	30704-64-4	3
Formaldehyde, polymer with 4-nonylphenol and oxirane	30846-35-6	1
Formaldehyde, polymer with ammonia and phenol	35297-54-2	2
Formamide	75-12-7	5
Formic acid	64-18-6	24
Fumaric acid	110-17-8	8
Furfural	98-01-1	1
Furfuryl alcohol	98-00-0	3
Glass fiber	65997-17-3	3
Gluconic acid	526-95-4	1
Glutaraldehyde	111-30-8	20
Glycerol (1,2,3-propanetriol, glycerine)	56-81-5	16
Glycol ethers	—[a]	9
Glycol ethers	9004-77-7	4
Glyoxal	107-22-2	3
Glyoxylic acid	298-12-4	1
Guar gum	9000-30-0	41
Guar gum derivative	—[a]	12
Haloalkyl heteropolycycle salt	—[a]	6
Heavy aromatic distillate	68132-00-3	1
Heavy aromatic petroleum naphtha	64742-94-5	45
Heavy catalytic reformed petroleum naphtha	64741-68-0	10
Hematite	—[a]	5
Hemicellulase	9025-56-3	2
Hexahydro-1,3,5-*tris*(2-hydroxyethyl)-*s*-triazine (triazine)	4719-04-4	4
Hexamethylenetetramine	100-97-0	37
Hexanediamine	124-09-4	1
Hexanes	—[a]	1
Hexylene glycol	107-41-5	5
Hydrated aluminum silicate	1332-58-7	4
Hydrocarbon mixtures	8002-05-9	1
Hydrocarbons	—[a]	3
Hydrodesulfurized kerosine (petroleum)	64742-81-0	3
Hydrodesulfurized light catalytic cracked distillate (petroleum)	68333-25-5	1
Hydrodesulfurized middle distillate (petroleum)	64742-80-9	1
Hydrogen chloride (hydrochloric acid)	7647-01-0	42
Hydrogen fluoride (hydrofluoric acid)	7664-39-3	2

Appendix. Chemicals Used in Hydraulic Fracturing

Chemical Component	Chemical Abstract Service Number	No. of Products Containing Chemical
Hydrogen peroxide	7722-84-1	4
Hydrogen sulfide	7783-06-4	1
Hydrotreated and hydrocracked base oil	—[a]	2
Hydrotreated heavy naphthenic distillate	64742-52-5	3
Hydrotreated heavy paraffinic petroleum distillates	64742-54-7	1
Hydrotreated heavy petroleum naphtha	64742-48-9	7
Hydrotreated light petroleum distillates	64742-47-8	89
Hydrotreated middle petroleum distillates	64742-46-7	3
Hydroxyacetic acid (glycolic acid)	79-14-1	6
Hydroxyethylcellulose	9004-62-0	1
Hydroxyethylethylenediaminetriacetic acid, trisodium salt	139-89-9	1
Hydroxylamine hydrochloride	5470-11-1	1
Hydroxypropyl guar gum	39421-75-5	2
Hydroxysultaine	—[a]	1
Inner salt of alkyl amines	—[a]	2
Inorganic borate	—[a]	3
Inorganic particulate	—[a]	1
Inorganic salt	—[a]	1
Inorganic salt	533-96-0	1
Inorganic salt	7446-70-0	1
Instant coffee purchased off the shelf	—[a]	1
Inulin, carboxymethyl ether, sodium salt	430439-54-6	1
Iron oxide	1332-37-2	2
Iron oxide (ferric oxide)	1309-37-1	18
Isoamyl alcohol	123-51-3	1
Iso-alkanes/n-alkanes	—[a]	10
Isobutanol (isobutyl alcohol)	78-83-1	4
Isomeric aromatic ammonium salt	—[a]	1
Isooctanol	26952-21-6	1
Isooctyl alcohol	68526-88-0	1
Isooctyl alcohol bottoms	68526-88-5	1
Isopropanol (isopropyl alcohol, propan-2-ol)	67-63-0	274
Isopropylamine	75-31-0	1
Isotridecanol, ethoxylated	9043-30-5	1
Kerosene	8008-20-6	13
Lactic acid	10326-41-7	1
Lactic acid	50-21-5	1
L-Dilactide	4511-42-6	1
Lead	7439-92-1	1
Light aromatic solvent naphtha	64742-95-6	11
Light catalytic cracked petroleum distillates	64741-59-9	1
Light naphtha distillate, hydrotreated	64742-53-6	1
Low toxicity base oils	—[a]	1
Maghemite	—[a]	2
Magnesium carbonate	546-93-0	1
Magnesium chloride	7786-30-3	4
Magnesium hydroxide	1309-42-8	4
Magnesium iron silicate	1317-71-1	3
Magnesium nitrate	10377-60-3	5

Appendix. Chemicals Used in Hydraulic Fracturing

Chemical Component	Chemical Abstract Service Number	No. of Products Containing Chemical
Magnesium oxide	1309-48-4	18
Magnesium peroxide	1335-26-8	2
Magnesium peroxide	14452-57-4	4
Magnesium phosphide	12057-74-8	1
Magnesium silicate	1343-88-0	3
Magnesium silicate hydrate (talc)	14807-96-6	2
Magnetite	—[a]	3
Medium aliphatic solvent petroleum naphtha	64742-88-7	10
Metal salt	—[a]	2
Metal salt solution	—[a]	1
Methanol (methyl alcohol)	67-56-1	342
Methyl isobutyl carbinol (methyl amyl alcohol)	108-11-2	3
Methyl salicylate	119-36-8	6
Methyl vinyl ketone	78-94-4	2
Methylcyclohexane	108-87-2	1
Mica	12001-26-2	3
Microcrystalline silica	1317-95-9	1
Mineral	—[a]	1
Mineral filler	—[a]	1
Mineral spirits (Stoddard solvent)	8052-41-3	2
Mixed titanium ortho ester complexes	—[a]	1
Modified alkane	—[a]	1
Modified cycloaliphatic amine adduct	—[a]	3
Modified lignosulfonate	—[a]	1
Monoethanolamine (ethanolamine)	141-43-5	17
Monoethanolamine borate	26038-87-9	1
Morpholine	110-91-8	2
Mullite	1302-93-8	55
N,N'-Dibutylthiourea	109-46-6	1
N,N-Dimethyl-1-octadecanamine-HCl	—[a]	1
N,N-Dimethyloctadecylamine	124-28-7	3
N,N-Dimethyloctadecylamine hydrochloride	1613-17-8	2
N,N'-Methylenebisacrylamide	110-26-9	1
n-Alkyl dimethyl benzyl ammonium chloride	139-08-2	1
Naphthalene	91-20-3	44
Naphthalene derivatives	—[a]	1
Naphthalenesulphonic acid, bis(1-methylethyl)-methyl derivatives	99811-86-6	1
Natural asphalt	12002-43-6	1
N-Cocoamidopropyl-N,N-dimethyl-N-2-hydroxypropylsulfobetaine	68139-30-0	1
N-Dodecyl-2-pyrrolidone	2687-96-9	1
N-Heptane	142-82-5	1
Nickel sulfate hexahydrate	10101-97-0	2
Nitrilotriacetamide	4862-18-4	4
Nitrilotriacetic acid	139-13-9	6
Nitrilotriacetonitrile	7327-60-8	3
Nitrogen	7727-37-9	9
N-Methylpyrrolidone	872-50-4	1
Nonane, all isomers	—[a]	1
Nonhazardous salt	—[a]	1

Appendix. Chemicals Used in Hydraulic Fracturing

Chemical Component	Chemical Abstract Service Number	No. of Products Containing Chemical
Nonionic surfactant	—ᵃ	1
Nonylphenol ethoxylate	—ᵃ	2
Nonylphenol ethoxylate	9016-45-6	2
Nonylphenol ethoxylate	9018-45-9	1
Nonylphenol	25154-52-3	1
Nonylphenol, ethoxylated and sulfated	9081-17-8	1
N-Propyl zirconate	—ᵃ	1
N-Tallowalkyltrimethylenediamines	—ᵃ	1
Nuisance particulates	—ᵃ	2
Nylon fibers	25038-54-4	2
Octanol	111-87-5	2
Octyltrimethylammonium bromide	57-09-0	1
Olefinic sulfonate	—ᵃ	1
Olefins	—ᵃ	1
Organic acid salt	—ᵃ	3
Organic acids	—ᵃ	1
Organic phosphonate	—ᵃ	1
Organic phosphonate salts	—ᵃ	1
Organic phosphonic acid salts	—ᵃ	6
Organic salt	—ᵃ	1
Organic sulfur compound	—ᵃ	2
Organic titanate	—ᵃ	2
Organiophilic clay	—ᵃ	2
Organo-metallic ammonium complex	—ᵃ	1
Other inorganic compounds	—ᵃ	1
Oxirane, methyl-, polymer with oxirane, mono-C10–C16-alkyl ethers, phosphates	68649-29-6	1
Oxyalkylated alcohol	—ᵃ	6
Oxyalkylated alcohols	228414-35-5	1
Oxyalkylated alkyl alcohol	—ᵃ	1
Oxyalkylated alkylphenol	—ᵃ	1
Oxyalkylated fatty acid	—ᵃ	2
Oxyalkylated phenol	—ᵃ	1
Oxyalkylated polyamine	—ᵃ	1
Oxylated alcohol	—ᵃ	1
Paraffin wax	8002-74-2	1
Paraffinic naphthenic solvent	—ᵃ	1
Paraffinic solvent	—ᵃ	5
Paraffins	—ᵃ	1
Perlite	93763-70-3	1
Petroleum distillates	26	
Petroleum distillates	64742-65-0	1
Petroleum distillates	64742-97-5	1
Petroleum distillates	68477-31-6	3
Petroleum gas oils	—ᵃ	1
Petroleum gas oils	64741-43-1	1
Phenol	108-95-2	5
Phenol-formaldehyde resin	9003-35-4	32
Phosphate ester	—ᵃ	6
Phosphate esters of alkyl phenyl ethoxylate	68412-53-3	1

Appendix. Chemicals Used in Hydraulic Fracturing

Chemical Component	Chemical Abstract Service Number	No. of Products Containing Chemical
Phosphine	—[a]	1
Phosphonic acid	—[a]	1
Phosphonic acid	129828-36-0	1
Phosphonic acid	13598-36-2	3
Phosphonic acid (dimethlamino(methylene)	29712-30-9	1
Phosphonic acid, [nitrilotris(methylene)]*tris*-, pentasodium salt	2235-43-0	1
Phosphoric acid	7664-38-2	7
Phosphoric acid ammonium salt	—[a]	1
Phosphoric acid, mixed decyl, octyl and ethyl esters	68412-60-2	3
Phosphorous acid	10294-56-1	1
Phthalic anhydride	85-44-9	2
Pine oil	8002-09-3	5
Plasticizer	—[a]	1
Poly(oxy-1,2-ethanediyl)	24938-91-8	1
Poly(oxy-1,2-ethanediyl), alpha-(4-nonylphenyl)-omega-hydroxy-, branched (nonylphenol ethoxylate)	127087-87-0	3
Poly(oxy-1,2-ethanediyl), alpha-hydro-omega-hydroxy	65545-80-4	1
Poly(oxy-1,2-ethanediyl), alpha-sulfo-omega-(hexyloxy)-, ammonium salt	63428-86-4	3
Poly(oxy-1,2-ethanediyl), alpha-(nonylphenyl)-omega-hydroxy-, phosphate	51811-79-1	1
Poly(oxy-1,2-ethanediyl), alpha-undecyl-omega-hydroxy	34398-01-1	6
Poly(sodium-*p*-styrenesulfonate)	25704-18-1	1
Poly(vinyl alcohol)	25213-24-5	2
Polyacrylamides	9003-05-8	2
Polyacrylamides	—[a]	1
Polyacrylate	—[a]	1
Polyamine	—[a]	2
Polyanionic cellulose	—[a]	2
Polyepichlorohydrin, trimethylamine quaternized	51838-31-4	1
Polyetheramine	9046-10-0	3
Polyether-modified trisiloxane	27306-78-1	1
Polyethylene glycol	25322-68-3	20
Polyethylene glycol ester with tall oil fatty acid	9005-02-1	1
Polyethylene polyammonium salt	68603-67-8	2
Polyethylene-polypropylene glycol	9003-11-6	5
Polylactide resin	—[a]	3
Polyoxyalkylenes	—[a]	1
Polyoxyethylene castor oil	61791-12-6	1
Polyphosphoric acid, esters with triethanolamine, sodium salts	68131-72-6	1
Polypropylene glycol	25322-69-4	1
Polysaccharide	—[a]	20
Polyvinyl alcohol	—[a]	1
Polyvinyl alcohol	9002-89-5	2
Polyvinyl alcohol/polyvinylacetate copolymer	—[a]	1
Potassium acetate	127-08-2	1
Potassium carbonate	584-08-7	12
Potassium chloride	7447-40-7	29
Potassium formate	590-29-4	3
Potassium hydroxide	1310-58-3	25
Potassium iodide	7681-11-0	6

Appendix. Chemicals Used in Hydraulic Fracturing

Chemical Component	Chemical Abstract Service Number	No. of Products Containing Chemical
Potassium metaborate	13709-94-9	3
Potassium metaborate	16481-66-6	3
Potassium oxide	12136-45-7	1
Potassium pentaborate	—[a]	1
Potassium persulfate	7727-21-1	9
Propanol (propyl alcohol)	71-23-8	18
Propanol, [2(2-methoxy-methylethoxy)methylethoxyl]	20324-33-8	1
Propargyl alcohol (2-propyn-1-ol)	107-19-7	46
Propylene carbonate (1,3-dioxolan-2-one, methyl-)	108-32-7	2
Propylene glycol (1,2-propanediol)	57-55-6	18
Propylene oxide	75-56-9	1
Propylene pentamer	15220-87-8	1
p-Xylene	106-42-3	1
Pyridinium, 1-(phenylmethyl)-, ethyl methyl derivatives, chlorides	68909-18-2	9
Pyrogenic silica	112945-52-5	3
Quaternary amine compounds	—[a]	3
Quaternary amine compounds	61789-18-2	1
Quaternary ammonium compounds	—[a]	9
Quaternary ammonium compounds	19277-88-4	1
Quaternary ammonium compounds	68989-00-4	1
Quaternary ammonium compounds	8030-78-2	1
Quaternary ammonium compounds, dicoco alkyldimethyl, chlorides	61789-77-3	2
Quaternary ammonium salts	—[a]	2
Quaternary compound	—[a]	1
Quaternary salt	—[a]	2
Quaternized alkyl nitrogenated compound	68391-11-7	2
Rafinnates (petroleum), sorption process	64741-85-1	2
Residues (petroleum), catalytic reformer fractionator	64741-67-9	10
Resin	8050-09-7	2
Rutile	1317-80-2	2
Salt of phosphate ester	—[a]	3
Salt of phosphono-methylated diamine	—[a]	1
Salts of oxyalkylated fatty amines	68551-33-7	1
Secondary alcohol	—[a]	7
Silica (silicon dioxide)	7631-86-9	47
Silica, amorphous	—[a]	3
Silica, amorphous precipitated	67762-90-7	1
Silicon carboxylate	681-84-5	1
Silicon dioxide (fused silica)	60676-86-0	7
Silicone emulsion	—[a]	1
Sodium (C14–C16) olefin sulfonate	68439-57-6	4
Sodium 2-ethylhexyl sulfate	126-92-1	1
Sodium acetate	127-09-3	6
Sodium acid pyrophosphate	7758-16-9	5
Sodium alkyl diphenyl oxide sulfonate	28519-02-0	1
Sodium aluminate	1302-42-7	1
Sodium aluminum phosphate	7785-88-8	1
Sodium bicarbonate (sodium hydrogen carbonate)	144-55-8	10
Sodium bisulfite	7631-90-5	6

Chemical Component	Chemical Abstract Service Number	No. of Products Containing Chemical
Sodium bromate	7789-38-0	10
Sodium bromide	7647-15-6	1
Sodium carbonate	497-19-8	14
Sodium chlorate	7775-09-9	1
Sodium chloride	7647-14-5	48
Sodium chlorite	7758-19-2	8
Sodium cocaminopropionate	68608-68-4	2
Sodium diacetate	126-96-5	2
Sodium erythorbate	6381-77-7	4
Sodium glycolate	2836-32-0	2
Sodium hydroxide (caustic soda)	1310-73-2	80
Sodium hypochlorite	7681-52-9	14
Sodium lauryl-ether sulfate	68891-38-3	3
Sodium metabisulfite	7681-57-4	1
Sodium metaborate	7775-19-1	2
Sodium metaborate tetrahydrate	35585-58-1	6
Sodium metasilicate, anhydrous	6834-92-0	2
Sodium nitrite	7632-00-0	1
Sodium oxide (Na_2O)	1313-59-3	1
Sodium perborate	1113-47-9	1
Sodium perborate	7632-04-4	1
Sodium perborate tetrahydrate	10486-00-7	4
Sodium persulfate	7775-27-1	6
Sodium phosphate	—[a]	2
Sodium polyphosphate	68915-31-1	1
Sodium salicylate	54-21-7	1
Sodium silicate	1344-09-8	2
Sodium sulfate	7757-82-6	7
Sodium tetraborate	1330-43-4	7
Sodium tetraborate decahydrate	1303-96-4	10
Sodium thiosulfate	7772-98-7	10
Sodium thiosulfate pentahydrate	10102-17-7	3
Sodium trichloroacetate	650-51-1	1
Sodium tripolyphosphate	7758-29-4	2
Sodium xylene sulfonate	1300-72-7	3
Sodium zirconium lactate	174206-15-6	1
Solvent refined heavy naphthenic petroleum distillates	64741-96-4	1
Sorbitan monooleate	1338-43-8	1
Stabilized aqueous chlorine dioxide	10049-04-4	1
Stannous chloride	7772-99-8	1
Stannous chloride dihydrate	10025-69-1	6
Starch	9005-25-8	5
Steam-cracked distillate, cyclodiene dimer, dicyclopentadiene polymer	68131-87-3	1
Steam-cracked petroleum distillates	64742-91-2	6
Straight run middle petroleum distillates	64741-44-2	5
Substituted alcohol	—[a]	2
Substituted alkene	—[a]	1
Substituted alkylamine	—[a]	2
Sucrose	57-50-1	1

Appendix. Chemicals Used in Hydraulic Fracturing

Chemical Component	Chemical Abstract Service Number	No. of Products Containing Chemical
Sulfamic acid	5329-14-6	6
Sulfate	—a	1
Sulfonate acids	—a	1
Sulfonate surfactants	—a	1
Sulfonic acid salts	—a	1
Sulfonic acids, petroleum	61789-85-3	1
Sulfur compound	—a	1
Sulfuric acid	7664-93-9	9
Sulfuric acid, monodecyl ester, sodium salt	142-87-0	2
Sulfuric acid, monooctyl ester, sodium salt	142-31-4	2
Surfactants	—a	13
Sweetened middle distillate	64741-86-2	1
Synthetic organic polymer	9051-89-2	2
Tall oil (fatty acids)	61790-12-3	4
Tall oil, compound with diethanolamine	68092-28-4	1
Tallow soap	—a	2
Tar bases, quinoline derivatives, benzyl chloride-quaternized	72480-70-7	5
Tergitol	68439-51-0	1
Terpene hydrocarbon byproducts	68956-56-9	3
Terpenes	—a	1
Terpenes and terpenoids, sweet orange-oil	68647-72-3	2
Terpineol	8000-41-7	1
tert-Butyl hydroperoxide	75-91-2	6
Tetracalcium-alumino-ferrite	12068-35-8	1
Tetraethylene glycol	112-60-7	1
Tetraethylenepentamine	112-57-2	2
Tetrahydro-3,5-dimethyl-2H-1,3,5-thiadiazine-2-thione (Dazomet)	533-74-4	13
Tetrakis (hydroxymethyl) phosphonium sulfate	55566-30-8	12
Tetramethyl ammonium chloride	75-57-0	14
Tetrasodium 1-hydroxyethylidene-1,1-diphosphonic acid	3794-83-0	1
Tetrasodium ethylenediaminetetraacetate	64-02-8	10
Thiocyanate sodium	540-72-7	1
Thioglycolic acid	68-11-1	6
Thiourea	62-56-6	9
Thiourea polymer	68527-49-1	3
Titanium complex	—a	1
Titanium oxide	13463-67-7	19
Titanium, isopropoxy (triethanolaminate)	74665-17-1	2
Toluene	108-88-3	29
Treated ammonium chloride (with anticaking agent a or b)	12125-02-9	1
Tributyl tetradecyl phosphonium chloride	81741-28-8	5
Tricalcium silicate	12168-85-3	1
Tridecyl alcohol	112-70-9	1
Triethanolamine (2,2,2-nitrilotriethanol)	102-71-6	21
Triethanolamine polyphosphate ester	68131-71-5	3
Triethanolamine titanate	36673-16-2	1
Triethanolamine zirconate	101033-44-7	6
Triethanolamine zirconium chelate	—a	1
Triethyl citrate	77-93-0	1

Chemical Component	Chemical Abstract Service Number	No. of Products Containing Chemical
Triethyl phosphate	78-40-0	1
Triethylene glycol	112-27-6	3
Triisopropanolamine	122-20-3	5
Trimethylammonium chloride	593-81-7	1
Trimethylbenzene	25551-13-7	5
Trimethyloctadecylammonium (1-octadecanaminium, *N,N,N*-trimethyl-, chloride)	112-03-8	6
Tris(hydroxymethyl)aminomethane	77-86-1	1
Trisodium ethylenediaminetetraacetate	150-38-9	1
Trisodium ethylenediaminetriacetate	19019-43-3	1
Trisodium nitrilotriacetate	18662-53-8	8
Trisodium nitrilotriacetate (nitrilotriacetic acid, trisodium salt monohydrate)	5064-31-3	9
Trisodium orthophosphate	7601-54-9	1
Trisodium phosphate dodecahydrate	10101-89-0	1
Ulexite	1319-33-1	1
Urea	57-13-6	3
Wall material	—[a]	1
Walnut hulls	—[a]	2
White mineral oil	8042-47-5	8
Xanthan gum	11138-66-2	6
Xylene	1330-20-7	44
Zinc chloride	7646-85-7	1
Zinc oxide	1314-13-2	2
Zirconium complex	—[a]	10
Zirconium dichloride oxide	7699-43-6	1
Zirconium oxide sulfate	62010-10-0	2
Zirconium sodium hydroxy lactate complex (sodium zirconium lactate)	113184-20-6	2

Source: U.S. House of Representatives, *Chemicals Used in Hydraulic Fracturing*, U.S. Government Printing Office, Washington, DC, 2011.

[a] These components appeared on at least one MSDS without an identifying CAS number. The MSDSs in these cases marked the CAS as proprietary, noted that the CAS was not available, or left the CAS field blank. These components may be duplicative of other components on this list, but it was not possible to identify such duplicates without the identifying CAS number.

Index

2-butoxyethanol (2-BE), 67–68
29 CFR 1910, 66, 70, 211, 215–216, 219, 222–223, 232, 235–238, 243–245, 247–249, 255, 258, 259
29 CFR 1926, 215, 226, 229, 231, 235, 250, 251
40 CFR 405–469, 147–148

A

abandoned mines, water coming out of, 57
abandoned wells, 194–195
absolute temperature, 108–109, 118
accident prevention tags, color coding of, 237
acid treatment, 25, 31, 32–33, 35, 61, 125
acids, 53, 55, 59, 61, 233
    carcinogenic, 69
    defined, 55
    humic, 55, 57, 98, 124
    hydrochloric, 8, 31, 35, 55, 61, 134, 135, 233
    injection of to increase permeability, 25
    injection of to improve porosity, 34
    normality of, 113
    pH, and, 56–57
    relative strengths of in water, 56
    sulfuric, 55, 200, 233
    used to repair drilling mud damage, 34
action level, 150
additives, *see* chemical additives
administrative controls, 234, 235
    for noise, 230
adsorption, 84, 91
    criteria ratings, 91
    granular activated carbon, and, 95–96, 124
    ion exchange, and, 134
advanced oxidation process (AOP), 44
advanced separators, 84
advanced treated wastewater, 186
aeration, 93
Age of Fishes, 3, 6
air binding, 98–99
air-purifying respirator, 236
air quality, 9
air stripping, 93–94
air-supplying respirator, 236
albedo, 170
alkalinity, 56
    carbon dioxide, and, 54
    hardness, and, 58
    Langelier Saturation Index, and, 107
    nitrification, and, 88
    seawater, 57
    total hardness, and, 133–134
    vs. pH, 57
alternative desalination technologies, 101
anions, 50, 52, 94, 102, 111, 125, 135
antiscalants, 107, 125, 135, 136
Appalachian Basin, 6, 7–8, 146
apparent color, 54

applied stress, 180, 181, 183–184, 185
    vs. effective stress, 183
aquicludes, 178, 181
aquifers, 151, 177, 178–180
    change in applied stress in, 183, 184
    deep wells, and, 189
    defined, 9
    shallow wells, and, 189
    specific storage of, 181
    types of, 178
    water table, 179
aquitards, 178
    compaction of, 180–181
    effective stress in, 184
arc welding safety, 258
arsenic, 80, 155
artesian aquifers, 179–180
arthropods, 3
artificial lift, 21
assimilable organic carbon (AOC), 95, 96
asymmetric membrane, 119
atmospheric testing, confined space entry and, 242, 245
audit, OSHA, 212, 213, 216–217, 219, 221, 244
    hazard communication program, 265
automated external defibrillator (AED), 246
Avogadro's number, 112

B

background radiation, 65
bacteria, 15, 54, 58, 64, 95, 96, 98, 124, 135, 150, 157
    coliform, 157
    constructed wetlands, and, 94
    nitrifying, 88
barium, 24, 80, 125, 131
Barnett Shale, 28, 31
    hydrocyclones, and, 88–89
    reused wastewater used in, 43
    well pad near Trinitiy Trails, 206
bases, 53, 55–56, 112, 113, 135, 233
basin, defined, 9
bentonite clay, 24, 170
benzene, 60, 68, 70, 79, 91, 93, 155
best available technology (BAT), 147
best conventional technology (BCT), 147
best management practices (BMPs), 150, 206
best practicable technology (BPT), 147
best professional judgment (BPJ), 146
biocides, 34
biodegradation, 119
biogenic gas, 9
biological aerated filtration (BAF), 85–88
    criteria ratings, 88
    removal capabilities, 86
biological nitrification–denitrification, 88
biologically active carbon (BAC), 95
biosolids, 50, 102

303

304 Index

block and tackle, 22
blowout preventers, 23
bored wells, 189
borehole, 9, 28, 33, 190, 191
    Class I injection well, 197
    well log, and, 193
boron, 80, 101
    in treated water, 94
bottom slope, 172
Boyle's law, 108
brazing, OSHA compliance and, 254–259
breakthrough, 95, 96, 98
brine, 103, 126, 168, 169, 170, 200
    subsurface injection of, 195
    underground injection of, 197
bromide, 80
bromine, 101
BTEX compounds, 68, 70, 79, 91, 95

C

cadmium, 95, 155
calcium, 56, 79, 94, 125, 126, 128, 131–133, 134, 136
    equivalent weight, 131
    hardness, 131
    ions, 52, 58, 111, 125
    Langelier Saturation Index, and, 107
    scaling, and, 93
calcium carbonate ($CaCO_3$), 53, 58, 107, 131–133
    as encrusting agent, 194
    equivalent weight, 131
    ionization, 52, 111
    Langelier Saturation Index, and, 107
calcium chloride salt, 35
calcium hydroxide, 55
capacitive deionization, 130–131
carbonate hardness, 133–134
carbon dioxide, 51, 54
    alkalinity, and, 57
    Class VI wells, and, 201
    enhanced recovery wells, and, 197
    foamed gels, and, 61
    in fracturing fluids, 35
    in wastewater, 16, 79
    sequestration, Class VI injection wells and, 201
    shale gas emissions of, 8
carbonic acid (H_2CO_3), 55
carcinogens, 55, 58, 60, 66, 68–69, 70, 153, 155
cardiopulmonary resuscitation (CPR), 246
casing head, 23
casing, well, 9
cations, 50, 51, 52, 94, 102, 111, 125, 135
CDM treatment process, 135–136
cement evaluation log, 26, 27
chains, 250–252
Charles's law, 108–109
chemical additives, 33–36, 58–67
chemical mixing, 58–65
chemical oxidation treatment, 92
chemical spills, 223, 233
chemicals commonly used in hydraulic fracturing, 67–68
chlorination, 96, 97, 124, 155, 157, 192
    trihalomethanes, and, 155

chlorine, 35, 125
    reverse osmosis membranes, and, 119
citations, OSHA, 216–217
    Industry Group 138, 219–265
Clean Air Act, 66, 68
Clean Water Act (CWA), 143–149, 159, 164
clean water reform chronology, 143–144
closed-loop drilling systems, 28
coagulation, 56, 90, 91, 124
coalbed methane (CBM), 9, 77, 86
    produced water disposal, 142
coalbed natural gas (CBNG), 9
coefficient of volume compressibility, 180
coliform bacteria, 157
colligative properties, 114, 115
colloidal
    defined, 52
    fouling, 124
    solids, 51, 53, 88, 96, 98, 157
        silt density index, and, 107
        water color, and, 54
color-coded accident prevention tags, 237
color of water, 54
combustible liquids, 231, 232, 249
compaction, 180–181
    residual, 180
    specific, 181
    specific unit, 181, 185
    unit, 181, 185
    vs. consolidation, 181
completion, well, 24, 25, 26, 27, 29
composite membranes, 119
Comprehensive Environmental Response, Compensation,
        and Liability Act (CERCLA), 234
comprehensive stress, 180
compressed air, drilling with, 28
computer modeling, 30, 44
concentrate, membrane output, 103
concentration factor, 104
concentration, mass, 50, 103
concentration, of a solution, 110–112, 116
concentration polarization, 106
confined aquifer, 178, 179–180
confined space entry
    OSHA standards for, 237–245
    permit required, 236
        OSHA standards for, 237–243
confined spaces, hot work and, 257
confining layer, 195
consolidation, 181, 182
    one-dimensional, 180
constituents of water, 53–58
constructed wetlands, 94–95
    criteria ratings, 95
Consumer Confidence Report (CCR), 150
contaminants, 46, 53, 83, 147
    air stripping, and, 93–94
    breakthrough, and, 95
    defined, 150
    dissolved air flotation, and, 90
    evaporation pond, 168
    formation water, 141
    forward osmosis, and, 126

Index

granular activated carbon, and, 95–96
groundwater, cleaning up, 200
ideal gas law, and, 109
inorganic chemicals, 155–156
microbial, 119, 150, 157
not subject to national primary drinking water
regulations, 70
organic, 79, 86, 153–157
oxidation, and, 92
primary, categories of, 153–154
radionuclides, 157
reference dose (RfD), 153
rejection of, 118, 121, 126
reverse osmosis, and, 116
Safe Drinking Water Act, and, 149–159
USEPA safety standards for, 152
water color, and, 54
water well, 194
contamination, 13
dug wells, 189
environmental, due to fracking slurries, 8
groundwater, 26, 27, 187, 198, 200
surface water, 24
water well, 193–194
conventional drive, 21
conversion rate, reverse osmosis, 104
copper mining, Class III injection wells and, 200
copper sulfate, 56
coral reefs, 6
corrosivity, 153
hazardous waste, 232
cross-linked gels, 60
crown block, 22
Cryptosporidium, 157, 159
crystallization, 168
cup muff, 230–231
cutting safety, OSHA standards for, 255–259

D

DeBruin–Keijman equation, 175
deep natural gas, 77
deep wells, 180, 188, 189
deionization, capacitive, 130–131
density, specific gravity and, 111
derrick, 22
desalination, 83, 101–136
alternative processes, 130–135
commercial processes, 135–136
technologies, 101–107
thermal, 129–130
zeolites, and, 94
Devonian Period, 3–7
dew point, 170
diesel fuel, 60, 64, 70, 198
constituents of, 68
diffusivity, hydraulic, 181–182
directional drilling, 9, 21, 24–25, 218
disclosure requirements, 65
disinfection, 83, 95, 125
turbidity, and, 53, 151
ultraviolet, 84, 96–98, 135, 136
disinfection byproducts (DBPs), 55, 70, 80, 95–96

disposal well, 9
Class II, 197
dissolved air/gas flotation, 89–91
criteria ratings, 91
dissolved gases, in produced wastewater, 79
dissolved organic compounds, 94
dissolved oxygen (DO), 54, 88
dissolved solids, 51, 53, 88, 120
drawdown, 191–192
draw solution, 126
drill bit, 23
drill rig, 9
drill string, 23
drilled wells, deep, 189
drilling, 21–28
directional, 9, 21, 24–25, 218
fluids, 28
mud, 23–24, 28, 33, 34
drinking water
hydraulic fracturing, and, 45–46, 58, 64, 70
membrane filtration, 98
pesticide residues in, 155
protection of, 152. *See also* Safe Drinking Water Act
(SDWA)
radionuclides in, 157
driven wells, 189
drop pipe, 191
dual reverse osmosis, 101
with chemical precipitation, 127
with seeded slurry precipitation, 128
with softening pretreatment and operation at high pH,
127–128
dug wells, shallow, 189
dynamic hydrocyclone, 88

E

effective stress, 180, 181, 184, 185
vs. applied stress, 183
effluent limitation guidelines (ELGs), 142
effluent limitations, 145, 146–147
egress, OSHA standards for, 225–226
elastic compaction, 180
electrodeionization (EDI), 101, 128
electrodialysis, 101, 125–126
high-efficiency (HEED®), 128
elevated falls, 227, 229
emergency response, OSHA standards for, 222–225
emergency response plan, 223–224
employee complaints, 213, 217
empty bed contact time, 95
emulsions, 50, 102
endangered species, 9
endothermic process, 114
energy budget model, lake evaporation, 174
engineering controls, 234, 235
for noise, 230
enhanced gravity separators, 88
enhanced recovery wells, Class II, 197–198
entry permit, confined spaces, 243
environmental controls, OSHA compliance and, 237–245
equivalents, moles and, 113
ethylene glycol, 67

306

evaporation ponds, 167–175
evapotranspiration, 174, 175
excess pore pressure, 181
exit routes, 225–226
exothermic process, 114
expansion, specific, 181
expansion, specific unit, 181
explosion hazards, 232
Exxon Valdez, 159

F

fall protection, 226–229
    measures, 229
Fayetteville Shale, 34
    fracturing fluid, 35
    seasonal water flow, and, 42–43
fecal contamination, 157
Federal Emergency Management Agency (FEMA), 224
Federal Water Pollution Control Act, 145
filtration, 95, 152
    media, 91
    vs. reverse osmosis, 113, 118
fire extinguishers, 246, 248, 255, 256
fire, OSHA standards for, 222
fire protection, OSHA compliance and, 246–249
fire watch requirements for hot work, 254–255, 256
first aid, 222
    OSHA compliance and, 245–246
fixed-film biological treatment, 85
flammable liquids, 232, 249
flotation, dissolved air/gas, 89–91
flowback, 9, 12, 77
    boron and bromide in, 80
    impoundment, volume of, 171
    iron and manganese in, 79–80
    leaks, 170
    organic contaminants in, 79
    salinity of, 79
    trace metals in, 80
fluorine, 101
flux, 105–106, 117, 129
foamed gels, 59, 61
foaming agents, 67
formation, geologic, 9
formation water, 26, 31, 61, 66, 141, 197
forward osmosis, 101, 126
fracking, *see* hydraulic fracturing
fracking water, composition of, 49–71
fracture design, 30–31
fracturing fluids, 10, 27, 29, 30, 33–35
    constituents of, 66–70
    diesel fuel in, 60, 64
    disclosure requirements, 65
    qualities of, 59
    toxic chemicals in, 58, 68–73
    types of, 58–67
Frasch process, 200
free water surface, 182
freeze–thaw cycle, 168–169
freeze–thaw/evaporation (FTE®) process, 168
friction, falls and, 227
friction reducers, 33, 34

G

gas drive, 21
gas laws, 107–109
gas welding and cutting, 258
gases; *see also* gas laws
    as hazardous materials, 231
    defined, 51
    dissolved in water, 54, 79, 98–99
    inert, 61
    miscibility, 50, 102
    pressurized, 108
    solubility of in liquids, 114
    toxic, 10, 236, 239, 242
    wastewater, 16, 79
gellants/gelling agents, 59–60, 61, 64
geological sequestration, 201
geologic time, 4–6
geopressurized zones, 77
geostatic stress, 184
Giardia, 157
Globally Harmonized System of Classification and
        Labeling of Chemicals, 259–260
granular activated carbon (GAC), 95–96
    criteria ratings, 96
gravitational stress, 184, 185
gravity, defined, 227
gravity injection well, 201
gravity, particle settling and, 50, 92, 102
gravity separation, 79, 84, 88
grease, 83–84, 90
    removal of by biological aerated filtration, 86
    removal of by dissolved air/gas flotation, 90
    removal of by filtration, 91
    removal of by hydrocyclones, 89
    removal of by separation, 83–84
Great Extinction, second, 3
greenhouse gases, 8
ground-fault circuit interrupters (GFCIs), 240
groundwater; *see also* underground source of drinking
        water (USDW)
    aquifers, 9, 25, 178, 179
    aquitards, and, 180
    as fracking water supply, 41, 45, 46, 61, 157, 186–187
    barium and strontium in, 80
    coalbed methane, and, 9
    contamination, 26, 27, 187, 198
        cleaning up, 200
    deep wells, 189
    defined, 10
    infiltration, 14
    injection wells, and, 26–27
    produced wastewater leaks, and, 170
    radionuclides in, 157
    salinization of, 197
    Safe Drinking Water Act, and, 152
    turbidity testing, and, 157
    use by hydraulic fracturing, 46
    water budget model, and, 173
grout, well, 190, 195
guardrails, 229
guar gum, 60
gypsum, 128

Index

307

H

Hampton Roads Sanitation District (HRSD), 186–187, 195, 201
hand tools, OSHA compliance and, 253–254
handrails, 228, 229
hardness, water, 49, 53, 58, 107, 131–134, 135, 153
hazard classification, per Hazard Communication Standard, 260, 261
hazard communication, OSHA compliance and, 259–265
  audit items, 265
  labeling requirements, 262–265
  worker training, 262
hazardous air pollutants (HAPs), 66, 67, 68, 69
Hazardous and Solid Waste Amendments (HSWA), 197
hazardous energy, control of, 244–245
hazardous materials, 233; *see also* hazardous wastes
  categories of, as defined by U.S. Department of Transportation, 231
  emergencies involving, 222–223
  OSHA compliance and, 231–234, 259–265
hazardous substances, *see* hazardous materials, hazardous waste
hazardous waste, 187
  categories of, 232
  Class I injection wells, and, 197
  Class IV injection wells, and, 200
  defined, 232–234
  disposal wells, 196
  sites, 222, 234
  vs. hazardous substances, 232
Hazardous Waste Operations and Emergency Response (HAZWOPER), 222–223
head loss, 184
hearing conservation standard, OSHA, 229–231
heavy metals, 54, 85, 91, 102, 168, 233
HEED®, 128
herbicides, 16, 155
HERO™, 127–128
Higgins Loop™, 134–135
high-efficiency electrodialysis (HEED®), 128
hollow-fiber membrane modules, 121–122
horizontal drilling, 8, 10, 21, 25, 27, 31–32, 33, 34
hot work permit procedure, 254
housekeeping
  fire prevention, and, 247
  sanitation, and, 237
  spills, and, 228
  stair falls, 228–229
  trips, and, 228
  work area, 220
humic acid, 55, 57, 98, 124
hybrid membrane processes, 101, 127–129
hydration factor, 114
hydraulic conductivity, 181
hydraulic diffusivity, 181–182
hydraulic fracturing, 8
  fracture design, 30–31
  history of, 8
  occupational deaths, 214
  OSHA standards applicable to, 215–219
  process, 21–36
    description of, 31–33

terminology, 8–16
wastewater, 43
  disposal, 141–164, 167–175, 177–202, 205–206
  treatment, 77–99
water, composition of, 49–71
water supply, 41–46
hydrocarbons, 16, 22, 35, 50, 56, 79, 84, 88, 101, 102
  storage wells for, 197, 198
hydrochloric acid (HCl), 8, 31, 35, 55, 61, 134, 135, 233
hydrocompaction, 182
hydrocyclones, 88–89
  criteria ratings, 89
hydrogen ions, 51, 55, 56, 57, 134
hydrological cycle, 41
hydrophilic acids, 55
hydrophilic membrane, 119
hydrophobic membrane, 119, 129
hydrostatic pressure, 10
hypertonic solution, 103, 104
hypotonic solution, 103, 104

I

ideal gas law, 109
ignitability, hazardous waste, 232
immediately dangerous to life or health (IDLH), 236
impoundment surface evaporation, 173–176
incrustation, 194
induced gas flotation (IGF), 90
Industry Group 138, Oil and Gas Field Services, 217–219
  OSHA standards cited most often in, 219–265
injection wells, 10, 16, 26–27, 78, 152, 187, 195–202
  Class I, 195–197
  Class II, 26–27, 197–198
  Class III, 198–200
  Class IV, 200–201
  Class V, 201–202
inorganic chemicals/compounds/matter, 13, 15, 16, 52, 53, 55, 56, 92, 94, 124, 153, 155–156, 193
in situ leaching (ISL), 199–200
ion exchange, 134–135
ionization, 51, 52, 111–112; *see also* capacitive deionization, electrodeionization
ions, 50, 51, 52, 101, 102, 111, 125, 126
  hydrogen, 51, 55, 56, 57, 134
iron, 53, 79–80
  bacteria, 194
  oxidation of, 135
  removal of
    by adsorbents, 91
    by air stripping, 93
    by biological aerated filters, 85, 86
    by oxidation, 92
  reverse osmosis membrane fouling, and, 107
  water color, and, 54
  water softening, and, 131
isopropyl alcohol, 67
isotonic solution, 103

J

jetted wells, 189

Index

L

labels, chemical, 260, 261–262
ladders, OSHA standards for, 221, 240
laminar flow, 106
lead, 155
    materials, use of, 152
leakance
    aquiclude, 178
    aquitard, 180
lighting, confined spaces, 239–240
linear gels, 60
lineshaft turbines, 192
liquid–gas solubility, 114
liquid–liquid deoilers, 88
liquid–liquid solubility, 114
liquid–solid solubility, 114
liquids, defined, 51
lithium, 101
lixiviant, 199–200
lockout/tagout, 244–245
lower flammable limit (LFL), 247

M

machine guarding, OSHA compliance and, 252–253
magnesium, 125, 126
    equivalent weight, 131
    hardness, 132
manganese, 79–80
    removal of by adsorbents, 91
    removal of by air stripping, 93
    removal of by biological aerated filters, 85
    removal of by oxidation, 92
    reverse osmosis membrane fouling, and, 107
manholes, 240
Marcellus Shale, 7, 21, 28, 29, 31, 146
    reused wastewater in, 43
    water use in, 42
mass concentration, 50, 103
Material Safety Data Sheets (MSDSs), 70–71; *see also* safety data sheets
materials handling and storage, OSHA compliance and, 249–252
maximum contaminant level goal (MCLG), 150, 153
    lead, 155
maximum contaminant level (MCL), 150, 152, 153
    coliform bacteria, 157
maximum residual disinfectant level goal (MRDLG), 150
maximum residual disinfectant level (MRDL), 150
mean depth, water body, 171–172
means of egress, OSHA compliance and, 221–226
means of retrieval, 241
media filtration, 91
medical aid, OSHA compliance and, 245–246
medical emergencies, OSHA standards regarding, 222
membrane desalination technology, 101
membrane distillation, 101, 129
membrane filtration, 98, 118, 125
membrane fouling, 107, 119, 124–125
membrane, reverse osmosis, 118–122
    modules, 119–122
    symmetric vs. asymmetric, 119

mercury, 155
metals, 15, 54–55, 56, 60, 131, 193
    heavy, 54, 85, 91, 102, 168, 233
    oxidation of, 135
    radioactive, 157
    trace, 80
methane, 79
    coalbed, 9, 77, 86
        produced water disposal, 142
    hydrates, 77
    shale gas emissions of, 8
methanol, 67
microbial inactivation/removal, 84–99
microfiltration, 98–99
    criteria ratings, 99
microorganisms, 7, 34, 53, 54, 55, 56, 85, 88, 96, 98, 124, 153, 157, 179
microseismic fracture mapping, 31
Middle Devonian era, 7
mining, Class III injection wells and, 198–200
miscibility, 50, 102
mitigation banking, 146
molality, 113
molarity, 112–113
mud balls, 99
mud pits/tanks, 23, 24
mud pump, 23
multiple-effect distillation (MED), 101, 130
multistage flash distillation, 101, 130
myriapods, 3

N

nanofiltration, 101, 125, 186
    two-pass, 101, 127
National Ambient Air Quality Standards (NAAQS), 153
national discharge standards, 142
National Drinking Water Standards, 153
National Pollutant Discharge Elimination System (NPDES), 145, 146, 147, 150, 164
National Primary Drinking Water Regulations, 153–157
    exemptions, 150
    variance, 151
National Secondary Drinking Water Regulations, 158
natural disasters, 224
natural gas, 7, 8, 9, 10, 12, 13, 14, 21, 28, 35, 66, 83, 197, 206
    unconventional, types of, 77
    usage, 42
    U.S. supply of, 64
natural organic matter (NOM), 55, 90, 95, 124
naturally occurring radioactive material (NORM), 10, 65–66
near-surface subsidence, 182
nephelometric turbidity units (NTUs), 157
neutral stress, 184
new source performance standards (NSPS), 147
nitrates, 153, 155
nitric acid (HNO_3), 55
nitrification, 86, 88
nitrifying bacteria, 88
noise exposure, occupational, 229–231
noncarbonate hardness, 133–134

Index

309

non-point-source pollution, 149
nonpolar substances, 53, 112, 114
nonvolatile solids, 53
normality vs. molarity, 113

O

occupational noise, OSHA compliance and, 229–231
Occupational Safety and Health Act
    compliance issues, 211–213
    employer responsibilities, 215–217
    violations, 217
Occupational Safety and Health Administration (OSHA),
    211–265
    citations, for Industry Group 138, 219
    standards applicable to fracking, 215–219
        lack of compliance or willful violation of, 219–265
offsite treatment and disposal, 205–206
oil, removal of
    by adsorption, 91
    by biological aerated filtration, 85–86
    by dissolved air/gas flotation, 90
    by filtration, 91
    by hydrocyclones, 88–89
    by separation, 83–84
oil-equivalent gas (OEG), 10
Oil Pollution Act, 143, 159–164
one-dimensional consolidation, 180
organic chemicals, 153–155
    removal of, 84–99
organic matter, 54, 55, 168
osmosis, defined, 103; *see also* dual reverse osmosis,
        forward osmosis, reverse osmosis
osmotic gradient, 104
osmotic pressure, 104, 114, 115, 116
    difference, 118
    gradient, 103
oxidation, 84, 92, 135
    advanced, 44
    biochemical, 85
    chemical, 85
    criteria ratings, 92
oxygen, dissolved, 54, 88
ozonation, 95
ozone, 80, 92, 97, 125

P

Paleozoic Era, 4
Papadakis equation, 175
particles, in suspension, 50, 102
particulates, 16
    defined, 10
    removal of, 83, 84–99
    reverse osmosis, and, 107, 119, 121, 125
    turbidity, and, 157
    ultraviolet disinfection, and, 96
Pascal's law, 108
pathogenic protozoans, removal of, 98
pay zone, 26
Penman equation, 174
performance standards, 215
permeability, 10–12, 13, 28, 59, 106, 178, 180

permeate, 80, 104–105, 116, 120–122, 136
    defined, 103
permit-required confined space entry, 236
    OSHA standards for, 237–243
personal protective equipment, 230
    classifications of, 235
    hot work, and, 258
    OSHA compliance, and, 234–237
        confined space entries, and, 239
personnel sanitation facilities, 237
pesticides, 14, 16, 70, 155, 193, 233
pH, 52, 54, 56–57, 80, 107
    dual reverse osmosis, and, 127–128
    fracking fluids, 170
    nitrification, and, 88
    pollution, and, 57
piezometer, 182–183
piezometric surface, 182–183, 184
piston pump, 24
plate-and-frame membrane modules, 119–120
plug inserts, for hearing protection, 230–231
point-source pollution, 149, 151
polar substances, 53, 112
pollutants, 9, 53, 57, 107, 143, 145, 146, 147, 149
    defined, 151
    discharge of, 150
    found in fish, 149
    hazardous air, 66, 67, 68, 69
    indicator, 143
    toxic, 151
pollution, point-source vs. non-point-source, 149
polyamide (PA) membranes, 119
pore pressure, excess, 181
pores, aquifer, 178
porosity, 10–12, 35, 59, 178
positive displacement pump, 24
potassium chloride (KCl), 12, 59, 61, 64
power tools, OSHA compliance and, 253–254
powered platforms, OSHA compliance and, 226–229
precipitate, defined, 52
preconsolidation stress, 180, 182, 185
pre-entry, confined spaces, 242
preload stress, 180, 182
presence/absence concept, 157
pressure injection well, 201
pretreatment, produced wastewater, 83, 84–99
Priestly–Taylor equation, 174
primacy, defined, 12
Process Safety Management (PSM), 223
produced water/wastewater, 8, 13–14; *see also* wastewater
    composition of, 16
    constituents, 79–80
    defined, 12
    desalination, 101–136
    disposal of, 141–164
        by evaporation, 167–175
        discharge to surface waters, 142
    electrodialysis, and, 126
    injection, 177–202
    leaks of from ponds, 170
    management costs, 78
    naturally occurring radioactive water (NORM), and, 66
    offsite treatment and disposal, 205–206

Index

quality, 79–80
reinjection, 78
salinity of, 56, 79
settling ponds, 92–93
spilled, 205
stabilization ponds, 171–175
treatment, 77–80
technology, 83–99
volume of, 141
productivity index, water well, 192
proppants/propping agents, 12, 30, 31, 33, 34, 35, 58, 61, 64, 67
Class II injection wells, and, 198
proprietary chemicals, 70–71
proved reserves, 12
Public Health Service Act of 1912, 150
publicly owned treatment works (POTW), 151
public water systems
contaminants that occur in, 70
deep wells, and, 189
defined, 151
maximum contaminant levels for, 150
number of in United States, 144
Safe Drinking Water Act provisions for, 152–153
shallow wells, and, 189
underground sources of drinking water, and, 13
pump test, 191–192
pump, well, 192

R

radiation, background, 65
radioactive wastes, Class IV injection wells and, 200
radionuclides, 80, 95, 153, 157
radium, 101, 157
radon, 10, 65, 66, 157
reactivity, hazardous waste, 232
Reasonable and Prudent Practices for Stabilization (RAPPS), 147
recharge zone, 151
reciprocating pump, 24
reclamation, 12
recordkeeping, per OSH Act, 213, 216
recovery rate, reverse osmosis, 104
reference dose (RfD), 153
regulations, OSH Act, 211–213
reinjected wastewater, 78, 197
reject stream, 103, 104, 130
reject, hydrocyclone, 88
rejection, of contaminants, 103, 104, 129
defined, 104
forward osmosis, 126
heavy metals, 102
nanofiltration, 125
reverse osmosis, 104–105, 118, 121
salt, 101, 104, 114, 129
relative humidity, 170
rescue and retrieval line, 241
rescue equipment, for confined spaces, 241
residual compaction, 180
residual solids, 50, 102
residual stream, 103
Resource Conservation and Recovery Act (RCRA), 197, 222, 233–234

respirators, types of, 236
respiratory protection, 234–237
retentate, 116, 118
membrane output, 103
retention pits, 28
reused hydraulic fracturing wastewater, 43
reverse osmosis (RO), 44, 80, 89, 94, 101, 102, 114–125, 186
concentration polarization, and, 106
dual, with chemical precipitation, 127
dual, with seeded slurry precipitation, 128
dual, with softening pretreatment and operation at high pH, 127–128
equipment, 118–122
flow rating, 104
flux, and, 105–106
high-pressure, 135, 136
low-pressure, 135, 136
membranes, 107, 118–122
posttreatment, 125
pretreatment, 124–125
process, 116–118
rejection of contaminants, 104–105, 118, 121
system configuration, 123–125
vs. filtration, 116, 118
reverse osmosis (RO), 44
rigging safety, OSHA standards for, 249–252
Risk Management Planning (RMP), 223
ropes, 250–252
rotary drill rigs, 21–22
rotating biological contactor (RBC), 85, 86
rule of capture, 83, 205

S

Safe Drinking Water Act (SDWA), 64, 65, 66, 67, 68, 143, 149–159, 186, 195, 197, 198
amendments to, 158–159
chemicals regulated, 70
groundwater, and, 152
implementation of, 159
provisions, 152–153
safety data sheets, 260, 262, 265
safety harness, 241
safety shoes, 228
safey and health considerations, 211–265
salinity, 56, 142; *see also* desalination, salts
feed water, 129
osmotic pressure, and, 114, 115
produced wastewater, 78, 79
reverse osmosis, and, 116, 125
salt passage, 104
salt solution mining wells, 200
salts, 50, 52, 55, 56, 57, 67, 80, 91, 101, 102, 112, 125, 128, 168, 199; *see also* desalination, salinity
alkalinity, and 57
as polar subtances, 53, 112
biological aerated filters, and, 85–86
freeze crystallization, and, 168
osmosis, and 103, 116
rejection of, 104, 114, 129
sanitary seal, 191
sanitation facilities, 237
saturated solution, 52, 53, 54

Index

311

scaling, membrane, 107
sediment, 50, 102
    runoff contaminated with, 147
sedimentation, 87, 88, 89, 93, 124, 147
seeded crystalline slurry, 128
seepage stress, 184, 185
self-contained breathing apparatus (SCBA), 236
self-ionization of water, 51
semipermeable membrane, 104, 105, 115, 121
separators, advanced, 84
septic system, 14, 193, 201
    leach fields, 201
sequestration, carbon dioxide, 201
setback, 12
settleable solids, 51, 53
settling ponds, 92–93
severed estate, 12, 205
sewage vs. wastewater, 14
sewerage, 14
shale gas, 6, 7–8, 9, 12, 61, 66, 77, 143, 145, 147, 164, 177, 205–206
    drilling for, 21, 27–36, 41–46; *see also* hydraulic fracturing
        safety and health considerations, 211–265
    water use, 42
shallow subsidence, 182
shallow wells, 188, 189
shoreline development index, 171, 172
significant hazard to public health, 151
silica sands, 35
silicon dioxide, 67
silt density index (SDI), 107
site emergency response plan, 223, 224
slickwater, 12, 29, 31, 33, 34
    pad, 33
sling safety, 250–252
slips, 227–228
slurries, 8, 33, 60, 89, 101, 128, 195
sodium absorption ratio (SAR), 126, 134, 136
sodium chloride, 50, 52, 67, 102, 135
    ionization, 112
sodium ion exchange, 134–135
softening, water, 131–134
solids, 51, 53
    colloidal, 51, 53, 54, 88, 96, 98, 107, 157
    disposal, 87
    dissolved, 51, 53, 88, 120
    nonvolatile, 53
    residual, 50, 102
    settleable, 51, 53
    suspended, 51, 53, 79, 124
    total, 51, 53
    total dissolved, 13, 56, 91, 101, 107, 125–126, 127, 134–135
    total suspended, 53, 95, 97, 98
    types of, 51, 53
    volatile, 53
solubility, 50, 102, 107, 113–114
solute, 52, 53, 54, 109–114, 116
    concentration polarization, and, 106
    tonicity, and, 103
solutions, 50, 51, 52, 102, 109–114
    calculations for, 110–112
    colligative properties, 114

    heavy metal, 233
    hydrochloric, 31, 35
    lixiviant, 199–200
    mass concentration, and, 50, 103
    osmosis, and, 103–104, 115, 118
    pH of, 56
    saturated, 53, 54, 90
    sodium chloride, 135
    sulfuric acid, 200
    temperature of, 107, 168
    tonicity, and, 103
    water, 52–53
solvents, 52, 109–114, 116
species, metal, 54
specific capacity, 192
specific compaction, 181
specific conductance, 57
specific expansion, 181
specific flux, 106
specific gravity, 111
specific standards, 215
specific unit compaction, 181
specific unit expansion, 181
Spill Prevention, Control, and Countermeasure (SPCC), 159, 164
spiral-wound membrane module, 120–121
split estates, 12, 205
springs, as type of aquifer, 178
spudding in, 26
stabilization ponds, wastewater, 171–175
stabilizers, 34
stages, fracturing, 31
stair falls, 227, 228–229
standard conditions (SC), 107
Standard Industrial Classification (SIC) Group 138, 217–219
standard temperature and pressure (STP), 107
static hydrocyclone, 88
steerable downhole motor, 24–25
stimulation, 13, 25, 27, 28–31, 33, 61, 65
Stokes' law, 84, 88
storage pits, 28
stormwater discharges, 147
stormwater drainage wells, 201
streamflow source zone, 151
streamlines, 106
stress
    applied, 180, 181, 183–184, 185
    comprehensive, 180
    effective, 180, 181, 183–184, 185
    geostatic, 184
    gravitational, 184, 185
    neutral, 184
    preconsolidation, 180, 182, 185
    preload, 180, 182
    seepage, 184, 185
stripper wells, 142
strontium, 80, 125, 131
submersible turbines, 192
subsidence, 182, 185
substages, fracturing, 31–33
sulfur mining, 200
sulfur dioxide (SO_2), 13
sulfuric acid (H_2SO_4), 55, 200, 233
Superfund, 234

surface water disposal, 141–164
surface water withdrawals, 46
surfactant-modified zeolite/vapor-phase bioreactor, 94
surfactants, 67
suspended solids, 51, 53, 124
    produced wastewater, 79
suspensions, 50, 102
Sustainable Water Initiative for Tomorrow (SWIFT),
        186–187, 195, 201
symmetric membrane, 119
synthetic organic chemicals (SOCs), 153–155

T

technically recoverable resources, 13
temperature, 57
    absolute, 108–109, 118
    air, 108, 169, 170
    air stripping, and, 93
    flotation, and 90
    oxygen saturation level, and, 54
    solubility, and, 53, 54, 112, 114
    water, 57, 107, 167, 168, 170
testing, well, 32
tetrapods, 3
thermal desalination technologies, 101, 129–130
thermal pollution, 114
thermogenic gas, 13
thin-film composite reverse osmosis membranes, 105, 124
thixotrophy, 13
tight gas, 13, 77
tonicity, 103
torch cutting safety, 258–259
total dissolved solids (TDS), 13, 56, 91, 101
    electrodialysis, and, 125–126
    ion exchange, and, 134–135
    Langelier Saturation Index, and, 107
    two-pass nanofiltration, and, 127
total hardness, 132–134
total organic carbon (TOC), 90, 91
total solids, 51, 53
total suspended solids (TSS), 53
    granular activated carbon, and, 95
    ultraviolet disinfection, and, 97, 98
toxic chemicals in fracturing fluids, 58, 68–73
toxic gases, 10, 236, 239, 242
toxic pollutants, 151
toxic substances, OSHA compliance and, 259–265
toxicity, hazardous waste, 232
trace elements, 101
trace metals, 80
trade secrets, 70–71
transmembrane pressure (TMP), 119
traveling block, 22
trickling filter, 85
trihalomethanes (THMs), 153–155
Trinity Trails, 206
trips, 227, 228
true color, 54
tubular membrane modules, 122
turbidity, 53, 151, 153, 157
    defined, 52
    microfiltration, and, 98
    ultraviolet disinfection, and, 96, 97

turbines, lineshaft and submersible, 192
turbulent flow, 106
two-pass nanofiltration, 101, 127

U

ultrafiltration, 84, 98–99, 135
    criteria ratings, 99
ultraviolet disinfection, 84, 96–98, 135, 136
    advantages and disadvantages, 97
    applicability, 97–98
    criteria ratings, 98
unconfined aquifer, 178, 179
unconventional natural gas, 77
underground injection, 12, 152, 177–202
    Energy Policy Act of 2005, and, 198
    wells, classes of, 195–201
Underground Injection Control (UIC) Program, 13, 164,
        195, 198, 201
underground source of drinking water (USDW), 26–27,
        195, 196, 197, 198, 200, 201
    defined, 13
unit compaction, 181
    to head decline ratio, 185
upper flammable limit (UFL), 247
uranium, 157, 199
    Class III injection wells, and, 199–200
urban water cycle, 43

V

vapor compression, 101, 130
vapor pressure, 167–168, 170
    saturation, 170, 175
vertical seepage stress, 185
virgin compaction, 180
volatile compounds, in produced wastewater, 79
    removal of, 90, 93–94
volatile organic chemicals (VOCs), 93, 153–155
volatile solids, 53
volume development, 173
volume, water body, 171

W

walking and working surfaces, OSHA compliance and,
        220–221
wastewater, 13, 43; *see also* produced water/wastewater
    composition of, 14–16
    defined, 14
    sources of, 14
    treatment of, 77–80
    vs. sewage, 14
wastewater stabilization ponds, 171–175
wastewater treatment plant effluent, 201
water budget model, pond and pit evaporation, 173
water chemistry, 49–58
    concepts and definitions, 49–52
    constituents, 53–58
    solutions, 52–53
Water Quality Act (WQA), 147
water quality regulations, 143–164
water retention time, 173
water storage pits, 28

Index

water supply, fracking, 41–46
water table, 10, 179
water temperature, 57, 107, 167, 168, 170
water use/consumption, defined, 41
water well, 21, 27–28, 187–195
watershed, 16
weak acid cation IX softener, 135–136
welding, OSHA compliance and, 254–259
well casing, 25–28, 190
   Class I injection well, 197
   vent, 191
well head, 23
   protection programs, 153
well log, 193
well pad, 191, 205
well problems, troubleshooting, 194
well pumps, 192
well screen, 191
well yield, 192
wells, 186–187
   conventional water, 187–195
      abandonment of, 194–195
      components of, 189–191

      development process, 188
      evaluation of, 191–192
      maintenance, 193–194
      operation and recordkeeping requirements, 192–193
      requirements for, 188
   injection, 195–201
wetlands, 146, 149, 152
   mitigation, 146
whipstock, 16
Williston Basin, 27
wire rope, 251
workover, 16

Y

yield, water well, 191

Z

zeolites, 94
zero liquid discharge (ZLD), 136
zone of saturation, 179